Purchased through a gift to the Library from
UNIVERSITY of GUELPH
the PARENTS PROGRAM
McLaughlin Library

Plasma Charging Damage

Springer
London
Berlin
Heidelberg
New York
Barcelona
Hong Kong
Milan
Paris
Singapore
Tokyo

Kin P. Cheung

Plasma Charging Damage

With 327 Figures

Springer

Kin P. Cheung, PhD
Lucent Technologies, Bell Laboratories, Room 1B217, 600 Mountain Avenue, Murray Hill, NJ 07974, USA

ISBN 1-85233-144-5 Springer-Verlag London Berlin Heidelberg

British Library Cataloguing in Publication Data
Cheung, Charles Kin P.
Plasma charging damage
1.Plasma (Ionized gases)
I.Title
621'.044
ISBN 1852331445

Library of Congress Cataloging-in-Publication Data
Cheung, Kin P.
Plasma charging damage / Charles Kin P. Cheung.
p. cm.
Includes bibliographical references and index.
ISBN 1-85233-144-5 (alk. paper)
1. Semiconductors--Effect of radiation on. 2. Metal oxide semiconductors--Defects. 3. Plasma radiation. I. Title
TK7871.85.C4784 1999
621.3815'2--dc21 99-34071

Printed in Great Britain

Typesetting: Camera ready by author
Printed and bound by Athenæum Press Ltd., Gateshead, Tyne & Wear
69/3830-543210 Printed on acid-free paper SPIN 10709949

Preface

In the 50 years since the invention of transistor, silicon integrated circuit (IC) technology has made astonishing advances. A key factor that makes these advances possible is the ability to have precise control on material properties and physical dimensions. The introduction of plasma processing in pattern transfer and in thin film deposition is a critical enabling advance among other things. In state of the art silicon IC manufacturing process, plasma is used in more than 20 different critical steps.

Plasma is sometimes called the fourth state of matter (other than gas, liquid and solid). It is a mixture of ions (positive and negative), electrons and neutrals in a quasi-neutral gaseous steady state very far from equilibrium, sustained by an energy source that balances the loss of charged particles. It is a very harsh environment for the delicate ICs. Highly energetic particles such as ions, electrons and photons bombard the surface of the wafer continuously. These bombardments can cause all kinds of damage to the silicon devices that make up the integrated circuits.

Very early on in the introduction of plasma processing in silicon IC manufacturing, people became aware of the potential danger due to plasma damage to the devices. Much effort had been devoted to understand the damage mechanisms and to find ways to avoid or fix them. There is a huge body of literature on the subject. Most of the damage studies focused on physical damage to crystalline silicon by energetic ions and electrical damage to SiO_2 and its interface by energetic photons.

More recently, a different kind of plasma damage has become important. This is charging damage, the subject of this book. The first reported observation of plasma charging damage was in 1983. A high electric field apparently developed across the gate-oxide of a Metal-Oxide-Semiconductor-Field-Effect-Transistor (MOSFET) during plasma processing. This high electric field leads to gate-oxide breakdown or wear-out (lifetime shortening). It was quite confusing at first on how can such an electric field develop. After 17 years of research and development, the basic mechanisms are largely understood, although there are still many mysteries that await further studies.

The importance of plasma charging damage grows as the IC industry migrates toward ever-finer geometry. This is due to the use of thinner gate-oxide and higher density plasma tools. Thinner gate-oxide breaks down more easily and therefore more susceptible to charging damage. The higher density plasma tools are introduced to handle the more stringent critical dimension (CD) control required to make the finer geometry devices. Increasingly, incidences of plasma charging damage that causes yield loss are hitting IC manufacturers. As a result, the number of engineers that are involved in solving the plasma charging damage problem increases rapidly. This book is written for these engineers.

The literature of plasma charging damage is not easy to digest, particularly for new comers. There are plenty of poorly executed experiments and poorly interpreted results that serve to confuse the issues. Furthermore, plasma charging damage originated from plasma processing but damaged the devices and circuits. Thus the engineers who have the responsibility to fix the process are not well equipped to characterize the outcome of the damage and the engineers who have to fix the technology are poorly equipped to understand the cause of the damage. This inherently multi-disciplinary subject requires an unusual mix of knowledge and experiences to understand. It is my hope that this book serves to provide many of this knowledge in a manner that can be easily understand.

The idea of writing this book grew out of the experience of teaching the subject of plasma charging damage in the 1996 VLSI Technology Symposium, the 1997 International Reliability Physics Symposium, the 1998 International Electron Device Meeting and the 1999 International Symposium on Plasma process Induced damage. The goal is to provide a uniform way of understanding the plasma charging damage mechanisms and measurement methods. It is not an exhaustive review of the subject. The selection of materials is based on how it may contribute to the understanding of the subject. While one can get a quick assessment of certain aspects of the subject by diving right into the section that covers the topic of interest, it is much better if the reader would read through the book as a whole first. The later chapters do rely on the basic understanding developed in Chapters 2 and 3.

Chapter 1 is included to provide the relevant information about gate-oxide wear-out and breakdown. For those who are already familiar with the topic, it can be skipped. I choose to cover the gate-oxide basics in Chapter 1 instead of starting with the plasma charging damage mechanism as the first chapter for two reasons. One is that plasma charging damage is mostly about gate-oxide degradation and therefore should be emphasized immediately. The other is that to understand the plasma charging mechanisms requires the various conduction modes in the oxide under a high electric field.

Some of the materials covered here are still not fully understood. The discussion is thus necessary influenced by my own bias. It may later on be proven incorrect. However, to omit them would be worse. The approach is to make the interpretation in a consistent framework. Hopefully, by going through the arguments, the reader can form an opinion of their own.

The following figures are reproduced or slightly adapted from other publications: Figures 1.25, 1.29, 1.33, 1.35, 2.25, 2.31, 2.32, 2.34, 3.50, 4.4, 4.5, 4.16, 4.17, 4.18, 4.26, 4.27, 5.23, 5.39, 6.6, 6.7, 6.24, 6.28, 6.30, 6.62, 6.63, 6.66, 6.67 and 6.68 are copyright of the IEEE and reproduced with permission; Figures 2.16 and 4.14 are reproduced by permission of The Electrochemical Society, Inc. and Figure 4.9 is reprinted from DiStefano, T.H. and D.E. Eastman, *The band edge of amorphous SiO2 by photoinjection and photoconductivity measurements.* Solid State Communications, 1971. volume 9: p. 2259 with permission from Elsevier Science.

I am in debt to many of my colleagues from whom I have learned much of the subject. One person in particular taught me a lot when I first started in this field. This person was Paul Aum, who is currently the president of SPIDER Systems Inc. More recently, the close interaction with Philip Mason, who is my partner in dealing with the plasma charging damage problems of our own company has been of great importance. Phil. has proof-read most of the book and has provided a lot of suggestions for improvement. Wes Lukascek of Wafer Charging Monitors, Inc. provided valuable feedback to me on the chapter on Damage Measurement I. Kay Bogart and Avi Kornblit, both colleagues of mine in Bell Laboratories, read some chapters and provided helpful suggestions.

Kin P. Cheung
Murray Hill, NJ, USA
July, 2000

Table of Contents

Chapter 1

Thin Gate-oxide Wear-out and Breakdown

Plasma charging damage is mainly a phenomenon of high-field stressing of thin gate-oxides during plasma processing. Any discussion of plasma charging damage cannot avoid the discussion of gate-oxide breakdown or wear-out. A good background knowledge of thin gate-oxide wear-out and breakdown under high field stress is a prerequisite for anyone who wishes to understand plasma charging damage. As the IC industry continues its relentless advance toward ever-smaller devices, the reliability of thin gate-oxide becomes increasingly more difficult to maintain. Somewhere around the 0.25μm generation of CMOS technology, the intrinsic breakdown lifetime of gate-oxide changes from being far longer than the expected service time (10 years) of a circuit to being the limiting factor. When the intrinsic lifetime of the thin oxide is barely long enough to meet reliability requirements of IC's, it is obvious that they are also least able to tolerate some of that lifetime being consumed by plasma charging damage. The challenge of measuring damage when the gate-oxide is very thin requires anyone in the plasma charging damage field to keep up with the latest advances in the gate-oxide reliability research. This is a formidable task even for specialists focusing on gate-oxide research, l*et al.*one non-specialists who are the intended readers of this book. To help bridge the gap, this chapter collects the relevant understanding of gate-oxide wear-out and breakdown and presents it in a manner suitable for non-specialists. The presentation will be from the prospective of charging damage. It is not intended to be a complete or in-depth discussion of all gate-oxide reliability related subjects.

1.1 The MOSFET

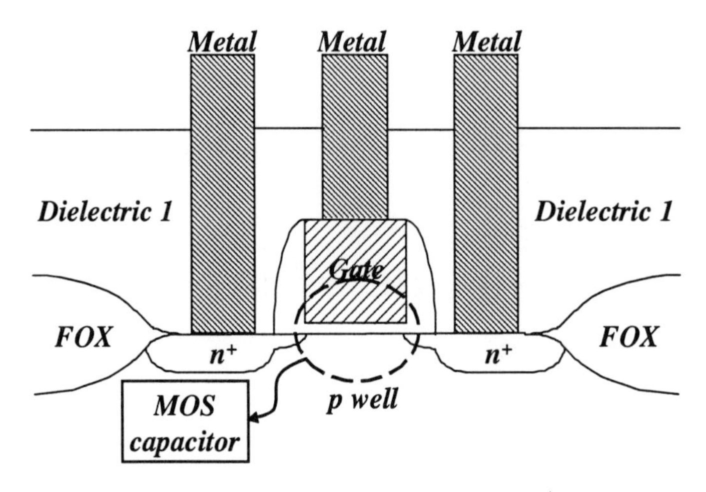

Figure 1.1. The structure of an n-MOSFET. The area within the broken circle is a MOS capacitor, which is the main concern of plasma charging damage.

The Metal-Oxide-Semiconductor-Field-Effect Transistor (MOSFET) (Figure 1.1) dominates the IC industry. When we talk about plasma charging damage, we are mostly concerned about the degradation of the MOSFET, both as individual devices and as integrated circuits. While the fabrication and operation of a MOSFET is rather complex, we can understand most of the plasma charging damage effect by concentrating on the MOS capacitor that is at the heart of the MOSFET. In particular, we need to understand the high-field stress response of the MOS capacitor.

1.2 Tunneling Phenomena in Thin Oxide

Since plasma charging damage concerns high-field stressing of thin gate-oxides, the conductivity of the thin oxide at various stress voltages is an important characteristic that we need to know in order to estimate the damage quantitatively. When defect related leakage is negligible, which is usually true for good quality gate-oxides, the current conduction mode is by quantum mechanical tunneling.

1.2.1 Fowler-Nordheim (FN) Tunneling

The band diagram of the MOS capacitor under positive and negative gate bias is shown in Figure 1.2. Here the "metal" gate is taken to be n+ polysilicon, which is

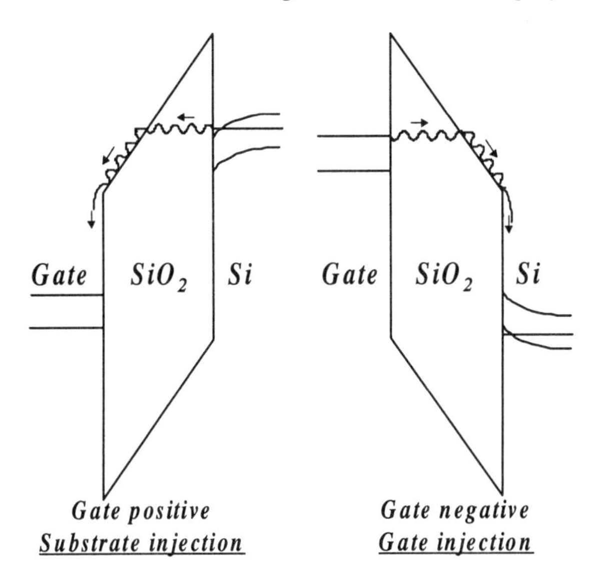

Figure 1.2. The band diagram of a MOS capacitor under bias. The gate is assumed to be n^+ poly-silicon.

the most common material for n-MOSFET. When the gate is biased positively, electrons tunnel from the substrate through the triangular barrier into the SiO_2 conduction band. When the gate is biased negatively, electrons tunnel from the gate into the SiO_2 conduction band. Electrons flowing from the substrate through oxide to the gate are called substrate injection, and the reverse is called gate injection. For a sufficiently large gate bias the barrier is triangular and the tunneling process is the well-known Fowler-Nordheim tunneling [1]:

$$I_{FN} = AE_{OX}^2 \exp\left(-\beta / E_{OX} \right) \tag{1.1}$$

where

$$A = \frac{q^3 m}{8\pi h m_{OX} \Phi_B} = 1.25 \times 10^{-6} \left(\frac{A}{V^2} \right)$$

$$\beta = \frac{-4\sqrt{2m_{OX}}}{3(h/2\pi) q E_{OX}} \Phi_B^{3/2} = 2.335 \times 10^8 \left(\frac{V}{cm} \right).$$

E_{OX} in Equation 1.1 is the oxide field, which is proportional to the potential drop across the SiO_2 layer. Φ_B is the 3.1eV barrier height that the electrons must tunnel through. Specifically it is the energy difference between the bottom of the silicon conduction band and the bottom of the SiO_2 conduction band (with some minor correction due to image force). m and m_{OX} are the electron effective masses in silicon and SiO_2 respectively. Note that the value of A and β are only approximate because experimental measurements are affected by trapped charges in the oxide.

For plasma charging damage studies, it is very useful to have in mind the quantitative current-voltage behavior of FN tunneling. Figure 1.3 plots the I-V characteristic in two popular ways. Marked on the left-hand-side plot are the oxide fields for 1μA/cm^2, 1mA/cm^2 and 1A/cm^2. These encompass the oxide fields encountered most often in charging damage.

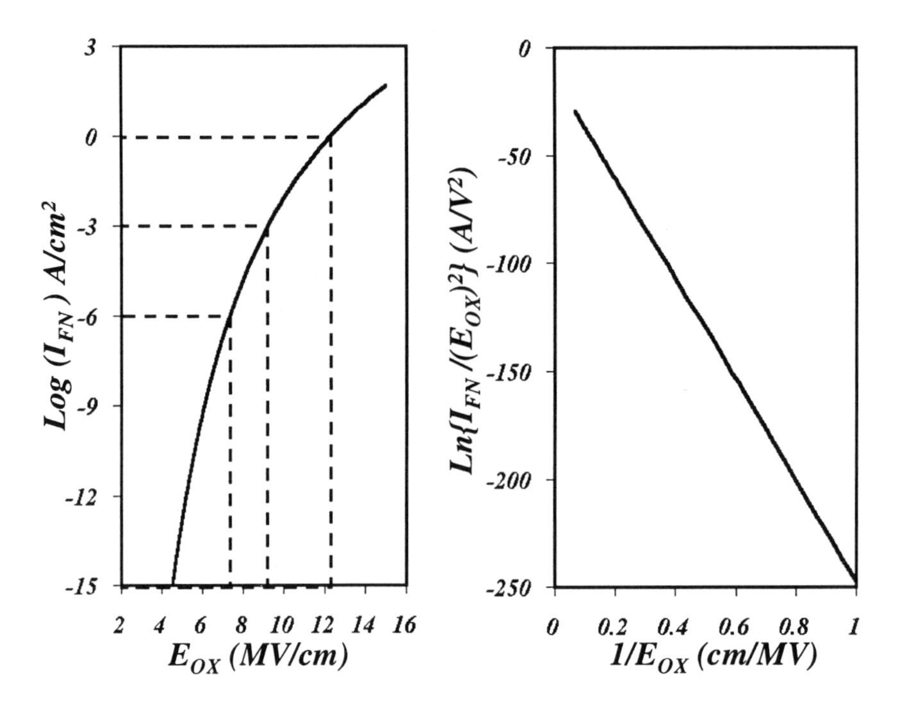

Figure 1.3. Current-voltage characteristic of FN tunneling. The linear plot on the right is often used to show that the observed I-V characteristic is FN.

Note that to calculate the oxide field, one cannot just divide the difference between the gate voltage and substrate voltage by the gate oxide thickness. There are a number of corrections that one must take to obtain an accurate oxide field. The largest factor is Fermi level difference between gate and substrate. This difference can be accurately determined from the flat band voltage of the CV curve of the gate-oxide-substrate MOS capacitor. The next factor is band bending. As can be seen from Figure 1.2, band bending in the substrate side absorbs some of the applied voltage. In addition to absorbing some of the applied voltage, band bending also confines the charges in a very thin region at the SiO_2/Si interfaces. This tight confinement has strong quantum effects that prevent the electron density from peaking right at the SiO_2/Si interfaces. The quantum confinement effect therefore adds an effective offset thickness to the insulator physical thickness. The correction factor for calculating the oxide field is dependent on the applied voltage. At high field, the correction becomes voltage independent, but it is still dependent on the doping level both in the gate and in the substrate. In general, the high-field thickness correction ranges from 5 to 9 Å [2]. That is, a physically 50Å thick gate-oxide is 55 to 59Å thick electrically. To calculate the oxide field, one must first subtract the flat band voltage out of the applied voltage and then divide the resulting voltage by the electrical thickness of the gate-oxide.

An additional factor that must also be taken into account is the larger than normal band bending in the polysilicon gate. For polysilicon doped to degeneration, this band bending should contribute about 2Å to the electrical thickness of the oxide. If it is not doped to degeneration, it can be much larger, up to 6Å is often observed. In advanced MOSFET fabrication, due to the limitation in thermal budget, poly-depletion (band bending) effect is a rather common problem. The problem is most severe for p-MOSFET using p^+-polysilicon gate.

1.2.2 Tunneling Current Oscillation

Equation 1.1 is the result of the Wentzel-Kramers-Brillouin (WKB) approximation. It works remarkably well for thicker oxides. However, when the gate-oxide is thin enough, electrons reaching the SiO_2 conduction band can traverse the rest of the distance ballistically. This phenomenon introduces an additional factor due to coherent interference. Gundlach [3] showed that such interference should exist in his analysis of tunneling current by solving the Schrödinger equation exactly. The result of coherent interference is oscillation in the I-V characteristic of tunneling. Almost 10 years later, Lewicki *et al.* [4] confirmed the existence of such oscillation experimentally.

Figure 1.4 shows some examples of oscillation in the I-V characteristic [5]. From the left-hand-side of Figure 1.4, oscillation is not observed in the I-V curve of the thicker sample. This is consistent with Fischetti *et al.*'s [6] theory of ballistic electron coherent interference. Note that the interference is not with different electrons since their tunneling events are random in time. The oscillation is due to a single electron wave interfering with its own reflection from the potential discontinuity at the SiO_2/Si interfaces. This tunneling current oscillation can sometimes cause problems in the interpretation of the I-V characteristic of thin gate-oxide, particularly when comparing one sample to another at low oxide field.

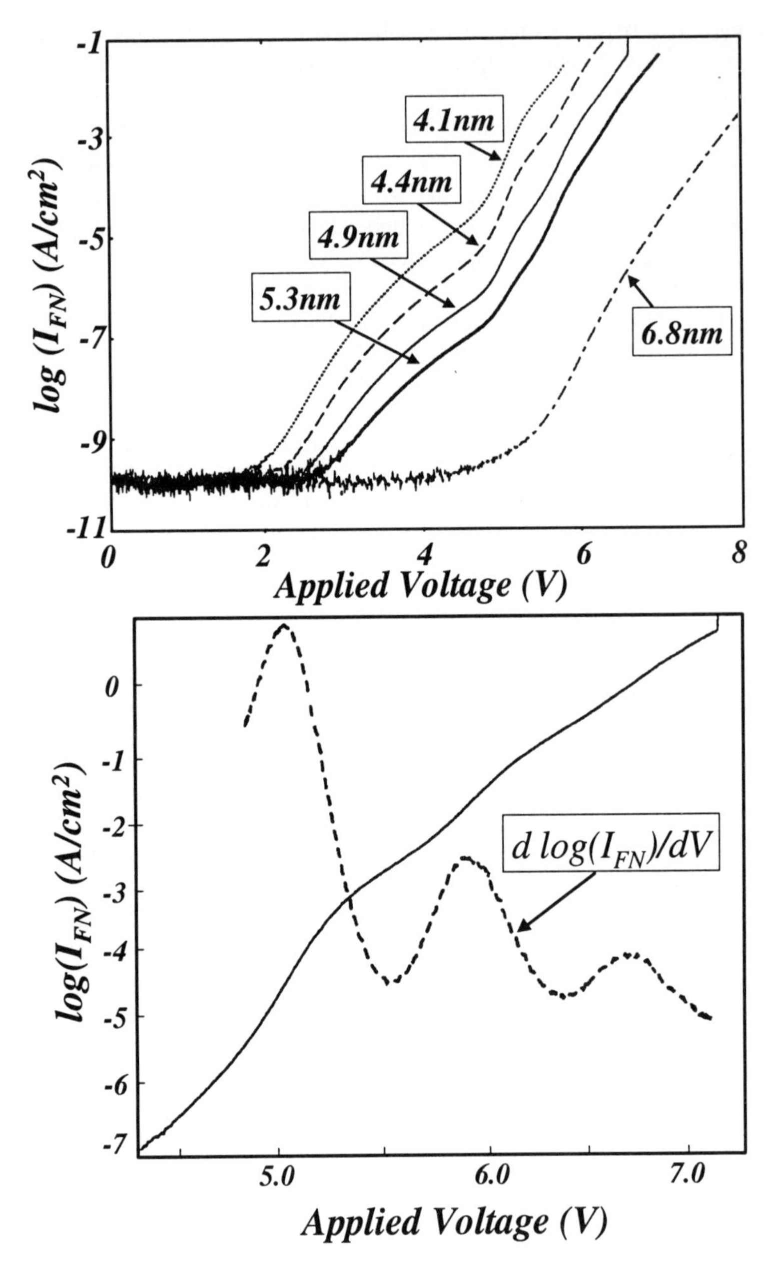

Figure 1.4. The tunneling currents of thin gate-oxides show oscillation (Upper figure). The lower figure shows the derivative of one of the curve to high-light the oscillation. Taken from [5].

1.2.3 Direct Tunneling

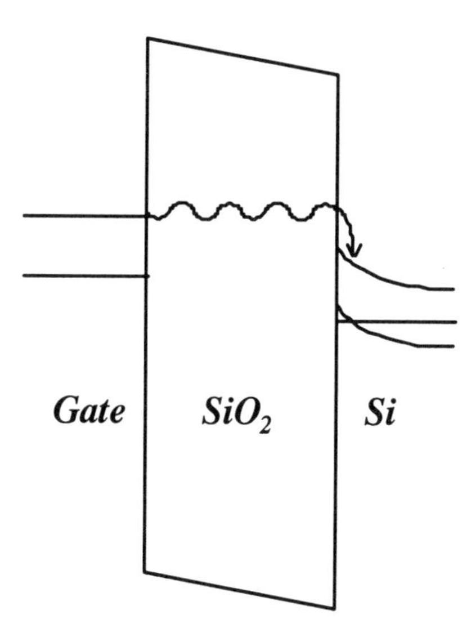

Figure 1.5. Direct tunneling through very thin gate-oxide under low voltage. The tunneling electrons never encounter the SiO_2 conduction band before reaching the anode.

When the gate-oxide is very thin, the probability for electrons to tunnel across the oxide at low voltages is enhanced. At low voltages, the potential drop across the oxide is smaller than the Si/SiO_2 barrier. The tunneling path is not through a triangular barrier, but rather a trapezoidal barrier as shown in Figure 1.5. The Fowler-Nordheim tunneling equation will no longer work for this barrier shape. Also quite naturally, no tunneling current oscillation is expected since the tunneling electrons never enter the SiO_2 conduction band.

The exact solution of the Schrödinger equation for direct tunneling was worked out by Gundlach [3] as well. The result is a very complicated equation that can only be dealt with numerically. For everyday use, a more convenient but less accurate analytical form was worked out by Schuegraf *et al.* [7]:

$$I_{tunnel} = \frac{AE_{OX}^2 \exp\left(-\beta/E_{OX}\right)\exp\left(P^3\beta/E_{OX}\right)}{(1-P)^2} \tag{1.2}$$

where

$$P = \sqrt{\frac{\Phi_B - E_{OX}}{\Phi_B}} \text{ is for } \Phi_B - E_{OX} \geq 1 \text{ , otherwise, } P = 0.$$

Similar to Equation 1.1, Equation 1.2 does not take into account the interference effect in the FN tunneling regime.

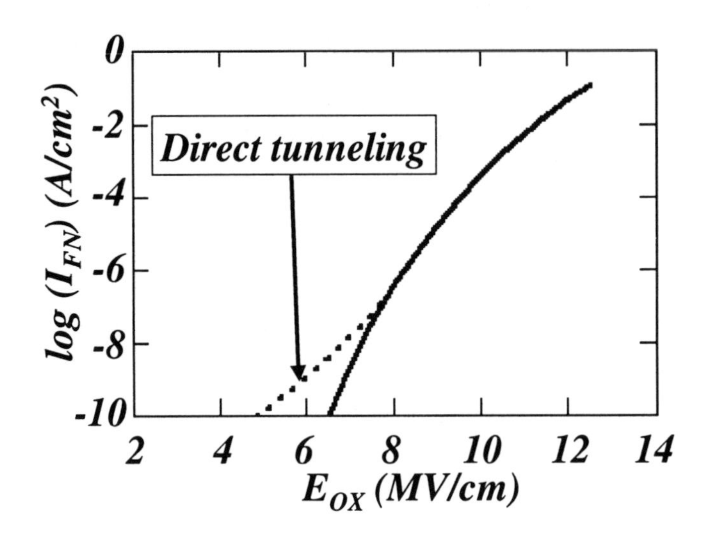

Figure 1.6. The I-V characteristics of tunneling current with direct tunneling correction. Significantly higher current is seen at low oxide field.

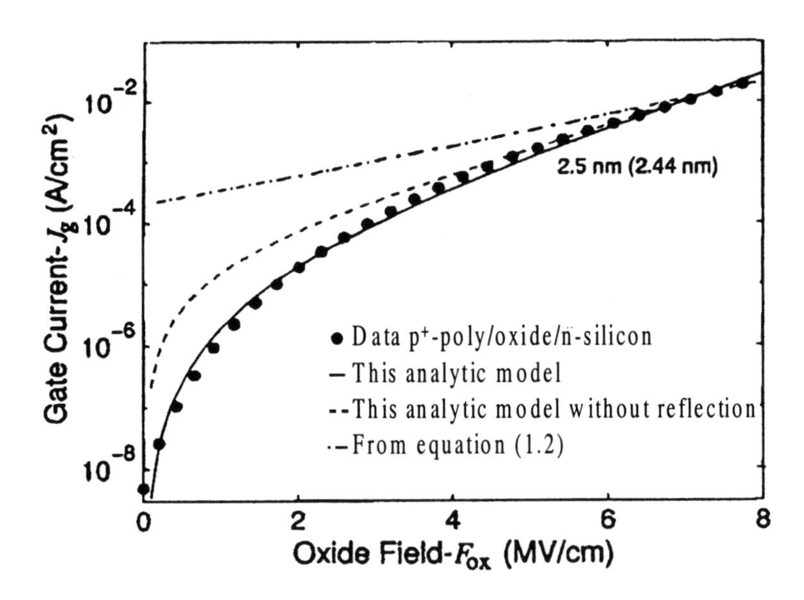

Figure 1.7. A more accurate analytic model fits the experimental data much better at low field than Equation 1.2. Quantum confinement as well as interference is taken into account in this model. Taken from [10].

Figure 1.6 shows the contribution of direct tunneling to tunneling current at low field. When a more accurate I-V characteristic of thin gate-oxide is needed, it is usually done through quantum mechanical simulation taking into account the band structure of the silicon [8, 9]. More recently, Register *et al.* [10] produced a far more accurate (Figure 1.7) analytic expression for tunneling current. Careful use of simplifying approximations without leaving out the quantum effects was the key to their success. However, the expression is still too complex to explain here, and the reader should consult the original reference.

Direct tunneling enhancement of the low field current is evident below 4.5 V of applied voltage for the thinner oxides in Figure 1.4. Note that although the barrier height is only 3.1eV, FN tunneling does not occur until the applied voltage reaches around 4.5 V because of the Fermi level correction plus the quantum effect correction for the thin oxide samples in Figure 1.4.

1.3 Thin Oxide Breakdown Measurements

When the thin gate-oxide is stressed under high field, defects are generated in the bulk of the oxide layer as well as at the SiO_2/Si interfaces. After a long enough period of time, dielectric breakdown occurs. The time duration required to break the oxide depends on a number of factors. Dielectric breakdown is the most common method for assessing gate-oxide quality. It is also the most extensively used method to detect plasma charging damage.

There are many widely employed methods to measure dielectric breakdown. They all yield similar results qualitatively. Subtle differences in the measurement details, however, can produce quite different results quantitatively. While some methods are considered better than others, almost all methods produce very useful information when used consistently. A comprehensive review on these various methods of dielectric reliability measurement has appeared recently [11].

1.3.1 Median Time to Breakdown (T_{BD})

T_{BD} is most often measured with the constant voltage time dependent dielectric breakdown (TDDB) method. The constant current TDDB method can do the job as well and is sometimes used. The device-under-test (DUT) is usually a large capacitor. A constant voltage that produces an oxide field of 5 to 9 MV/cm is commonly used. The wafer is usually maintained at an elevated temperature of 100 to 200°C. The capacitor is stressed until breakdown and the time for breakdown to occur is recorded. A large number of devices are needed because breakdown is a statistical process. The breakdown statistic is often shown in a probability plot (lognormal plot, Figure 1.8). The time for 50% of the devices to fail is the median time to failure.

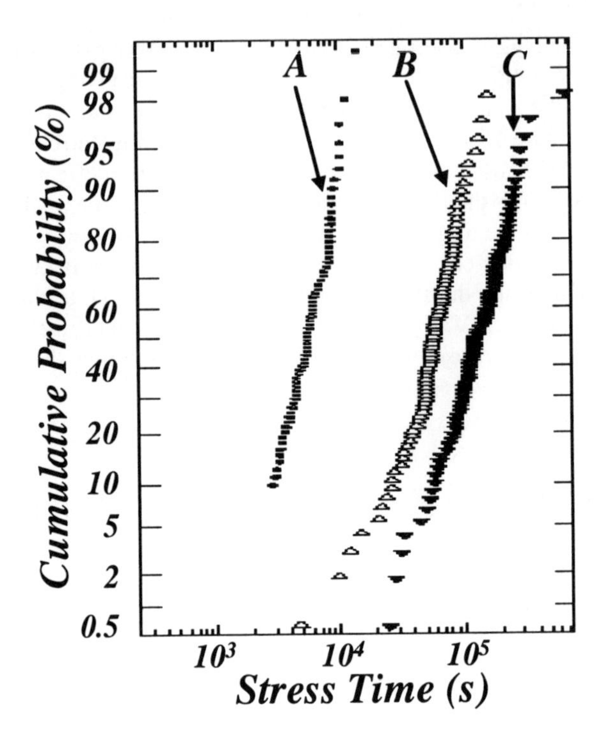

Figure 1.8. The probability plot of time to breakdown for 55Å gate oxides. The capacitor size was $0.2cm^2$.

A: 7.5MV/cm, 210°C

B: 7.0MV/cm, 210°C

C: 7.5MV/cm, 150°C

Notice that changing the field or temperature causes the whole distribution to shift in parallel.

As shown in Figure 1.8, the median time to failure is dependent on oxide field and temperature. It is also dependent on oxide thickness, capacitor size, capacitor design (length of isolation edge, for example), failure criteria (10μA of leakage current through the $0.2cm^2$ capacitor in this case), stress polarity (accumulation in this case) and the doping concentration and type of both the gate and substrate. Most of the time, these details are not provided when T_{BD} is reported in the literature. Without the details, the reported value is meaningless. In plasma charging damage measurements, one compares the measured T_{BD} of antenna testers to control samples tested under identical conditions.

The T_{BD} is a very important quantity. It is usually measured at a few accelerated fields (fields that are higher than the expected operation condition of the integrated circuit) and temperatures. The results are used for lifetime projection for large circuits under operating conditions. To project lifetime at a condition different from the test condition, one needs a model that describes the breakdown over the fields and temperature of interest for the accelerated test. The acceleration model will be discussed later in this chapter. For technology purposes, T_{BD} is not the only quantity of concern. The breakdown time distributions shown in Figure 1.8 often have "tails". In other words, small number of devices shows particularly short lifetime. These "tails" are not reflected in the values of T_{BD}, but they signal a high failure rate for large circuits. Plasma charging damage often increases these "tails" without significantly altering the T_{BD}.

1.3.2 Charge to Breakdown (Q_{BD})

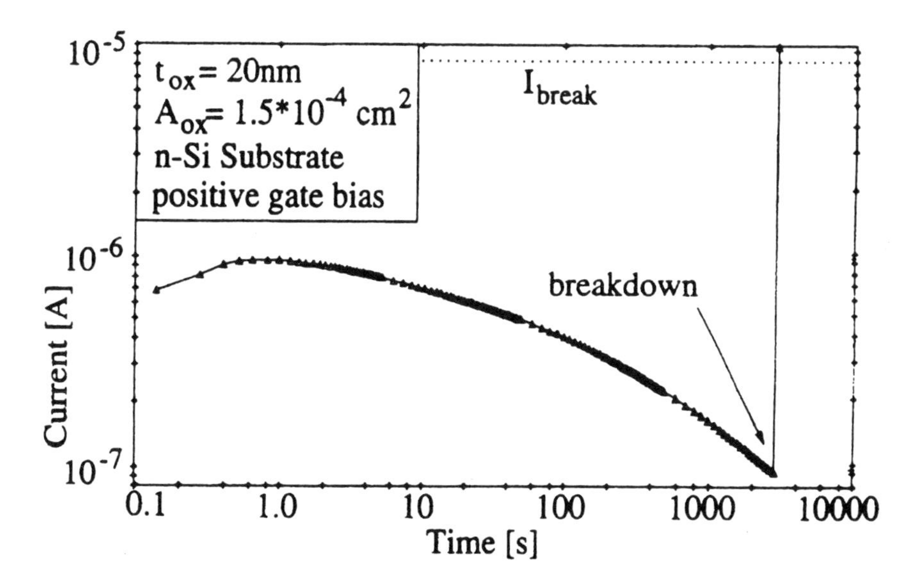

Figure 1.9. The current level is not constant in a constant voltage stress. It increases at first and then turns around to become smaller as stress continue. Taken from [11].

Charge to breakdown (Q_{BD}) is a measure of the total charge that flows across the oxide during high-field stress until breakdown occurs. It is a quantity extracted from the TDDB measurement. During the same constant voltage stress measurement that T_{BD} is obtained, one typically also monitors the current as a function of time. Q_{BD} is simply the integration of current over the stress duration. Like T_{BD}, Q_{BD} normally refers to the 50% point in the cumulative probability versus charge to breakdown plot. Even though it is not explicitly specified, it is understood that Q_{BD} is the median charge to breakdown. Due to the statistical nature of breakdown, measurement of charge to breakdown or time to breakdown on only a few samples does not produce a sufficiently accurate value for Q_{BD}. Note that one cannot simply measure the initial current and multiply it by the time to breakdown. Even though the applied voltage remains constant, the current level changes as a function of time (Figure 1.9). This current change is due to charge trapping in the oxide that modifies the effective oxide field.

Q_{BD} can also be obtained by using constant current stress. In this case, the current multiplied by the time to breakdown gives the Q_{BD} for each sample directly. During constant current stress, the applied voltage required to maintain the current level also changes with time due to charge trapping which modifies the oxide field. The changes in current as a function of time during constant voltage stress and the changes in voltage as a function of time during constant current stress are useful tools for the study of charge trapping in gate-oxide under high field. We will return to this phenomenon later.

1.3.3 Voltage to Breakdown (V_{BD})

Another breakdown measurement often encountered in the plasma damage literature is the voltage to breakdown (V_{BD}). V_{BD} is measured either by the current ramp or voltage ramp methods. In these methods, current or voltage is ramped from low to high values while monitoring the voltage or current to detect breakdown.

In the voltage ramp (V-Ramp) method, the voltage is ramped up linearly until breakdown occurs. In the current ramp method (J-Ramp), current density is ramped up exponentially until breakdown occurs. Using either method, one can get the V_{BD} directly. One can simultaneously get the charge to breakdown. As mentioned earlier, Q_{BD} is highly sensitive to the measurement method. Q_{BD} measured by the ramp method is very different from the constant stress method. In fact, V_{BD} is also very sensitive to the details of the ramp. Subtle differences such as dwell time at each step can have a strong influence on the result. It is for this reason that a standard methodology for V-Ramp or J-Ramp has been defined by the JEDEC to allow comparison of results from one measurement to another. Even with the adoption of JEDEC standard methodology, it is still not advisable to take the comparison of various reports too seriously.

Similar to T_{BD} and Q_{BD}, one should always use the median value of a distribution for V_{BD}. Also similar to T_{BD} and Q_{BD}, in regard to plasma charging damage, we are more often concern about the "tail" of the distribution than the median value. Note that the median value does change, but it takes a very severe damage to change it sufficiently to give a reliable signal.

While all the quantities that measure gate-oxide breakdown are highly sensitive to measurement conditions, they are still very useful indicators for intrinsic gate-oxide quality as well as degradation. For very thin gate-oxides they emerge as more important tools to characterize gate-oxides.

1.3.4 Soft Breakdown

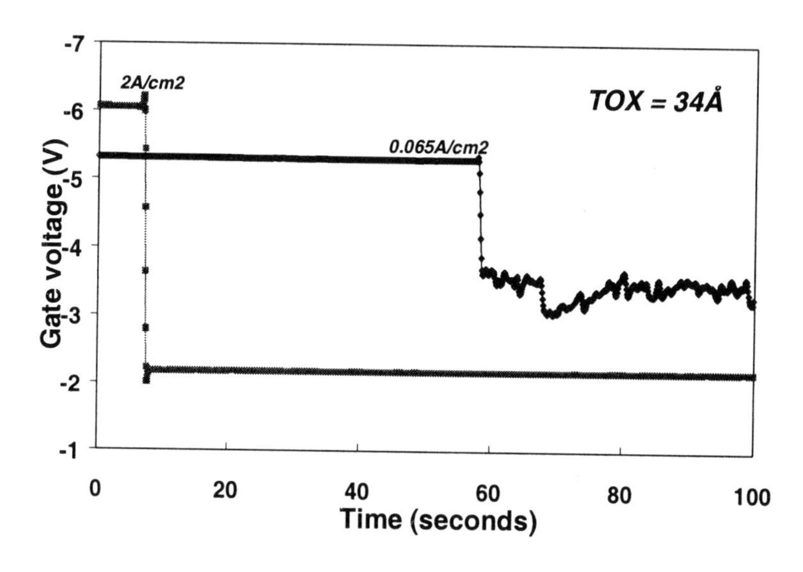

Figure 1.10. The voltage required to maintain a constant current stress drops sharply and cleanly to a low value when breakdown occurs during high current-density stress. The drop is much milder when breakdown is during lower current-density stress. The post breakdown voltage is also much noisier.

Typically, when one stresses an oxide to breakdown, the resistivity takes a sudden sharp drop from a very high value to almost zero. Figure 1.10 shows the voltage required to sustain constant current stress as a function of stress time at two current densities. Breakdown occurs earlier when stress is done at high current density and the post breakdown voltage is much lower and clean. Breakdown occurs at a later time when stress is done at low current density and the post breakdown voltage is higher and much noisier. This low current density stress-induced breakdown is very different from breakdown commonly observed in thicker oxides. The post breakdown voltage, although supporting a lower current density, is significantly higher than the post breakdown voltage of the higher stress current density case indicating the post breakdown resistance is much higher. The noisy behavior is very characteristic to this soft breakdown phenomenon.

Whether soft breakdown leads to device and circuit failure, and therefore, should be treated as normal breakdown is currently under debate. There is evidence that soft breakdown does not lead to device failure [12, 13]. If the devices don't fail, soft breakdown is not a real oxide failure from a technology viewpoint. On the other hand, there is evidence that under some circumstances, soft breakdown does lead to device failure [14]. In an integrated circuit, all devices have to work for the circuit to function, and the fear is that some of the soft breakdowns will lead to device failure and those failures will cause circuit failure. If that were the case, then one must treat soft breakdown as an oxide failure. Since soft breakdown tends to take place most often in thin oxide under constant current stress and that plasma charging damage is typically a constant current stress, we expect it to show a high probability of causing soft breakdown instead of hard breakdown. It is therefore important for the purpose of plasma charging damage study to identify soft breakdown failure.

Soft breakdown becomes softer as oxide thickness decreases. As oxide thickness drops to 25Å or less, soft breakdown becomes so soft that only the noise level of the gate current changes [13]. Most of the conventional breakdown measurement technique legacies from thicker oxide discussed in this chapter have difficulty detecting soft breakdown. Unfortunately, such difficulty has lead to much confusion about how plasma charging damage will impact ultra-thin gate-oxide [15-17]. We shall discuss this problem in more detail in Chapter 7.

1.3.5 Statistical Behavior of Oxide Breakdown

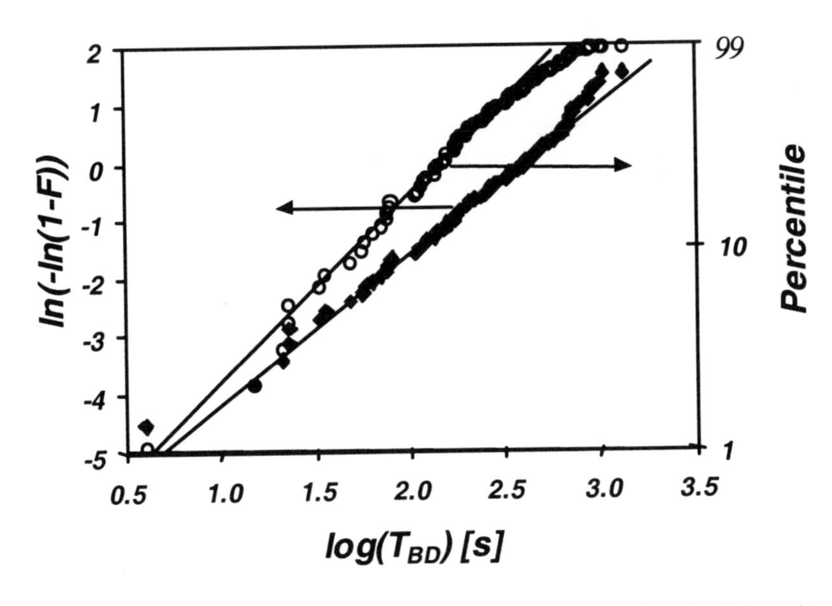

Figure 1.11. The time to failure data is plotted in both log-normal scale (right axis) and Weibull scale (left axis). The Weibull scale gives a better linear fit of the data.

As mentioned earlier, the measured breakdown events, such as time to breakdown, charge to breakdown or voltage to breakdown, are often plotted in a lognormal scale. This usually produces straight lines (Figure 1.8) that are convenient to use for extrapolation. However, given the limited number of tested samples, lognormal is not the only plotting method that can produce a straight line.

The choice of plotting method, and therefore the underlying probability distribution function, has important implications both in the breakdown mechanism and lifetime projection.

In an integrated circuit, there are millions of transistors and each will eventually fail given a long enough operational time. However, since the whole chip fails whenever one of the transistors fails, the median chip lifetime is much shorter than the median device lifetime. Similarly, when stressing a large capacitor, it can be considered as a large number of small capacitors being stressed in parallel. When one of the small capacitor breaks, the whole capacitor is broken. This type of failure from a large number of competing events is called the weakest link problem and is best described by extreme value statistics. Extreme value statistical events should have a Weibull distribution [18-22].

The Weibull distribution function is, by definition:

$$F(t) = 1 - e^{-\left(t/\eta\right)^{\beta}} \tag{1.3}$$

where $F(t)$ is the cumulative failure fraction at time t, η is the time to 63.2% failure (characteristic failure time) and β is the Weibull slope or the shape factor of the Weibull distribution. Typical value for β is in the range of 0.5 to 5. For intrinsic oxide lifetime during stress, β must be larger than 1. β larger than 1 implies that the instantaneous failure rate as a function of stress time is increasing and is fundamentally necessary for a wearout type event. Note that the function $F(t)$ describe the total failure fraction after a stress duration t has been applied to the samples. Obviously, shorter stress-time results in smaller failure fraction. As mentioned earlier in this chapter, the commonly reported failure fraction is 0.5 or 50% (median time to breakdown, for example). For those who are familiar with Weibull statistics, the preferred failure fraction to report is 0.632 or 63.2%, corresponding to the value of η. On the other hand, the industrial standard reliability criteria for chips is $F(10\ year)<0.0001$, or less than 0.01% failure in 10 years of operation. Equation 1.3 thus allow one to link the experimentally measure median time to failure to the time to very low failure fraction required by reliability standard.

From the $F(t)$ definition, we get:

$$ln[-ln(1 - F(t))] = \beta ln(t) - \beta ln(\eta) \tag{1.4}$$

In other words, if we plot $ln[-ln(1-F)]$ versus $ln(t)$, we should get a straight line with slope β. Figure 1.11 is an example of a time to breakdown Weibull plot. The same data are also plotted in lognormal scale in the same figure. In this case, the Weibull plot is clearly a better fit to a straight line than the lognormal plot. More importantly, the two plots have very different slopes and therefore will have very different lifetime projection when extrapolated to low-percentile range (or low failure fraction range). The projection based on Weibull plot is far more pessimistic than the projection based on lognormal plot. The distinction between Weibull plot and lognormal plot become much clearer when the number of devices is larger. Figure 1.12 (a) and (b) [23] show much more clearly that Weibull scale is the right scale for oxide breakdown plot.

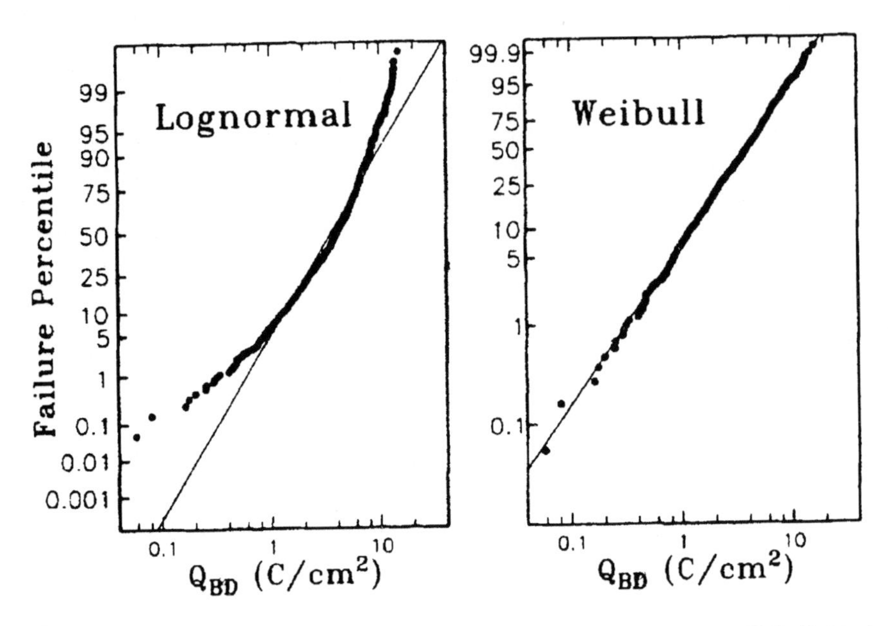

Figure 1.12. When the number of devices is larger, the difference between Weibull (b) plot and lognormal (a) plot is very clear. Taken from [23].

From Equation 1.4, a scaling relationship between stress time and failure fraction can be worked out:

$$\frac{\ln(1-F_1)}{\ln(1-F_2)} = \left(\frac{t_1}{t_2}\right)^{\beta} \tag{1.5}$$

If $F \ll 1$, Equation 1.5 becomes:

$$\frac{F_1}{F_2} \cong \left(\frac{t_1}{t_2}\right)^{\beta} \tag{1.6}$$

From the weakest link model, stressing a large capacitor is known to be equivalent to the stressing of a large number of small capacitors. If the large capacitors were 100 times larger than the small equal sized capacitors, the stressing of 100 large capacitors can be viewed as the equivalent of stressing 10,000 small capacitors. Since when one of the small capacitors that make up the large capacitor breakdown, the large capacitor is broken, 10 out of 100 large capacitor failure is clearly equivalent to 10 out 10,000 small capacitor failure. The stress time required to achieve both is the same. This is in contradiction with Equation 1.5 and 1.6. Obviously, Equation 1.5 and 1.6 do not apply when the failure fractions are from different sample populations. Another way to look at this is that implicit in Equation 1.5 and 1.6, the failure fractions are from the same distribution or from distributions that have the same β and η. Since η of the large capacitors is different from η of the small capacitors, their failure fractions cannot be linked simply by Equations 1.5 or 1.6.

The fact that smaller capacitors have much lower failure fraction after the same amount of stress time means that they are far more difficult to break. Usually, the question to ask is, for the same failure fraction, how much shorter will the stress time be when the area increases. The answer can be found in Equation 1.7, the derivation of which can be found in a recent paper by Nigam *et al.* [24] and will not be reproduced here.

$$\frac{\eta_1}{\eta_2} = \left(\frac{A_2}{A_1}\right)^{1/\beta} \tag{1.7}$$

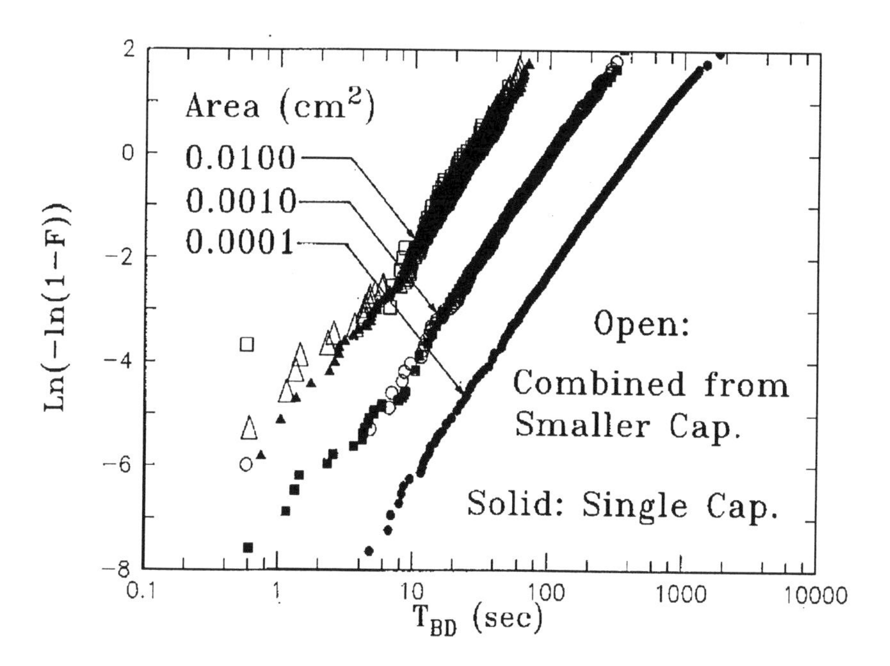

Figure 1.13. Large area capacitors' time to failure plot can be reproduced from random grouping of smaller area capacitors, demonstrating the validity of extreme value statistics applying to oxide breakdown. Taken from [25].

Equation 1.7 clearly shows that η for different area capacitors are different. Note that η is the time to 63.2% failure. If median time to failure T_{BD} is used instead of η, the equation is the same. In fact, one can choose a failure fraction and write Equation 1.7 as the time ratio to reach the same failure fraction. In a recent report, Wu *et al.* [25] demonstrated the weakest link model for ultra-thin gate-oxides. Figure 1.13 [25] shows three T_{BD} distributions (solid symbols) with three different areas and two other constructed T_{BD} distributions (open symbols). The open symbol distributions are constructed by randomly dividing the total population of a given capacitor size into equal size subgroups and by selecting the minimum T_{BD} from each subgroups. For example, from the 4000 $10^{-4}cm^2$ capacitors, 40 subgroups of 100 capacitors are formed. The total area of each subgroup is $10^{-2}cm^2$. The distribution of the minimum T_{BD} from each subgroup would be the same as the T_{BD} distribution from a group of $10^{-2}cm^2$ capacitors if the weakest link model were the correct model for oxide breakdown. As can be seen from Figure 1.13, this is indeed the case.

The firm experimental and theoretical foundation of Weibull statistics describing oxide breakdown provides a powerful scaling tool to link measurements with yield and reliability projections of finished products. It is one of the most important properties of gate-oxide breakdown behavior. When the sensitivity issues of damage measurement is discussed later in chapter 7, the scaling property of Weibull statistics will be relied upon heavily. In a nutshell, the question one must ask is, when the small area, limited sample size antenna device does not show any sign of damage, can one conclude that the large product chips will meet the 0.01% reliability failure fraction specification?

1.4 Gate-oxide Breakdown Models

Gate-oxide breakdowns, in general, are classified into two modes. One is intrinsic breakdown and one is extrinsic. Extrinsic breakdowns are due to defects introduced or incorporated during the growth of the thin gate-oxide. These breakdowns tend to happen much earlier than intrinsic types. Figure 1.14 is an example of a time to breakdown distribution that has significant number of extrinsic failure.

Extrinsic failures, while important, are not the focus of plasma charging damage. The only point to keep in mind is that plasma charging damage tend to turn the marginal gate-oxides with extrinsic defects into a dead-on-arrival extrinsic failure. In that sense, plasma charging damage serves to highlight hidden problems in the processes that affect gate-oxide quality.

It is generally recognized that the intrinsic gate-oxide breakdown mechanism in thick oxide is different from thin oxide. However, there is no reason to believe that nature would make a clean transition from one mechanism to another at some specific oxide thickness. The most reasonable way to think about the transition is that the relative importance of each mechanism is gradually shifting as the oxide thickness changes. While one usually discuss only the dominant mechanism for a given oxide thickness, one must also keep in mind that the other mechanism may be the reason why the model does not fit the data as well as it could be.

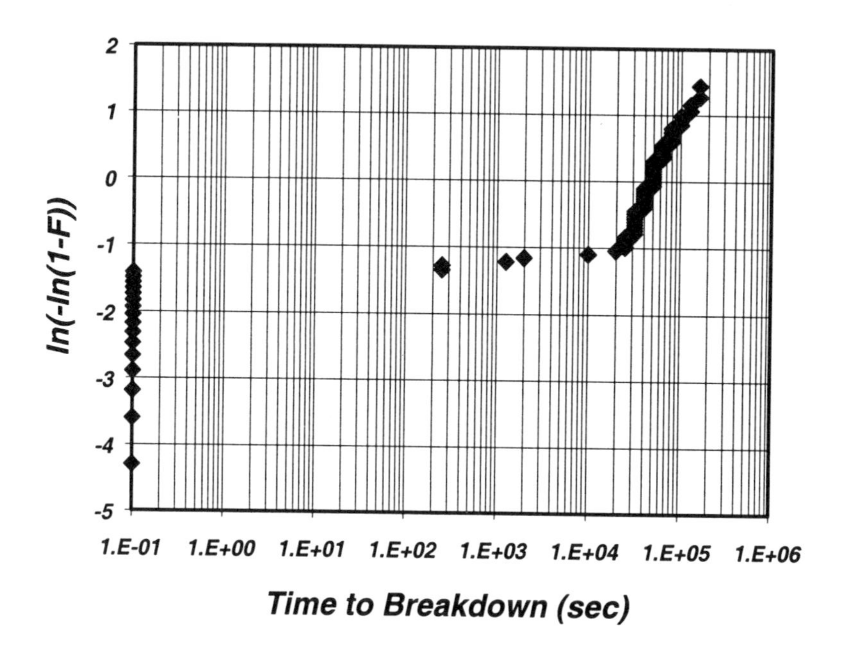

Figure 1.14. The time to breakdown distribution that shows bimodal behavior. The early failures are extrinsic failures.

Sometimes plasma charging damage is detected in test devices that have thick gate-oxide but not in test devices with thin gate-oxide, or vice versa. These seemingly confusing results can be understood by the fact that different breakdown mechanisms dominate at different oxide thickness. Additional insight into the damaging process can be gained from these types of test results.

1.4.1 Oxide-field Runaway Model

When the oxide is thick, impact ionization in the bulk of the oxide is thought to be the dominant mechanism [26-28] for charge creation and trapping. Under a high applied-field, electrons tunnel across the triangular barrier into the SiO_2 conduction band (Figure 1.15). After they reach the SiO_2 conduction band, electrons can gain enough energy from the electric field to cause excitation of another electron across the SiO_2 band gap by collision. This impact ionization process creates electron-hole pairs in the SiO_2. In the presence of the applied field, the created electrons are quickly swept away toward the anode while the created holes will drift toward the cathode. Due to their low mobility, holes have a high probability to be trapped in the oxide.

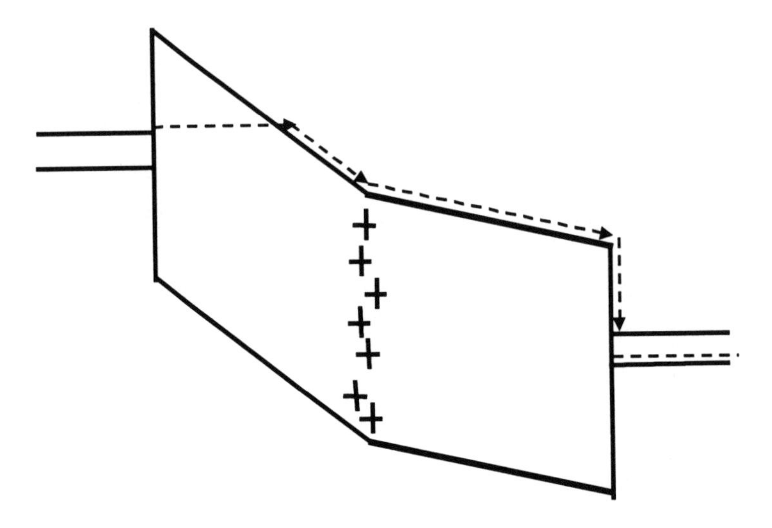

Figure 1.15. Band diagram showing the reduction of the oxide field near the anode and the enhancement of the oxide field near the cathode due to hole trapping.

By definition, the tunneling process does not gain or lose energy. The triangular barrier region is therefore a "dead zone" in which acceleration of the electron is not possible. After electrons reach the SiO_2 conduction band, they gain energy from the electric field resulting in an increase in electron mean energy. However, the mean-free-path for electrons in the oxide is very small. Inelastic scattering with optical phonons limits the energy gain between collision. In this sense, the electrons are being "heated" up gradually as they propagate toward the anode. The electron temperature and thus the population of the high-energy tail will increase with distance from the injecting electrode. The electron energy distribution is important since only those electrons with energy above 9eV (the SiO_2 band gap) can produce electron-holes pairs in the oxide via impact ionization.

Since electron temperature increases with distance from the cathode, impact ionization rate increases with distance as well. As impact ionization increases, more and more holes are trapped in the oxide. These positive charges modify the electric field in the oxide such that the field strength near the anode is reduced (Figure 1.15). A reduced field would reduce the heating of electrons. As a result, the trapped charge density continues to build only to a certain distance, beyond which the impact ionization rate will decrease. This distance was determined to be around 200Å to 300Å from the injecting electrode [29] for very thick oxides.

Trapping of positive charges near the cathode also serves to increase the field near the cathode. Since the FN tunneling process is very sensitive to the field in this region, the hole-trapping induced stronger field causes tunneling current to increase. The increase in tunneling current in turn causes more impact ionization and therefore more holes to be trapped. This feedback effect eventually leads to a run away situation, which causes breakdown to occur.

For gate-oxide thinner than the distance to reach peak impact ionization efficiency, the total impact ionization rate is decreased. This happens for gate-oxides 200Å or thinner. The reason for impact ionization rate to decrease in thinner oxide is that electrons do not have enough distance to fully "heat" up. An additional effect diminishes the importance of impact ionization as a breakdown mechanism in thin oxide. Specifically, the positive charge distribution no longer peaks in the bulk of the oxide layer. As a result, the run away mechanism due to feedback is also diminishing. At around 100Å, impact ionization efficiency is too low to be the dominant breakdown mechanism.

1.4.2 The Percolation Model

As the IC industry has advanced, researchers have studied in detail thinner and better quality gate-oxides. Out of these studies, another breakdown model has begun to emerge [30-32]. The central idea of this model is that, neutral traps (defects that would trap charges) are created at random location in the oxide during stress. When the neutral trap density reaches a critical value, breakdown occurs. Unlike the impact ionization breakdown model where the feedback mechanism induces a runaway situation so that a breakdown path can be "burnt" through the oxide layer by thermal energy, the exact mechanism of breakdown in this critical trap density model has not yet been made clear.

On the other hand, there is ample evidence for the existence and build-up of neutral traps. The most important observation is the growth of trap-assisted tunneling current. For thin oxides, the leakage current across the oxide increases far beyond normal tunneling levels after high-field stress. This additional stress-induced leakage current (SILC) component has been observed by many groups [33-35]. The commonly accepted explanation for SILC is trap-assisted tunneling due to neutral traps in the oxide. As stress time increases, SILC continues to grow (Figure 1.16), indicating neutral trap density also increases.

The critical neutral-trap-density breakdown model has been a tentative model for a while. One difficulty of the model was that it lacked any clear-cut experimental evidence that a critical neutral trap density has been reached at breakdown. It addition, it couldn't reconcile with Chen *et al.*'s [36] observation that the critical hole-fluence to breakdown is independent of stress field (Figure 1.17). In their experiment, Chen *et al.* measured the electron-current that flows across the gate-oxide and the hole-current that flows back in the opposite direction separately using the carrier separation technique during stress. They found that while the total electron fluence (Q_{BD}) is a strong function of stress field, the total hole-fluence (Q_P) remains quite constant across a wide range of stress field.

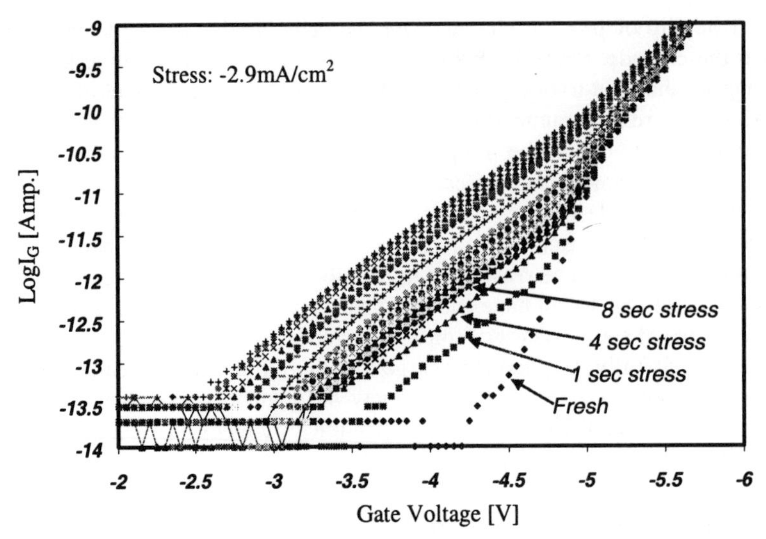

Figure 1.16. The leakage current at low oxide field increases as a function of stress time.

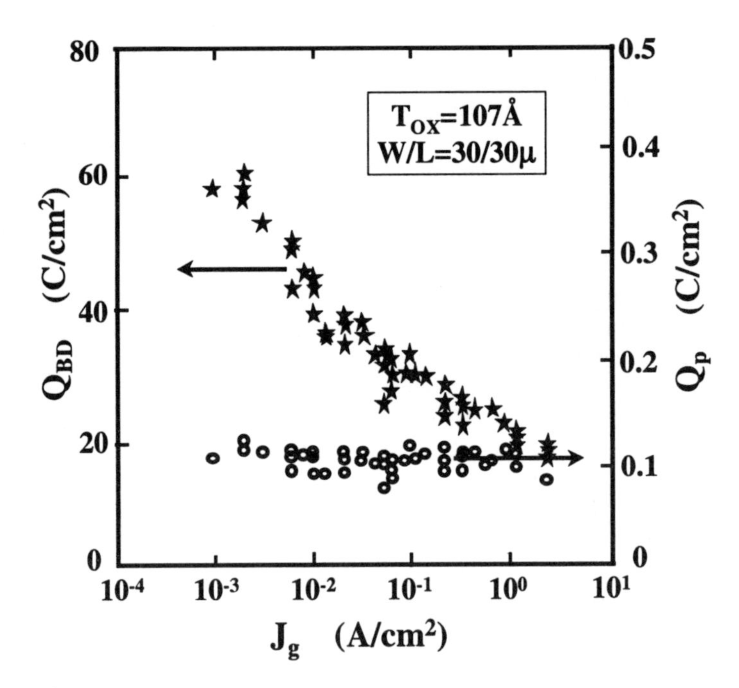

Figure 1.17. While the Q_{BD} varies as a function of stress current density in a constant current stress, the Q_p remains constant. Q_p is the integrated substrate current, which is interpreted as the hole current flowing through the gate-oxide from the poly-silicon gate. Taken from [36].

Both difficulties were resolved by Degraeve *et al.* [37], who found that the neutral trap density is related to Q_P (Figure 1.18). Note that Figure 1.18 plots the total trapped charges rather than neutral trap density. These total trapped charges were measured using substrate hot electron injection under a fixed oxide field to fill the neutral traps with electrons. While not all the neutral traps are filled, it is expected that the filling factor is constant if the field is constant [30]. The measured total trapped charge is thus directly proportional to the neutral trap density. The relationship shown in Figure 1.18 says that the critical hole-fluence for breakdown simply translates to a critical neutral trap density for breakdown. Chen *et al.*'s observation turns out to be the missing clear-cut evidence for the critical neutral-trap-density model for which everyone was looking.

Why should the oxide breakdown when the neutral trap density reaches a critical value? What happens when the trap density reaches the critical value? Note that when we say that the oxide would breakdown at a critical trap density, we are referring to a specific probability of breakdown. In other words, the trap density created by stressing the oxide for a duration equals the median time to breakdown (T_{BD}) would be the critical trap density for a 50% breakdown probability. One way to think about the critical trap density to breakdown model is to think of breakdown

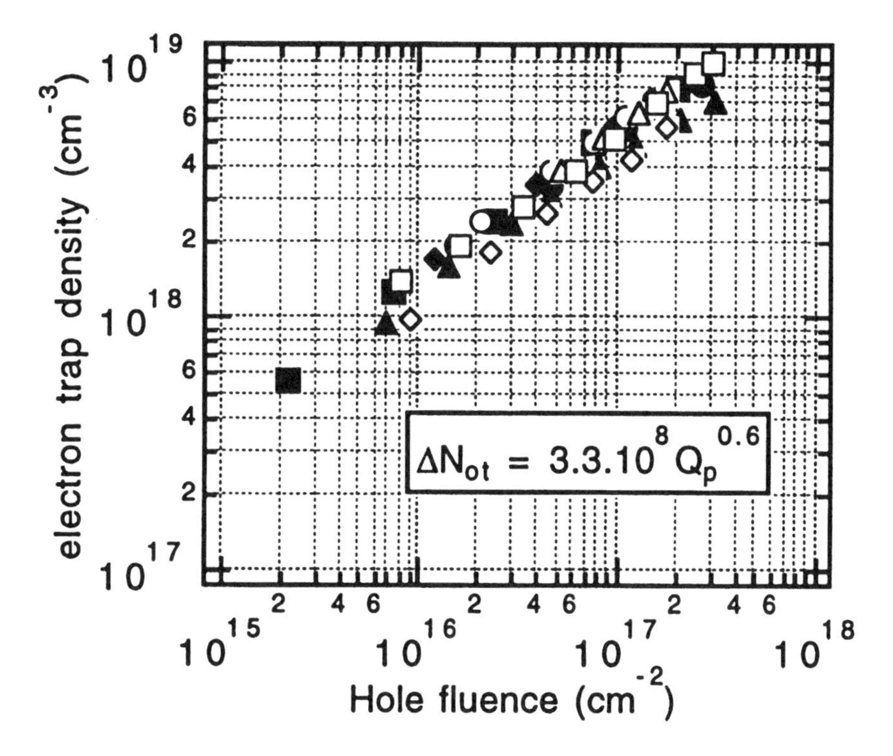

Figure 1.18. The total trapped electron density measured by filling the neutral traps with substrate hot electron injection at fixed oxide field shows that the neutral trap density is related to the total hole fluence during stress. Taken from [37].

as the formation of a continuous path in which the traps are so close to each other that electrons jumping from one trap to the next is extremely efficient. Then the theoretical problem is to find the minimum number of randomly created traps for a conduction path to form from one electrode to another. Such a problem is, by definition, a percolation problem. Suñé *et al.* [32] was the first to try a percolation modeling approach, but the most successful percolation model was by Degraeve *et al.* [37, 38]. In this model, traps are treated as uniform spheres with an effective radius. Breakdown occurs when a continuous link (the spheres touching each other) of the traps from the cathode to the anode (Figure 1.19).

This "sphere based" percolation model is remarkably successful. Using this model, Degraeve *et al.* [38] came up with an effective trap radius of 4.5Å. Not only can it fit the Weibull distribution nicely for a large range of oxide breakdown data (Figure 1.20a); it can also predict quantitatively how the Weibull slope changes with oxide thickness (Figure 1.20b). The success of the "sphere based" percolation model is probably one of the most important advances in gate-oxide reliability research in recent years. It enjoys widespread acceptance as being the best model for gate-oxide breakdown. Note that while the "sphere based" percolation model derives its strongest support from the experimentally observed Weibull distributions of oxide breakdown, it also provides a good conceptual foundation for the Weibull statistical behavior of oxide breakdown. Many of the scaling behaviors, the failure fraction and area scaling discussed above and the voltage and temperature scaling discussed later in this chapter can be understood from the standpoint of the percolation model. The percolation model of Degraeve *et al.* is not perfect, but good enough for our purpose. We shall not dwell on the various "improvements" that may be applied to the model here.

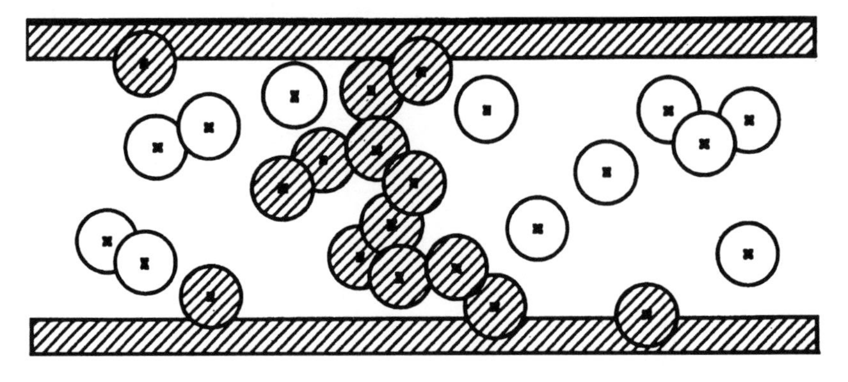

Figure 1.19. The sphere model for intrinsic breakdown. Shaded spheres are in electrical contact with the electrodes. Shaded path connecting the electrodes is the breakdown path. Taken from [37].

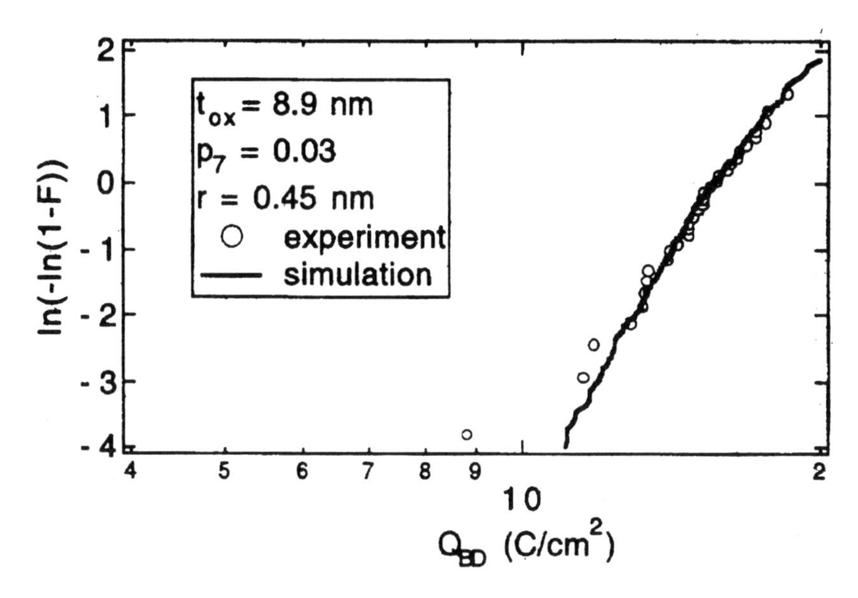

Figure 1.20a. The computer simulated Q_{BD} distribution fits the experimental data very well when the effective trap radius is 0.45nm. Taken from [37].

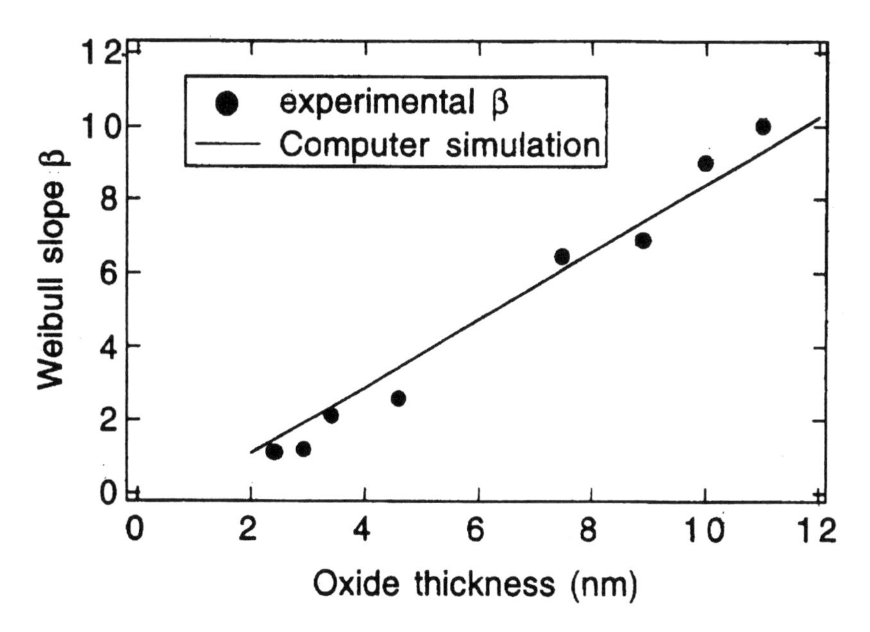

Figure 1.20b. Using the same 0.45nm effective trap radius, the simulated Q_{BD} distributions for a variety of oxide thickness have Weibull slopes agree with experiments. Taken from [37].

According to the model, once the effective trap radius is known, the Weibull slope for different oxide thickness can be obtained by simulation. While Degraeve *et al.* were able to reproduce the Weibull slope for all their oxide breakdown distributions, others [39] have found Weibull slopes that are not in agreement with their results. Since the Weibull slope is of critical importance for scaling, it is unfortunate that the theoretical values cannot be relied upon.

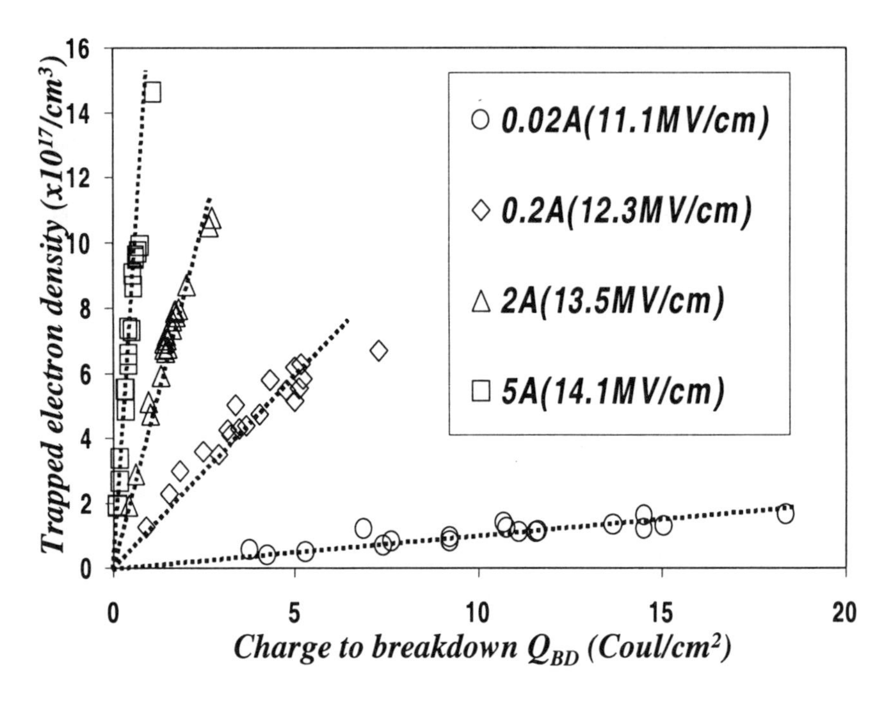

Figure 1.21. The total trapped electron densities at breakdown were found to increase with the stress field. The trapped holes have been subtracted from the result.

In the percolation model, the nature of the trap as well as the effective trap radius are assumed to be field independent. As a result, the critical trap density for breakdown is expected to be field independent as well. A recent study [40] measured the total amount of trapped electrons at breakdown for a large number of oxide samples. The median total trapped electron density at breakdown was found to increases significantly with stress field (Figure 1.21), suggesting that the critical trap density to breakdown increase with stress field.

To explain this observation, a modification of the definition of breakdown in the "sphere based" percolation model was proposed [40]. Instead of defining breakdown as when the connecting path is established, the new definition regard the establishment of the connection path as the onset point of a feedback process that leads to breakdown. In the modified model, at the moment the last link for the connection path is created, the phonon assisted hopping conduction along the connection path increases suddenly and significantly. The larger current flux causes more holes to be created at the anode. As a result, more hot holes will be back injected into the oxide right at the point where the connection path is formed. More traps are created along that path and more electron current will flow. This feedback mechanism tends to enlarge the conduction path so that the magnitude of the conductivity jump at the moment of breakdown is much larger.

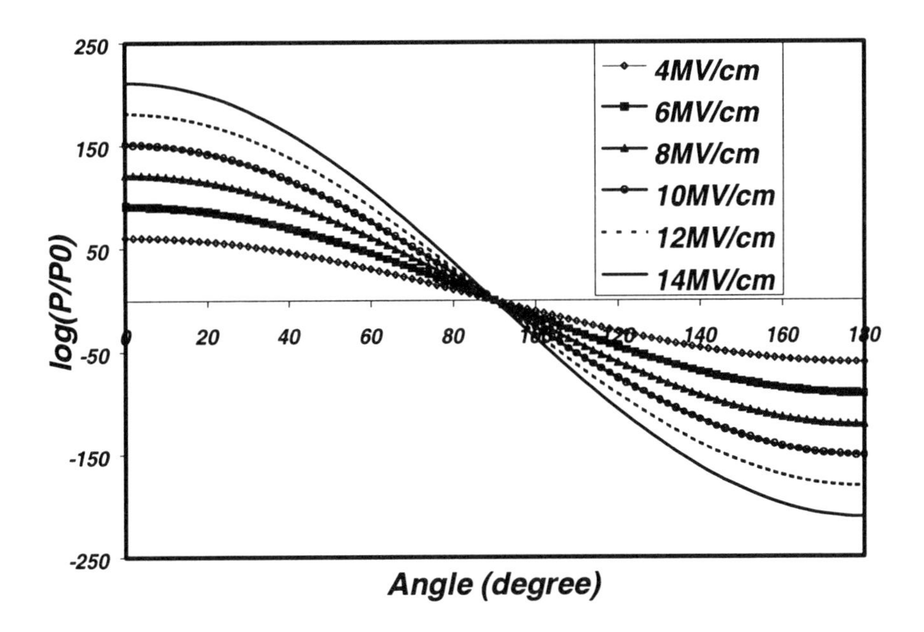

Figure 1.22. The hopping probability, P, normalized to the field-free hopping probability, P_0, is plotted as a function of the angle between the hopping direction and the direction of the applied field.

The key point in the modified model is the sudden increase in hopping current when the connection is made. Under the influence of an applied field, the hopping probability is strongly dependent on the hopping direction. Figure 1.22 shows the hopping probability P, normalized to field-free hopping probability P_0, as a function of angle between the hopping direction and the direction of the applied field for a range of field strength.

The directional hopping probability of Figure 1.22 suggests that only hopping along the forward direction is allowed. Sideways or backward hops are strongly suppressed (by many orders of magnitude). Since conduction is limited by the weakest link, when the last link on the connection path is created, hopping current may not jump. This situation is illustrated in Figure 1.23.

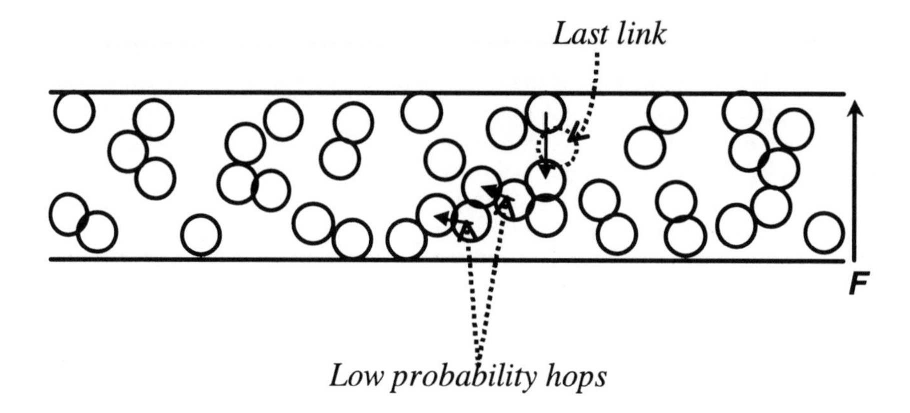

Figure 1.23. The sideways hops have such a low probability that they are the weakest link in the connection path as shown. Such a connection path does not produce breakdown that causes the conductivity to have a sudden jump.

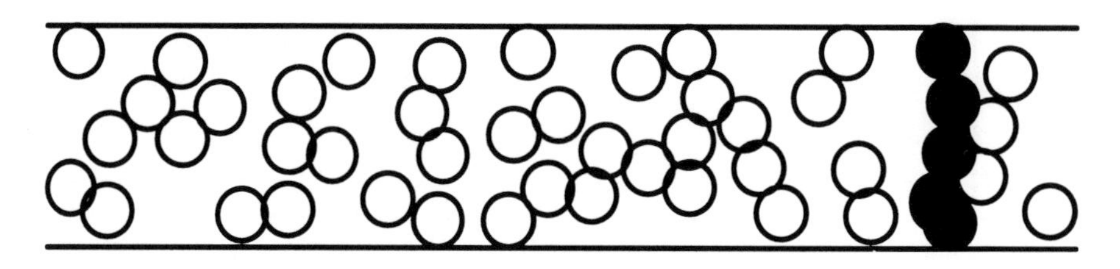

Figure 1.24. Under the influence of an applied field, only the connection path that is in line with the field will produce breakdown.

When the last link in Figure 1.23 is created, a connection path is formed but no breakdown can occur without the current jump that drives the feedback process. Only forward hops within a small angle from the field direction can be linked together and produce a breakdown (Figure 1.24).

Obviously, connections that are in line with the applied field are a subset of all possible connections. The critical trap density for breakdown for the modified model must be higher than the original "sphere based" model. As the applied field goes up, the allowed angle of deviation for hopping from the applied field direction goes down. This further reduces the probability of forming a breakdown path at a given trap density. As a result, the critical trap density for breakdown must go up with the applied field.

At moderate to low field, this field dependent critical trap density modification of the model is negligible. The field dependent effect can be ignored for the purpose of plasma charging damage discussion. It is included here only because it has important consequence on a popular damage measurement method used in ultra-thin gate-oxide. The issue at hand is the size of the conductivity jump at breakdown. With the modified breakdown model, it is clear that there are cases where the connection path is made but no sudden jump in conductivity. In fact, one should expect the post breakdown conductivity ranges continuously from no change in conductivity at all to a large sudden jump in conductivity. Such a lack of clear-cut signature for breakdown makes the precise determination of failure fraction very difficult, if not impossible. We will revisit this problem in Chapter 6.

An important result of the percolation model is that it explains why a thinner oxide has a shorter lifetime naturally. When the oxide is thinner, the number of traps needed to link together to form a breakdown path is fewer and the probability of forming such a path for a given trap density is higher. In other words, for a given probability to breakdown, the critical trap density is smaller. Under the same trap generation rate, the time to breakdown is shorter. Figure 1.25 shows [24] the critical trap density to breakdown as a function of oxide thickness. As oxide gets thinner, the critical trap density to breakdown drops faster.

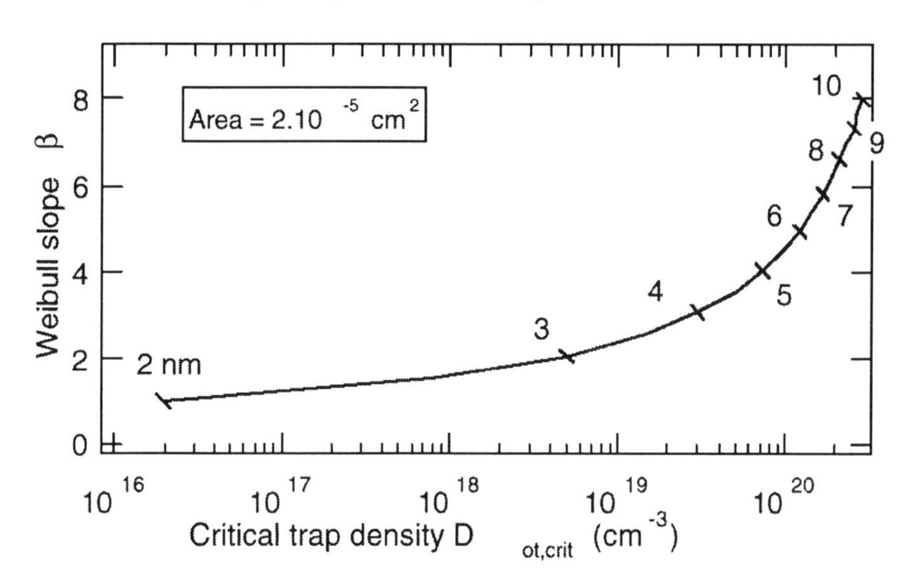

Figure 1.25. The simulated Weibull slopes and the critical trap densities to breakdown as a function of oxide thickness for the given oxide area. Taken from [24].

1.5 Trap Generation Model and Acceleration Factors

The percolation model of breakdown assumes traps are generated at a random location in the oxide during stress, but it does not deal with the trap generation mechanism or trap generation rate. For the trap generation mechanism, there are currently many models. Among these models, three enjoy the widest support and there is a lively debate in the literature on which one is correct. They are the Thermal-Chemical model [41, 42], Anode-Hole Injection model [43, 44] and Hydrogen model [45, 46]. It is likely that each of these models is partially correct and that the real mechanism involves all three and more [47]. The details of each model and why they are right or wrong will not be discussed here. At this moment, the subject is best left to the oxide reliability specialists. We will concentrate on experimentally known facts that will impact plasma charging damage.

According to the percolation model, a well-defined critical trap density is associated with a given probability of breakdown. Obviously, when traps are generated at a higher rate during stress, we will reach the critical trap density faster. Naturally, it is very important for us to know the factors that accelerate trap generation. Two of the most important acceleration factors are stress voltage and stress temperature. Both of these factors are well known to affect the lifetime of gate-oxide and are commonly employed in accelerated testing of oxide reliability (see Section 1.3.1).

1.5.1 Voltage Acceleration Factor

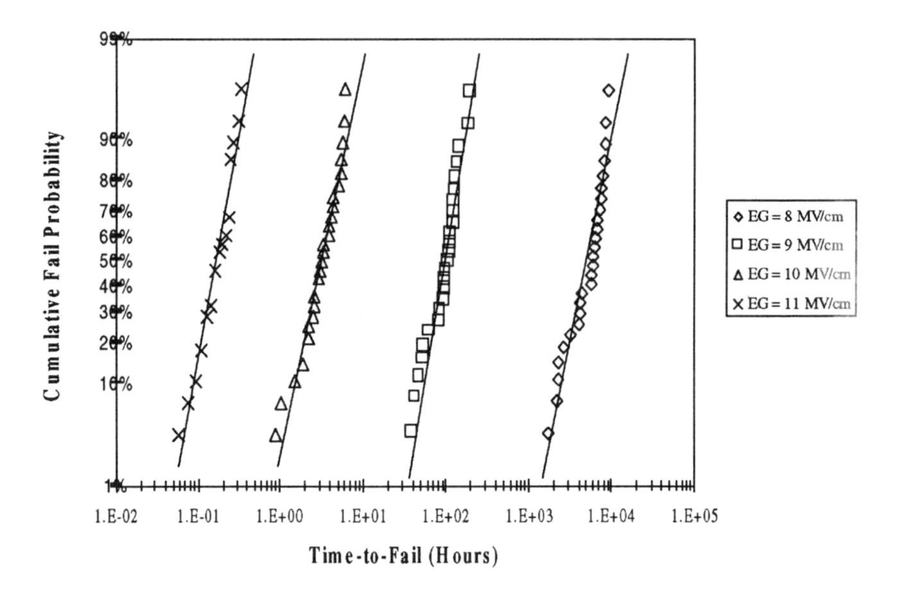

Figure 1.26. The breakdown time distribution of a 90Å gate-oxide at 4 different stress fields. Taken from [42].

As mentioned in Section 1.3.1, in order to characterize the reliability of thin oxide, median lifetimes at a few different oxide-fields are measured. These oxide-fields are chosen so that all the oxide samples in a population will be broken in a reasonable stress time period. This means that these oxide-fields are significantly higher than the operational oxide-field for the circuit built with the same gate-oxide thickness. From each stress-field population, T_{BD} is extracted. A stress-field acceleration factor is obtained from the slope of the T_{DB} versus stress-field plot. Figure 1.26 shows [42] the Weibull distribution of oxide breakdown measured at four different stress-fields. From these distributions, T_{BD} are extracted and plotted in log-scale against stress-field [42] as shown in Figure 1.27. As can be seen, for the stress field range plotted, $\log(T_{BD})$ is a linear function of stress-field. From the slope of the least-square fitted line, the stress-field acceleration factor is obtained, which is -1.50decade per MV/cm (the -3.45cm/MV value marked on the figure is for exponential scale). This acceleration factor means that for every 1 MV/cm increase in stress field, the median time to failure of the oxide samples is shortened by 1.5 decade. In terms of trap generation rate, it is also increased by 1.5 decade.

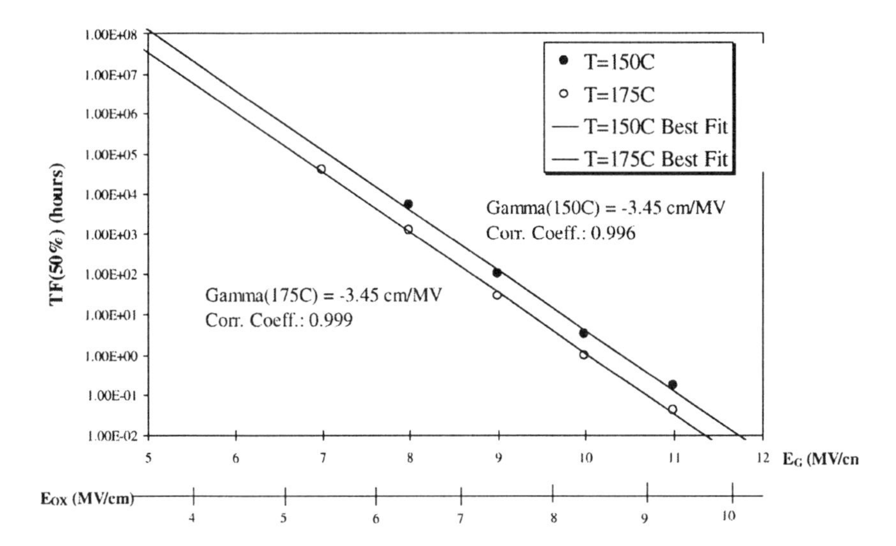

Figure 1.27. The median breakdown time as a function of the stress voltage and stress field at two different temperatures. The slopes of these least square fitted lines are field acceleration factors. Taken from [42].

Assuming that the stress-field acceleration factor remains unchanged as oxide thickness is decreased, the impact of stress voltage on oxide lifetime or trap generation rate will increase. This is simply because each 1 volt change in stress voltage will translate to a larger stress-field change in thinner oxide. Figure 1.28 shows the acceleration factor as a function of physical thickness of the oxide (assuming the electrical thickness is 8Å thicker than physical thickness as discussed in Section 1.2.1). The experimentally measured voltage acceleration factors [25, 42, 48-50] are in reasonable agreement with the calculated value based on the stress-field acceleration factor reported by McPherson *et al.* It should be pointed out that the stress-field acceleration factor is not universal. Its value depends on the oxide quality. There are other considerations that further complicate the comparison of measured acceleration factors to the calculated values. Nevertheless, the trend both theoretically and experimentally is very clear; the stress voltage acceleration factor increases rapidly when the oxide thickness is reduced.

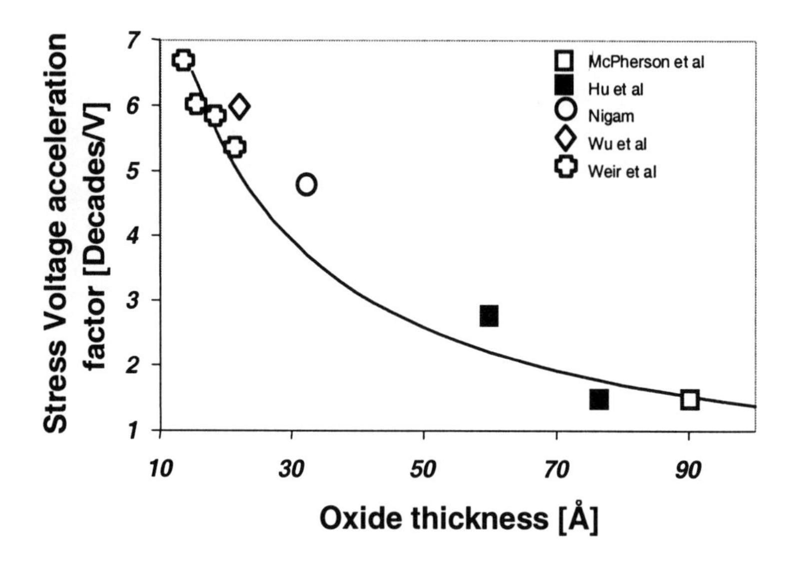

Figure 1.28. The stress voltage acceleration factor as a function of oxide thickness. Solid curve is calculated from 1.5decade per MV/cm.

The impact of this increasing voltage acceleration factor as oxide thins down is that, even though the difference between the voltage during charging stress and the operation voltage is smaller in thinner oxide, the level of traps generated by charging damage is not diminishing. Coupled with the much lower critical trap density to breakdown in thinner oxide, the impact of charging damage to gate-oxide reliability becomes more serious for thinner oxide.

The stress voltage acceleration factor is one of the important parameters routinely measured during technology development stage for the gate-oxide thickness to be used in each technology generation. This parameter is used for gate-oxide reliability projection. Expressing this in a quantitative form, we have:

$$\frac{T_{BD1}}{T_{BD2}} = 10^{\alpha(V_2 - V_1)} \tag{1.8}$$

where α is the voltage acceleration factor in decades per volt for the oxide thickness in question. With Equation 1.8, once we know the T_{BD} for one stress voltage, we can calculate the T_{BD} at another stress voltage.

1.5.2 Temperature Acceleration Factor

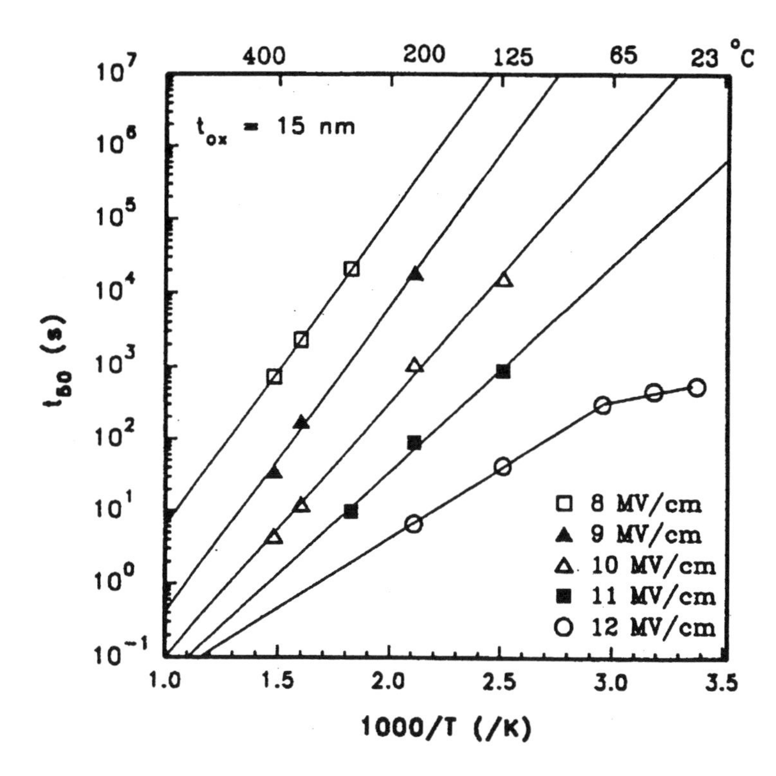

Figure 1.29. The temperature dependent T_{BD} in Arrhenius plot shows the stress-field dependent activation energy. Taken from [51].

As shown in Section 1.3.1, gate-oxide lifetime (T_{BD}) is shorter at higher temperature. The temperature acceleration factor, or activation energy, is usually obtained by measuring T_{BD} at several elevated temperatures and then plotting log(T_{BD}) versus 1/T (Arrhenius plot) to evaluate the activation energy. Figure 1.29 shows a few of these plots [51]. As can be seen, the temperature dependent T_{BD} is well described as an activated process and that the activation energy is a function of the stress-field.

Plasma processing is frequently carried out at elevated temperatures. Plasma charging damage at elevated temperatures is more serious than at room temperature simply because the gate-oxide is easier to break. The stress field dependence of activation energy makes it difficult to predict how bad the temperature effect will be on damage. In a recent report, Nojiri *et al.* [52] clearly demonstrated this problem. Figure 1.30 plots the antenna device breakdown failure rate as a function of 1/T for different antenna ratio and plasma density. For plasma charging damage, both higher antenna ratio as well as higher plasma density would lead to higher stress current and higher stress field. As can be seen, the slopes of the Arrhenius plots are larger for the higher stress field cases.

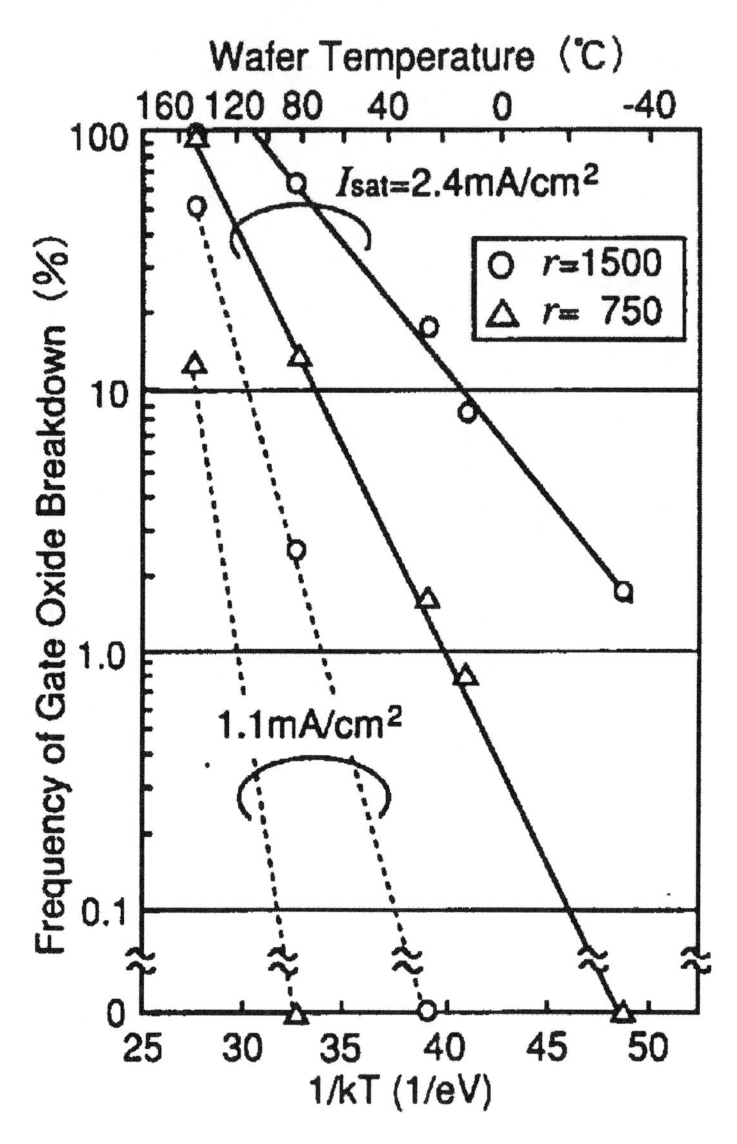

Figure 1.30. The temperature dependent antenna device yield in Arrhenius plot for a 4.5nm thick gate-oxide. The slope is clearly both antenna ratio r and ion current density I_{SAT} (and therefore plasma density) dependent. Taken from [52].

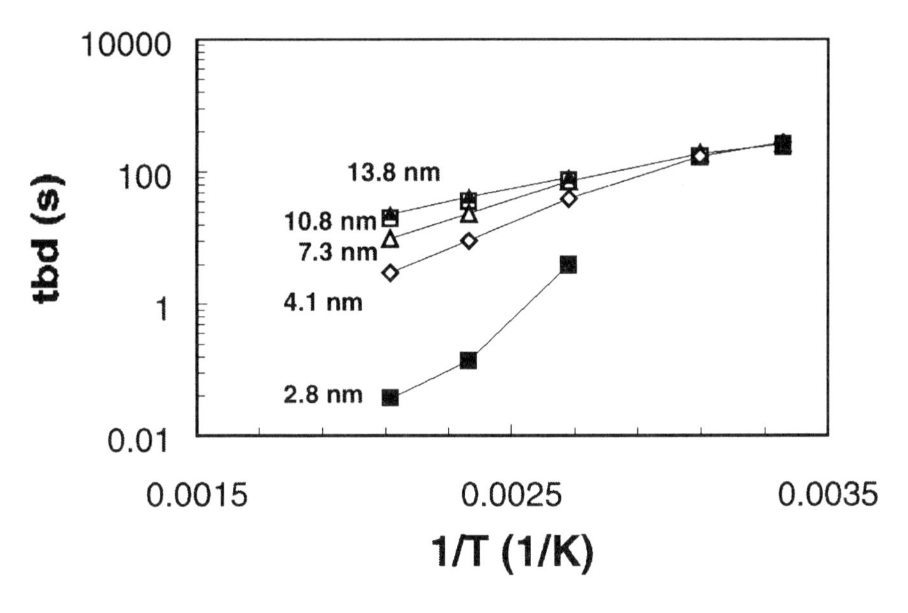

Figure 1.31. T_{BD} versus 1/T does not produce a straight line, an indication of non-Arrhenius behavior. Taken from [54].

The activation energy for temperature acceleration is also one of the important parameters necessary for gate-oxide reliability projection. For plasma charging damage studies, it is often convenient to express the activation energy in degree per decade of lifetime rather than in electron volts. When using the deg/decade as energy unit, the activation energy is more straight forwardly named temperature acceleration factor (γ). To calculated the gate-oxide lifetime at the same stress field but different temperature, one can use the following expression:

$$\frac{T_{BD1}}{T_{BD2}} = 10^{(T_2 - T_1)/\gamma} \tag{1.9}$$

Recently, the temperature dependent T_{BD} for ultra-thin gate-oxide has been reported to be non-Arrhenius (see Figure 1.31) [53], [54]. Rather than plotting T_{BD} versus 1/T, these authors found that plotting T_{BD} versus T is more linear (see Figure 1.32). While this result has important implication for trap generation models, the more important message from the data shown in Figure 1.32 is that the temperature acceleration factor is increasing rapidly as oxide thickness is reduced. This characteristic of thin oxide makes plasma charging damage at elevated temperature particularly troublesome. When combining the effect of low critical trap density to breakdown, very high voltage acceleration factors and very large temperature acceleration, any charging damage will be fatal.

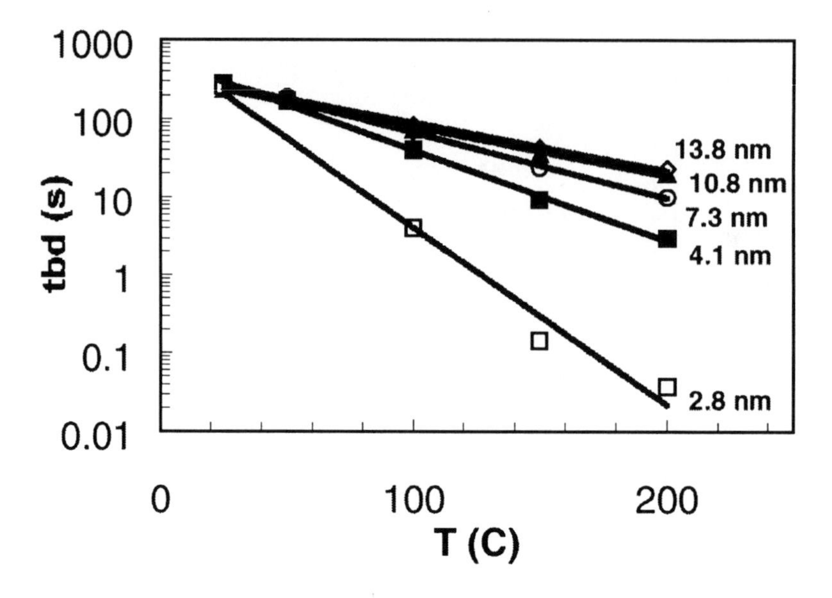

Figure 1.32. T_{BD} versus T is closer to a straight line. Taken from [54].

The non-Arrhenius dependence of T_{BD} on temperature seems to be most prominent in ultra-thin oxides. Thicker oxides are still thought to be better fit by an Arrhenius relationship. Thus, for ultra-thin oxides, a new temperature acceleration factor (γ') should be defined corresponding to the slope of the log(T_{BD}) versus T plot. Equation 1.9 also should be rewritten as:

$$\frac{T_{BD1}}{T_{BD2}} = 10^{\gamma'(T_2 - T_1)} \tag{1.10}$$

1.6 Defects, Traps and Latent Defects

1.6.1 Defects and Traps

The terminology of traps and defects in the oxide network can be quite confusing. In the previous section, the question of trap generation model was not addressed because the subject is still under intense debate. One may wonder why after so many years of intense research the subject is still being debated. The main reason is that no one really knows what constitute traps and defects in the oxide network. All we know is that traps can be detected electrically and that they affect the electrical behavior of the MOS capacitor and therefore the performance of the MOSFET. When certain types of traps are accumulated to a certain concentration, the oxide breaks down. Since few spectroscopic methods exist that have the sensitivity to detect traps or defects in thin oxides, little is known about them. The only exception is Electron-Spin-Resonance (ESR) spectroscopy, which barely has enough sensitivity to measure defect sites that have an unpaired electron. Thus, those types of defects are known in greater detail. The rest are all based on modeling using quantum molecular dynamic calculation. Since theoretical calculations are not yet able to handle the effect of an electric field, all we know are possible defect states in the absence of an electric field. The dynamic of trapping and detrapping is all based on electrical measurements that give no specific information on defect structure. For a review of the defects in the SiO_2 system, a good place to start is Helms *et al.*'s paper [55].

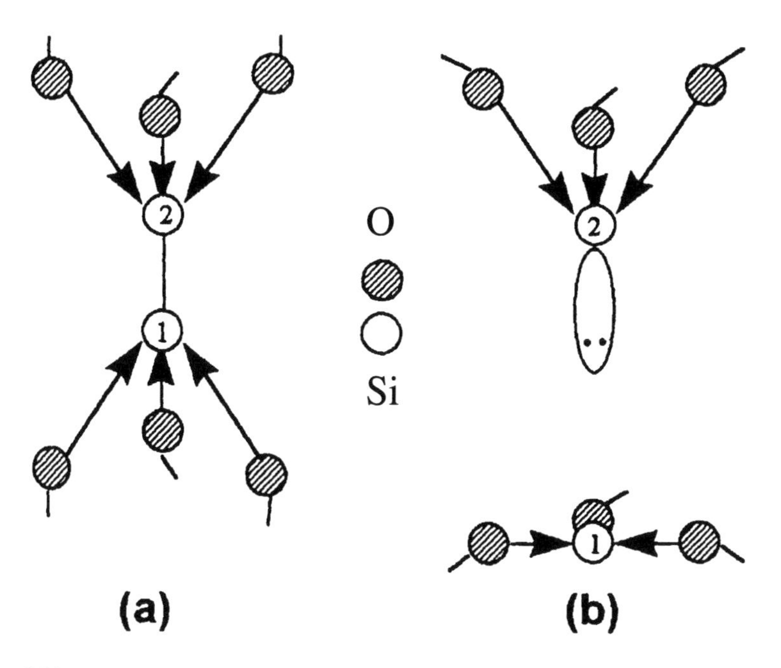

Figure 1.33. An example of breaking the Si-Si bond in an oxygen vacancy defect (a) by electrical stress to form a neutral trap (b). After [56].

Simplistically, traps are defect sites that are ready to capture charges during a collision event (trapping of charge). Defect sites that capture electrons are called electron traps while those that capture positive charges are called hole traps. Some defect sites are amphoteric, which means that it can capture either type of charges. In the discussion of percolation model of oxide breakdown, we speak of the accumulation of neutral traps. Obviously, a neutral trap is a defect site that is ready to trap charge. No one has specifically discuss whether this neutral trap is an electron trap, hole trap or amphoteric. Experimentally, the supporting evidence that a critical neutral trap density must be reached for breakdown to occur is from measuring the trapped electrons under certain applied field [38]. Thus we expect the neutral traps involved at least trap electrons.

Why isn't the neutral traps simply be called defects? In the long history of oxide study, it is well accepted that degradation is related to pre-existing defects sites. Since the pre-existing defects do not cause breakdown or measurable SILC, they must be different from the defects formed after electrical stress. The neutral traps are, in effect, transformed defects. One example of such transformation is turning an oxygen vacancy defect (Si-Si bond in the SiO_2 network) into a defect containing Si dangling bond by breaking the Si-Si bond [56] (Figure 1.33). Although it is still an uncharged defect, it is different from those pre-existing ones. Figure 1.33 represents only one of the possible structure of neutral traps. Since no one is sure at this point what the real structure is, we cannot call the stress created defect by its proper name. For no better way to distinguish them, the uncharged defects created by electrical stress are called neutral traps.

It is assumed that neutral traps exist in virgin (before any electrical stress has been applied), good quality thermal oxide in negligible concentration only [57]. The concentration increases with stress time. By the time 63.2% of the oxide samples are broken, the concentration can be as high as $10^{20}/cm^3$ [38] depending on oxide thickness and capacitor size. Obviously, most pre-existing defects are not at such high concentrations. Thus we can expect the defect conversion process to involve more than one type of pre-existing defect. The conversion rate will be dominated by the easiest to convert type at first. When this type is depleted, the conversion rate will be dominated by the next easiest type of pre-existing defect [58], and so on. This raises an obvious question: should one expect the neutral trap be all the same regardless of its origin? If not, can one expect various types of neutral trap contribute to breakdown equally? There is no clear-cut answer to these questions.

In addition to classifying defects into electron traps and hole traps, they are further classified into bulk traps, near interface traps (border traps, slow states) and interface traps. Theoretically, interface traps are those defects that are associated with the interface between SiO_2 and silicon, near interface traps and bulk traps are away from the interface. Practically, it is all determined by response time.

Interface states are typically measured by using capacitance-voltage (CV) curves. By subtracting the high-frequency CV curve from the quasi-static CV curve, one can measure all the traps that can exchange charge with silicon between the frequency of the two measurements. Typical high-frequency curves are measured at 100kHz and low frequency curves are measured at ~10Hz. Thus this classifies all traps that have a time constant of 0.1s or shorter for charge exchange as interface states.

Another method of measuring interface states is charge pumping (CP) method which measures only those traps that can exchange charges at pulse repetition rate or faster. Since it is typical to use pulse repetition rate of 100kHz or higher in CP method, this classifies only those traps that can exchange charges at 100kHz or faster as interface states.

Border traps typically refer to traps that exchange charge with a time constant from a fraction of a second to a few days. Charge exchange between border traps and the electrode is thought to be by the tunneling process. For example, tunneling of valence electron into the oxide to neutralize a trapped hole [59]. The distance of the trap location from the interface dictates the time constant. According to the tunneling front model [59], [60], the time for a trapped hole to be neutralized by an electron tunneling from the electrode is:

$$t/t_0 = \exp(2\beta x)$$

where t_0 is roughly 10^{-13} s and $\beta = 0.572$ Å^{-1}. Figure 1.34 plots the hole detrapping time as a function of distance from the electrode. Marked on Figure 1.34 is the distance of 26Å reached by the tunneling front in 1 second. In other words, trapped-hole density at 26Å distance will decrease to 1/e times the original value in 1 second.

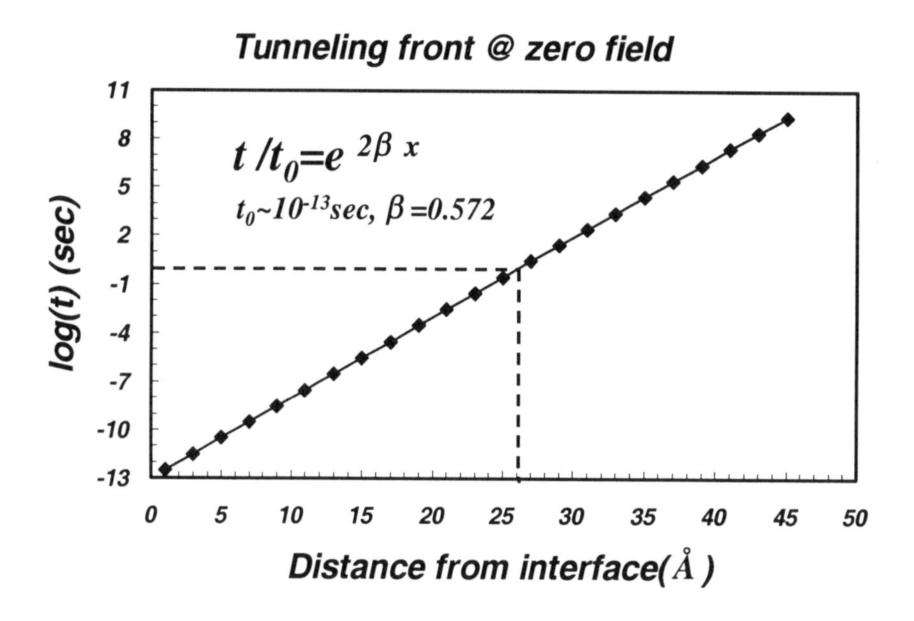

Figure 1.34. The time required for the tunneling front to reach positions in the oxide from the electrode.

Obviously, border traps in this picture include traps that are less than 35Å from the interface. From this picture, it is also clear that assigning all fast-responding traps to interface states is not correct. For example, Figure 1.34 shows that traps at ~12Å away from the interface can exchange charges with silicon at a time constant of 1μs. Thus, the measured interface state density in most reports actually include significant amount of border traps.

1.6.2 Latent Defects

The fact that there are multiple types of neutral traps means that we can expect them to have different anneal response. Some neutral traps are annealed completely while other are only become hidden, and ready to reappear rapidly when the oxide is stressed again. Those that become hidden upon anneal are called latent defects. When a defect become hidden, it may mean that it has been passivated by some chemical species such as hydrogen and rendered non-responsive to measurement. The attached chemical species can easily be detached when the oxide is under stress again. It may also simply lose the charge and become invisible to most measurement methods that rely on detecting the trapped charges.

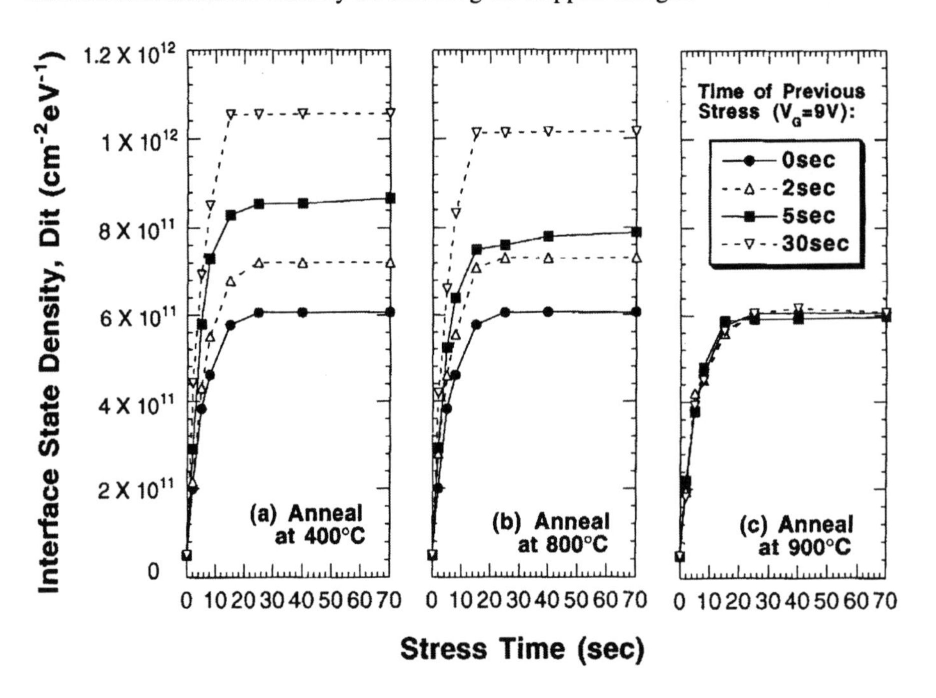

Figure 1.35. The growth of interface state density as a function of stress time shows that pre-anneal stress remains clearly seen after anneal. Only when the anneal temperature is at 900°C does the difference disappear. Taken from[61].

During the course of IC manufacturing, wafers go through many elevated temperature cycles. Some of these cycles are at high enough temperature that traps created by charging damage at an earlier step may be annealed or passivated. Trap annealing requires extended anneal time at very high temperature. King *et al.* [61]

studied the annealing of interface traps generated by electrical stress. Figure 1.35 shows the interface state density as a function of stress. Regardless of pre-anneal stress time, the post anneal interface state density is very low even for the lowest anneal temperature of 400°C. However, as soon as the samples are stressed, the interface state density rises rapidly and saturates at levels depending on the pre-anneal stress. This is a clear indication that anneal did not actually repair all the traps, at least a portion of it is only passivated and readily reappears under stress. As shown in Figure 1.35, the portion of passivated traps remains unchanged even at 800°C. Only the 900°C anneal can fully repair all the traps created by pre-anneal stress.

In plasma charging damage studies, transistors with attached antenna are the most common types of test device. These devices are manufactured the same way as any products. The common practice in manufacturing is to anneal the wafer in forming gas just before it is removed from the production line. As a result, the traps in the gate-oxide of the transistor are all empty. Since the transistor threshold voltage is sensitive to trapped charges in the gate-oxide, no threshold shift can be detected on as-fabricated devices. However, once the transistors are stressed electrically, the traps will be filled again and the damage in the gate-oxide reveal itself. An example of such phenomena has been demonstrated by Li *et al.* [[62] (Figure 1.36).

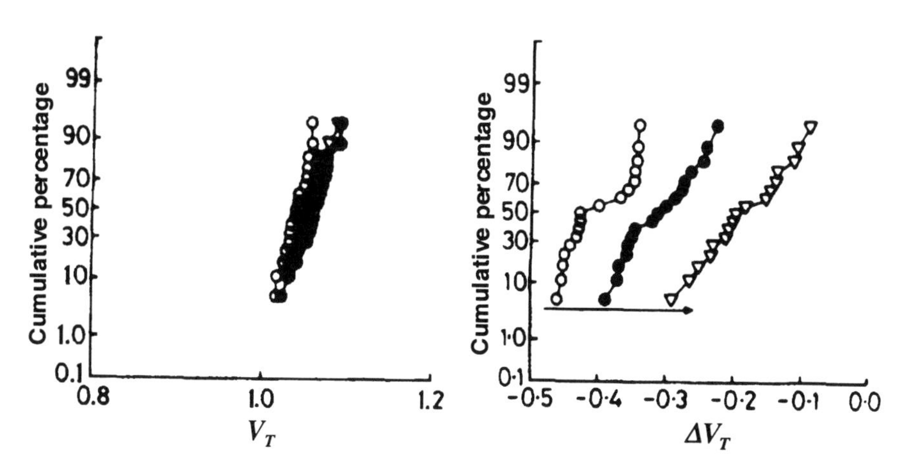

Figure 1.36. The transistor threshold voltages are independent of antenna ratio for wafers directly come out of the production line (left hand side). However, after a negative FN stress, the threshold voltage shifts due to the stress is highly dependent on the antenna ratio, indicating the existence of plasma charging damage (right hand side). Taken from [62].

Some defects can be annealed at moderate temperature. For example, Lu *et al.* [58] reported that annealing at 450°C in nitrogen for 60 minutes could remove the traps that contribute to SILC. They found that this removal is permanent so that re-stressing the oxide does not bring a fast reappearing of these traps. However, annealing at lower temperature only turns these traps into hidden defects that will reappear rapidly upon stress.

Defects that reappear rapidly will cause serious reliability problem to circuits and thus need to be quantified. This latent damage phenomenon is now well recognized in the plasma charging damage field. Most people know enough to use stress to reveal the hidden defects.

References

1. Sanchez, J.J. and T.A. DeMassa, *Modeling gate emissions: A review - Part I.* Microelectronic Eng., 1993. **20**: p. 185.
2. Krisch, K.S., J.D. Bude, and L. Manchanda, *Gate Capacitance Attenuation in MOS Devices with Thin Gate Dielectrics.* IEEE Electron Dev. Lett., 1996. **17**(11): p. 512.
3. Gundlach, K.H., *Zur Berechnung des Tunnelstroms Durch Eine Trapezförmige Potentialstufe.* Solid-State Electronics, 1966. **9**: p. 949.
4. Lewicki, G. and J. Maserjian, *Oscillations in MOS tunneling.* J. Appl. Phys., 1975. **46**(7): p. 3032.
5. Kramer, T.A., K.S. Krisch, and D.P. Monroe, *Characterization of Thin SiO_2 Dielectrics Using Room-Temperature Tunneling Current Oscillation.* Technical Memorandum, Lucent Technology, 1996.
6. Fischetti, M.V., *et al.*, *Ballistic electron transport in thin silicon dioxide films.* Phys. Rev. B, 1987. **35**(9): p. 4404.
7. Schuegraf, K.F., C.C. King, and C. Hu, *Ultra-thin Silicon Dioxide Leakage Current and Scaling.* in *Symp. VLSI Technology*. 1992. p. 18.
8. Bowen, C., *et al.*, *Physical Oxide Thickness Extraction and Verification using Quantum Mechanical Simulation.* in *IEDM*. 1997. p. 869.
9. Wu, E., *et al.*, *Determination of Ultra-Thin Oxide Voltages and Thickness and The Impact on Reliability Projection.* in *IRPS*. 1997. p. 184.
10. Register, L.F., E. Rosenbaum, and K. Yang, *Analytic model for direct tunneling current in polycrystalline silicon-gate metal-oxide-semiconductor devices.* Appl. Phys. Lett., 1999. **74**(3): p. 457.
11. Martin, A., P. O'Sullivan, and A. Mathewson, *Dielectric Reliability Measurement Methods: A Review.* Microelectron. Reliab., 1998. **38**(1): p. 37.
12. Okandan, M., *et al.*, *Soft-Breakdown Damage in MOSFET's Due to High-Density Plasma Etching Exposure.* IEEE Electron Dev. Lett., 1996. **17**(8): p. 388.
13. Weir, B.E., *et al.*, *Ultra-Thin Gate Dielectrics: They Break Down, But Do They Fail?* in *International Electron Device Meeting (IEDM)*. 1997. p. 73.
14. Wu, E., *et al.*, *Structural dependence of dielectric breakdown in ultra-thin gate oxides and its relationship to soft breakdown modes and device failure.* in *International Electron Device Meeting (IEDM)*. 1998. p. 187.

15. Cheung, K.P., *et al.*, *Charging Damage in Thin Gate Oxides - Better or Worse?* in *1998 International Symp. Plasma Process Induced Damage (P2ID)*. 1998. p. 34.
16. Cheung, K.P., P. Mason, and J.T. Clemens, *Measuring plasma charging damage in ultra-thin gate-oxide*. in *European Symp. Plasma Process Induced Damage (ESPID)*. 1999. p. 88.
17. Cheung, K.P., P. Mason, and D. Hwang, *Plasma charging damage of ultra-thin gate-oxide -- The measurement dilemma*. in *International Symp. Plasma Process Induced Damage (P2ID)*. 2000. p. 10.
18. Hill, R.M. and L.A. Dissado, *Theoretical basis for the statistics of dielectric breakdown*. J. Phys., 1983. **C16**: p. 2145.
19. Yamabe, K. and K. Taniguchi, *Time-dependent dielectric breakdown of thin thermally grown SiO2 films*. IEEE Trans. Electron Dev., 1985. **32**: p. 423.
20. Hokari, Y., T. Baba, and K. Kawamura, *Reliability of 6-10nm thermal SiO2 films showing intrinsic dielectric integrity*. IEEE Trans. Electron Dev., 1985. **32**: p. 2485.
21. Wolters, D.R. and J.F. Verwey, *Instabilities in Silicon Devices*. 1986: Elsevier Science Publishers. Ch. 6, p332.
22. Vollertsen, R.P. and W.G. Kleppmann, *Dependence of dielectric time to breakdown distributions on test structure area*. in *IEEE Int. Conf. Microelectronic Test Structure*. 1991. p. 75.
23. Wu, E.Y., J.H. Stathis, and L.K. Han, *Ultra-thin oxide reliability for ULSI application*. Semicond. Sci. technol., 2000. **15**: p. 462.
24. Nigam, T., *et al.*, *Constant Current Charge-to-Breakdown: Still a Valid Tool to Study the Reliability of MOS Structures?* in *IEEE Internation Reliability Physics Symposium (IRPS)*. 1998. p. 62.
25. Wu, E.Y., *et al.*, *Challenges for Accurate Reliability Projections in the Ultra-Thin Oxide Regime*. in *IEEE Internation Reliability Physics Symposium (IRPS)*. 1999. p. 57.
26. SiStefano, T.H. and M. Shatzkes, *Impact ionization model for dielectric instability and breakdown*. Appl. Phys. Lett., 1974. **25**(12): p. 685.
27. Klein, N. and P. Solomon, *Current runaway in insulators affected by impact ionization and recombination*. J. Appl. Phys., 1976. **47**(10): p. 4364.
28. DiMaria, D.J., D. Arnold, and E. Cartier, *Impact ionization and positive charge formation in silicon dioxide films on silicon*. Appl. Phys. Lett., 1992. **60**(17): p. 2118.
29. Nissan-Cohen, Y., J. Shappir, and D. Frohman-Bentchkowsky, *High-field and current-induced positive charge in thermal SiO_2 layers*. J. Appl. Phys., 1985. **57**(8): p. 2830.
30. Nissan-Cohen, Y., J. Shappir, and D. Frohman-Bentchkowsky, *Trap generation and occupation dynamics in SiO_2 under charge injection stress*. J. Appl. Phys., 1986. **60**(6): p. 2024.
31. Avni, E. and J. Shappir, *A model for silicon-oxide breakdown under high field and current stress*. J. Appl. Phys, 1988. **64**(2): p. 743.

32. Suñé, J., *et al.*, *On the breakdown statistics of very thin SiO_2 films.* Thin Solid Films, 1990. **185**: p. 347.
33. Nguyen, T.N., P. Olivo, and B. Riccó, *A New Failure Mode of very Thin (<50Å) Thermal SiO_2 Films.* in *IEEE Internation Reliability Physics Symposium (IRPS).* 1987. p. 66.
34. Moazzami, R. and C. Hu, *Stress-Induced Current in Thin Silicon Dioxide Films.* in *International Electron Devices Meeting (IEDM).* 1992. p. 139.
35. Takagi, S., N. Yasuda, and A. Toriumi, *Experimental Evidence of Inelastic Tunneling and New I-V Model for Stress-induced Leakage Current.* in *International Electron Devices Meeting (IEDM).* 1996. p. 323.
36. Chen, I.C., S. Holland, and C. Hu, *Oxide breakdown dependence on thickness and hole current-enhanced reliability of ultra thin oxide.* in *IEDM.* 1986. p. 660.
37. Degraeve, R., *et al.*, *A consistent model for the thickness dependence of intrinsic breakdwon in ultra-thin oxides.* in *IEDM.* 1995. p. 863.
38. Degraeve, R., *et al.*, *New Insights in the Relation Between Electron Trap Generation and the Statistical Properties of Oxide Breakdown.* IEEE Trans. Electron Dev., 1998. **45**(4): p. 904.
39. Stathis, J.H. and D.J. DiMaria, *Reliability Projection for Ultra-Thin Oxide at Low Voltage.* in *IEDM.* 1998. p. 167.
40. Cheung, K.P., *et al.*, *Field Dependent Critical Trap Density for Thin Gate Oxide Breakdown.* in *IRPS.* 1999. p. 52.
41. McPherson, J.W. and D.A. Baglee, *Acceleration factor for thin gate oxide stressing.* in *IEEE Internation Reliability Physics Symposium (IRPS).* 1985. p. 1.
42. McPherson, J., *et al.*, *Comparison of E and 1/E TDDB Models for SiO_2 under lond-term/low-field test conditions.* in *International Electron Devices Meeting (IEDM).* 1998. p. 171.
43. Chen, I.-C., S. Holland, and C. Hu, *Oxide breakdown dependence on thickness and hole current - enhanced reliability of ultra thin oxides.* in *IEDM.* 1986. p. 660.
44. Bude, J.D., B.E. Weir, and P.J. Silverman, *Explanation of Stress-Induced Damage in Thin Oxides.* in *IEDM.* 1998. p. 179.
45. Weinberg, Z.A., *et al.*, *Exciton or hydrogen diffusion in SiO_2?* J. Appl. Phys., 1979. **50**(9): p. 5757.
46. DiMaria, D.J. and J.W. Stasiak, *Trap creation in silicon dioxide produced by hot electrons.* J. Apl. Phys., 1989. **65**(6): p. 2342.
47. Cheung, K.P., *A Physics-Based, Unified Gate-Oxide Breakdown Model.* in *International Electron Device Meeting.* 1999. p. 719.
48. Hu, C. and Q. Lu, *A Unified Gate Oxide Reliability Model.* in *IRPS.* 1999. p. 47.
49. Nigam, T., Ph.D. thesis, Katholieke Universiteit Leven, 1999.
50. Weir, B.E., *et al.*, *Gate oxide reliability projection to the sub-2nm regiem.* Semicond. Sci. Technol., 2000. **15**: p. 455.

51. Suehle, J.S. and P. Chaparala, *Low Electrical Field Breakdown of Thin SiO2 Films Under Static and Dynamic Stress.* IEEE Trans. Electron Devices, 1997. **44**(5): p. 801.
52. Nojiri, K., K. Kato, and H. Kawakami, *Evaluation and Reduction of Electron Shading Damage in High Temperature Etching*. in *International Symposium on Plasma Process Induced Damage (P2ID)*. 1999. p. 29.
53. DiMaria, D.J. and J.H. Stathis, *Non-Arrhenius temperature dependent of reliability in ultrathin silicon dioxide films.* Appl. Phys. Lett., 1999. **74**(12): p. 1752.
54. Degraeve, R., *et al.*, *Temperature acceleration of oxide breakdown and its impact on ultra-thin gate oxide reliability*. in *VLSI Technology Symp.* 1999. p. 59.
55. Helms, C.R. and E.H. Poindexter, *The silicon-silicon-dioxide system: its microstructure and imperfections.* Rep. Prog. Phys., 1994. **57**: p. 791.
56. McPherson, J.W. and H.C. Mogul, *Disturbed Bonding States in SiO2 Thin-Films and Their Impact on Time Dependent Dielectric Breakdown*. in *IEEE International Reliability Physics Symposium (IRPS)*. 1998. p. 47.
57. Nigam, T., *et al.*, *A Fast and Simple Methodology for Lifetime Prediction of Ultra-thin Oxides*. in *International Reliability Physics Symp. (IRPS)*. 1999. p. 381.
58. Lu, Q., *et al.*, *A Model of the Stress Time Dependence of SILC*. in *IRPS*. 1999. p. 396.
59. Oldham, T.R., A.J. Lelis, and F.B. McLean, *Spatial dependence of trapped holes determined from tunneling analysis and measured annealing.* IEEE Trans. Nucl. Sci., 1986. **33**: p. 1203.
60. Dumin, D.J. and J.R. Maddux, *Correlation of Stress-Induced Leakage Current in Thin Oxides with Trap Generation Inside the Oxides.* IEEE Trans. Electron Dev., 1993. **40**(5): p. 986.
61. King, J.C. and C. Hu, *Effect of Low and High Temperature Anneal on Process-Induced damage of Gate Oxide.* IEEE Electron Dev. Lett., 1994. **15**(11): p. 475.
62. Li, X.Y., *et al.*, *Evaluation of plasma damage using fully processed metal-oxide-semiconductor transistors.* J. Vac. Sci. Technol. B, 1996. **14**(1): p. 571.

Chapter 2

Mechanism of Plasma Charging Damage I

This chapter develops the basic mechanism of plasma charging damage. We start the discussion from basic plasma physics. Only the concepts required to understand plasma charging damage are introduced. For a more thorough discussion of the physics of processing plasma, the readers should consult books that are dedicated to the subject, such as those by Lieberman and Lichtenberg [1] and by Chapman [2].

2.1 Basic Plasma Characteristics

2.1.1 Quasi-neutral Character of Plasma

Plasma is a collection of charged and uncharged particles (Figure 2.1). Positively charged particles are ions. Negatively charged particles are either electrons or negative ions or both. Being a gaseous collection (usually, plasma is considered a fourth state of matter, not a gaseous state), the charged particles can easily move around and therefore shield the Coulombic force effectively. As a result, plasma is quasi-neutral at all times. In other words, owing to the gaseous nature charge fluctuation is always present, particularly in short time-scale. However, on the average, a charge balance is always maintained. Plasma is not a system at thermal equilibrium. Rather, it is at steady state very far from equilibrium. It is a balance between ionizing excitation (power coupling into the plasma) and loss due to particle recombination.

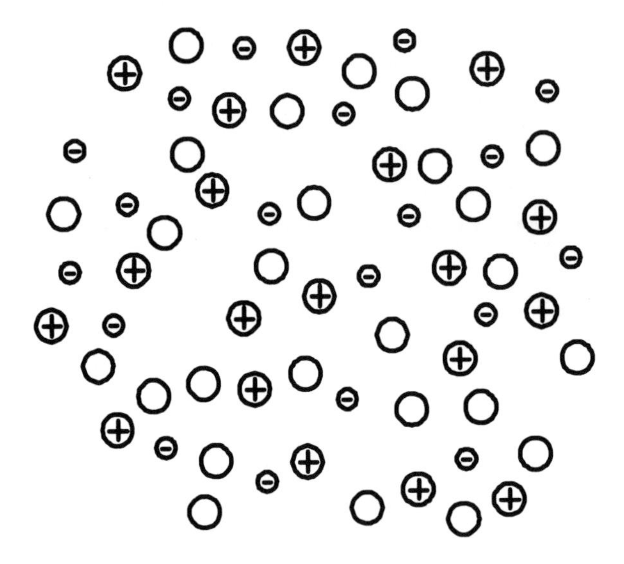

Figure 2.1. Plasma is a collection of charged and uncharged particles in a quasi-neutral state.

2.1.2 Particle Temperatures

While the plasma is in general not in thermal equilibrium, it is often described using statistic terms that apply, strictly speaking, only to equilibrium systems. One reasonable approximate method treats each type of particle within plasma as a separate equilibrium. In other words, while the velocity distribution of all particles is not Maxwellian, the velocity distribution of each type of particle is approximately Maxwellian. These approximations are often not very good, but much better than trying to approximate the whole system as an equilibrium state. Thus we often use terms like electron temperature, ion temperature and temperature of the neutrals. We need to make these approximations in order to develop models to describe plasma behaviors. However, we must always keep in mind that these approximations have strict limitations and thus are sometimes poor. In plasma damage studies, the concept of electron temperature is often used in modeling. It is not unusual to encounter situations where the model only holds if the velocity distribution is really Maxwellian. The validity of these models should therefore be taken with a big grain of salt.

Since each type of particle is not in thermal equilibrium with each other, each associated temperature can be quite different. In most plasma, energy is coupled-in via the acceleration of electrons by an external electromagnetic field. It is then natural that electrons are most energetic and have a much higher temperature than other species. It is common to have electron temperatures in processing plasma ranging from 20,000K to 100,000K (1.5eV to 8eV) while the ion temperature and neutral temperatures are only in the 2000K and 300K range, respectively.

2.1.3 Floating Potential and Plasma Sheath

Imagine an isolated object comes in contact with plasma. It is immediately bombarded by particles in the plasma. The bombardment rate is $nv/4$, where n is the number density and v is the average velocity of each type of particle (Figure 2.2). Since the electron is the lightest and hottest particle, its average velocity is far higher (by around 10,000 times) than either the ion's or neutral's. Very rapidly, a net negative charge will accumulate on the isolated object. A negative potential with respect to the plasma potential (V_p) rapidly builds up. This negative potential, which is called the floating potential (V_f) serves to repel the lower energy electrons coming from plasma. It continues to build up until the net flux of negative charge arriving at the isolated object is exactly the same as the net flux of positive charge. This buildup process is generally completed in microseconds thanks to the very high velocity of electrons.

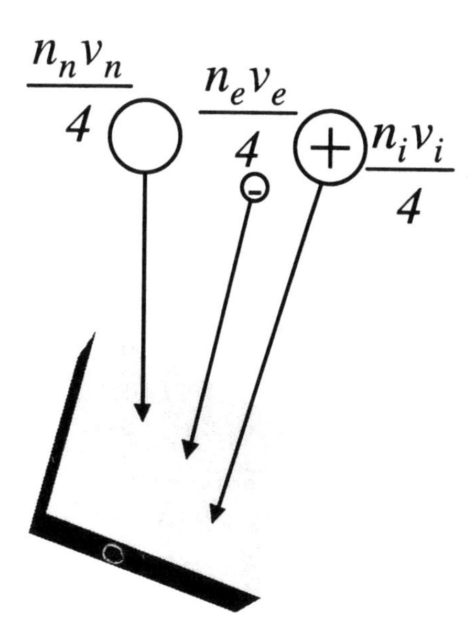

Figure 2.2. The bombardment rate to a surface is proportional to the number density and average velocity of each particle type.

Due to the repulsing potential, the immediate neighborhood of the isolated object will have fewer electrons than the rest of the plasma (Figure 2.3). This is the plasma sheath region, which is also called the dark region (lower electron density leads to lower light emission). The potential difference between plasma and the isolated object is the sheath potential ($V_{sh}= V_p-V_f$). While this sheath potential is a retarding potential for the electrons, it is an accelerating potential for the positive ions. The ability to accelerate ions so that they become highly collimated is the primary reason plasma processes are used in VLSI manufacturing. Notice that in Figure 2.3 the potential drop does not extend into plasma. This is because electrons in plasma are highly mobile, they can effectively screen the electric field. As a result, the ability of extracting positive ions from plasma by the electric field is limited. Only ions emerging from plasma into the sheath region due to random thermal motion are accelerated toward the isolated object.

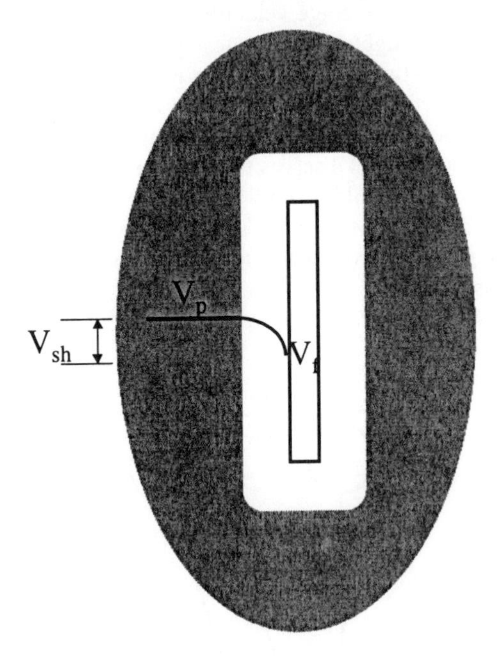

Figure 2.3. A "dark region" surrounds a floating object. Within it electrons are repelled. This region is called the plasma sheath. The potential difference between the floating object and the plasma drops almost entirely across this plasma sheath.

2.1.4 Pre-sheath and the "Bohm Criterion"

Actually, the potential does penetrate plasma by a short distance the size of an electronic Debye length (screening length). A value of $kT_e/2e$ potential is dropped across this distance. Here k is the Boltzmann constant, T_e is the electron temperature and e is the electronic charge. Note that no matter how large the sheath potential is, the potential drop across the screening length is not affected. Within this screening length (also called pre-sheath), ions are accelerated by the $kT_e/2e$ potential toward the sheath. This results in an ion flux significantly larger than just due to random thermal motion alone. This is the "Bohm Criterion" for a stable sheath formation, which gives the saturated ion current that is independent of the sheath potential.

The saturated ion current according to the "Bohm Criterion" is:

$$I_{ion} = 0.6n\sqrt{\frac{kT_e}{M}} \tag{2.1}$$

where n is the plasma density and M the positive ion mass. For simple plasma like an Ar discharge, plasma density equals ion density, which equals electron density. For a quantitative feel of the equation, we apply it to a typical high-density plasma as follows: n=1x10^{12}/cm^3, kT_e=4eV=6.4x10^{-19}K$_g$m^2/s^2, M=40u=6.64x10^{-26}K$_g$, I_{ion}=1.86x10^{17}/cm^{2}s=29.8mA/cm^2.

2.1.5 Independence of Sheath Potential

At steady state, the fraction of electrons in the Maxwellian velocity distribution that is energetic enough to overcome the sheath potential to reach the isolated object per unit time should equal the ion current (charge balance condition). Thus we have:

$$\frac{1}{4}n\sqrt{\frac{8kT_e}{\pi m}}\exp\left(\frac{eV_{sh}}{kT_e}\right) = 0.6n\sqrt{\frac{kT_e}{M}} \tag{2.2}$$

where m is the electron mass. From this we get:

$$\exp\left(\frac{eV_{sh}}{kT_e}\right) = \sqrt{\frac{M}{2.3m}} \tag{2.3}$$

or:

$$V_{sh} = \frac{kT_e}{2e}\ln\left(\frac{M}{2.3m}\right) \tag{2.4}$$

Equation 2.4 is a well-known relation [2]. It is also central to the mechanism of plasma charging. It says that the sheath potential is not a function of plasma potential. It can be determined entirely from the electron temperature and the mass ratio of the electron and ion. Physically, this result makes sense. Sheath potential develops because of the difference in bombardment rate of electrons and ions on the isolated object. Its magnitude should therefore be determined only by parameters that can affect the bombardment rate. Of these parameters, only the electron and ion densities are related to the plasma potential. Since the density of electrons and density of positive ions are exactly equal in simple positive plasma, they cancel each other. Plasma potential therefore does not play a role in the formation of sheath potential.

2.2 Charge Balance and Plasma Charging

2.2.1 Plasma Charging from Non-uniform Plasma

So far, we have established that any isolated object which comes in contact with plasma will be driven to a floating potential that is negative with respect to the plasma potential. We have also established that this potential difference is independent of the plasma potential. Both of these points are *direct consequences of the charge balance requirement* that is fundamental to any plasma. Yet, as we shall see below, just these two points alone can lead to plasma charging damage.

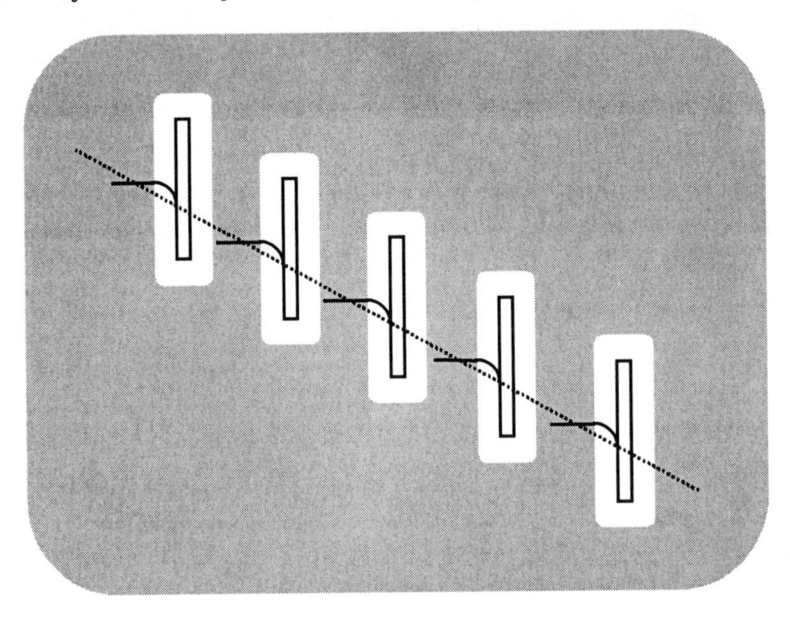

Figure 2.4. A series of isolated objects in a plasma with position dependent plasma potential V_P (dotted line). Each object develops a floating potential V_f according to the local plasma potential. However, V_{sh} remains the same for all of them.

If we have a series of isolated objects immersed in a plasma whose potential V_P is position dependent (Figure 2.4), each of these objects will develop a floating potential, V_f, with respect to the plasma. Since V_{sh} is independent of V_P, the floating potential, V_f, of each object will track the plasma potential and will also be position dependent.

Consider now a wafer with simple capacitors defined on a blanket layer of thin SiO_2 is exposed to highly non-uniform plasma (Figure 2.5). Each top electrode, because of the insulating SiO_2 layer, is an isolated object and, similar to the case in Figure 2.4, will develop a floating potential with respect to the plasma potential at the position of the electrode. For the substrate, it is shielded from the plasma by the insulating SiO_2 layer except at the edge. It therefore develops a floating potential dictated by the plasma potential at the edge of the wafer. Figure 2.6 illustrates the potential distribution of the top electrodes and the potential of the substrate. Since V_f tracks V_P, the floating potential distribution of the top electrodes mimic the plasma potential. On the other hand, the substrate potential has only one value. This is not because it is directly exposed to the plasma at only one point (it is exposed all around the wafer edge), but because it is a relatively good conductor compared to the insulating SiO_2 layer and therefore can only support one potential.

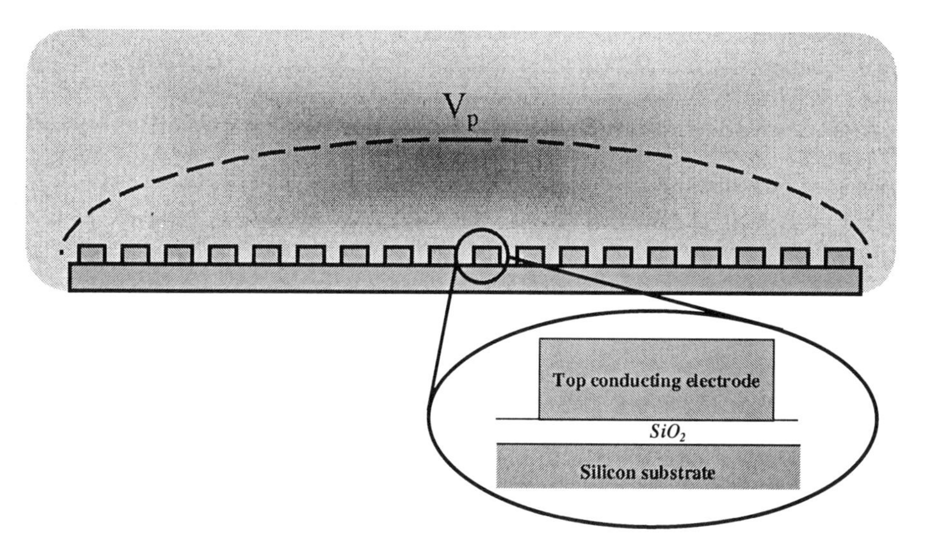

Figure 2.5. A wafer with simple capacitors submerged in highly non-uniform plasma. The broken line represents the plasma potential.

From Figure 2.6 we can see that at the center of the wafer, the top electrode is at a floating potential rather different from that of the substrate. This is but an example of how a high electric field can develop across the oxide. If this field is high enough, the oxide can be damaged. Therefore, simply by two very basic plasma physics principles, namely charge balance and sheath potential being independent of plasma potential, plasma charging damage can readily occur.

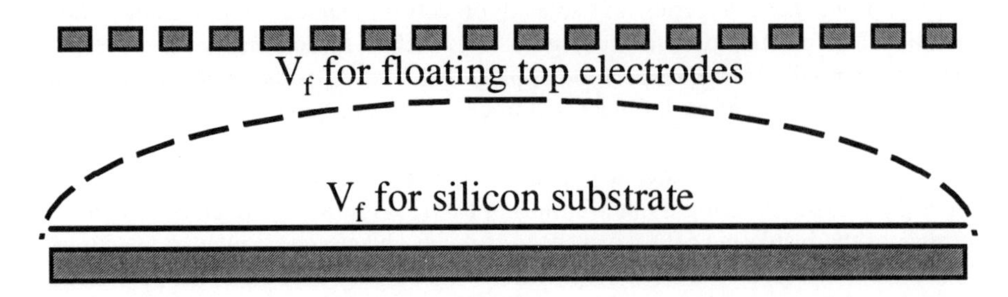

Figure 2.6. The resulting potential distribution for the top floating electrodes follows the potential distribution of the plasma. The potential for the silicon substrate can have only one value and is determined by the plasma potential at the edge of the wafer where it is exposed directly to the plasma.

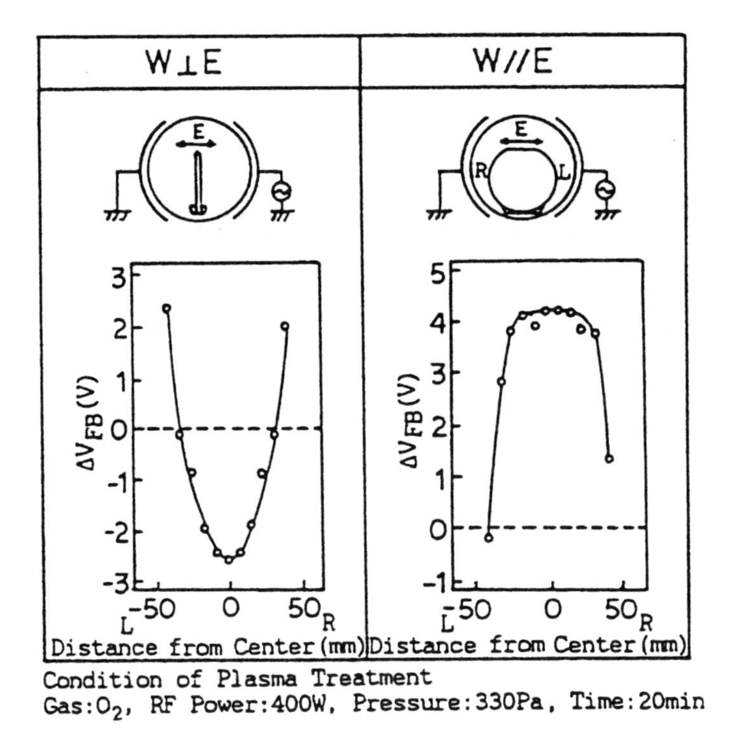

Figure 2.7. An example of charging damage from non-uniform plasma. The barrel etcher has a higher potential at halfway between the two electrodes. However, the halfway plane near the edge has higher potential than the center due to the curved electrodes. The shape of non-uniform plasma potential across the wafer is therefore dependent on the orientation of the wafer. The measured flat-band shift is from MNOS devices. Taken from [3].

A good early example of charging damage arising from non-uniform plasma is the work reported by Tsunokuni *et al.* [3]. Figure 2.7 shows the plasma potential non-uniformity of their barrel etcher measured using MNOS devices shown in Figure 2.8 (see also Chapter 5). To understand their data, we need to understand what the measured flat-band voltage shift in a MNOS device means.

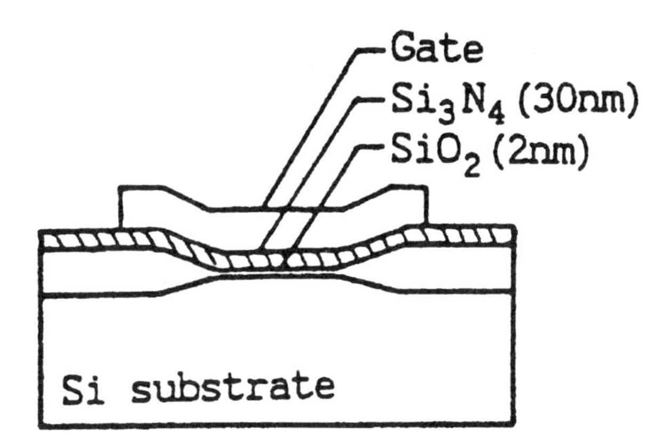

Figure 2.8. Structure of the MNOS device used to produce the data in Figure 5a. Tsunokuni *et al.* [3] did not provide the calibration curve in their paper. All we can say is that a relatively high voltage must exist when a flat-band shift is detected.

A MNOS device is a type of non-volatile memory structure. It is programmable by applying a high voltage. Kawamoto [4] was the first to use this device as a means to detect the high voltage developed across an insulating layer during plasma exposure. To use this device, a calibration curve must be established first by a controlled stress. The device structure and its reference curve used by Kawamoto are shown in Figure 2.9. As can be seen, the flat-band shifts are related to the stress voltage (and time). Tsunokuni *et al.* did not provide the necessary reference curve. One can, based on Kawamoto's result, expect a high voltage (not quite as high as those in Figure 2.9 because of a thinner nitride layer) must have been developed during plasma exposure to produce the kind of flat-band voltage shift shown in Figure 2.7.

Tsunokuni *et al.* examined two different geometries. One wafer is in parallel with the electrodes and the other wafer is perpendicular to the electrodes. In a dipole type reactor such as the one being studied here, it is expected that the plasma density and hence the plasma potential is the highest at mid-point between the two electrodes. This is exactly what is observed in the parallel case (right-hand side ΔV_{FB} curve of Figure 2.7). However, since the electrodes are curved, the plasma potential in the plane that is midway between the two electrodes is not uniform. It is expected to be higher where the electrodes are the closest. This also agrees with the MNOS result for the perpendicular case (left-hand side ΔV_{FB} curve of Figure 2.7). Clearly, the ΔV_{FB} curves obtained from the MNOS devices are reflecting a highly non-uniform plasma potential distribution.

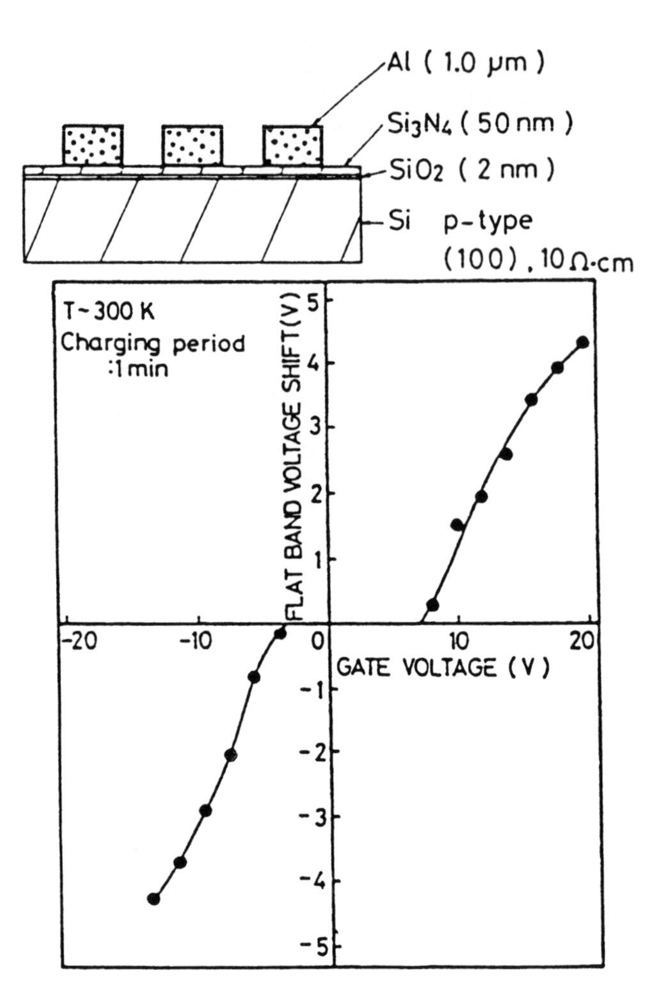

Figure 2.9. The MNOS device used by Kawamoto [4]. The calibration curve was constructed with a known stress and stress time. The flat-band voltage shift after stress has a well defined relation with the stress voltage.

Tsunokuni *et al.* then uses large capacitors (1mm^2 gate area) on the wafer to detect plasma charging damage to gate-oxide. Figure 2.10 shows the results of capacitors in the negative ΔV_{FB} region of the plasma. The breakdown (dead-on-arrival) frequency after plasma exposure of 20 minutes is dependent on both the gate-oxide thickness and the ΔV_{FB} value (i.e. location on the wafer) as measured using MNOS devices. Clearly, the higher the voltage developed across the gate-oxide due to plasma potential non-uniformity, the more damage results and, for a given voltage, thinner oxide suffers more.

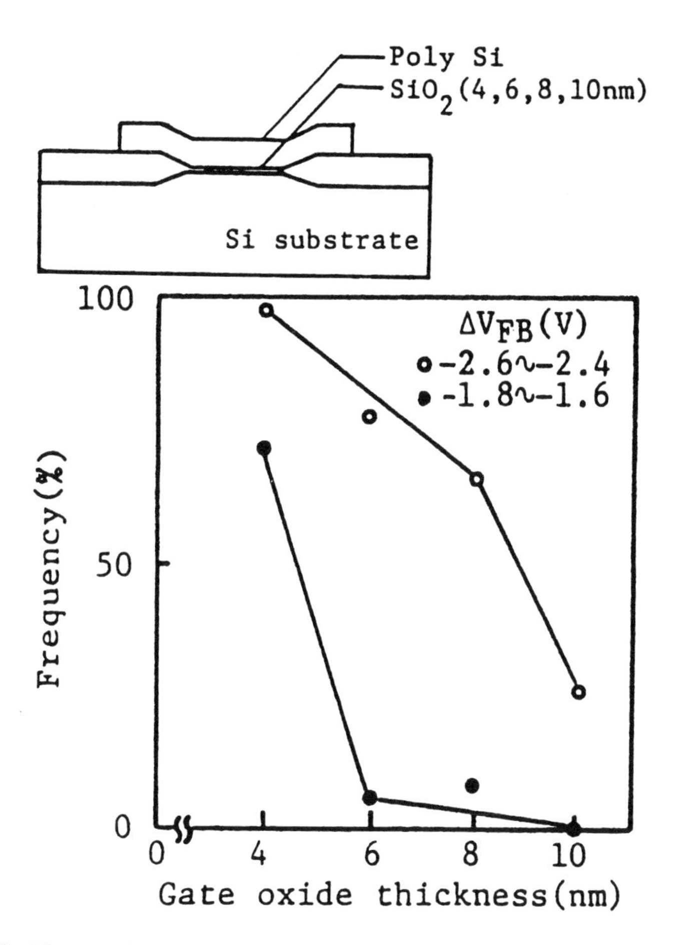

Figure 2.10. The capacitor structure used for plasma charging damage detection. The gate area is 1mm². Gate-oxide thickness varies. Breakdown frequency is shown to be a function of gate-oxide thickness as well as how negative the flat-band voltage is as measured using MNOS devices. Taken from [3].

Note that the left hand side ΔV_{FB} curve of Figure 2.7 covers both negative and positive shifts, implying that the gate potential is both negative and positive with respect to the substrate on the same wafer. This is not the situation depicted in Figure 2.6 where all the gate potentials have the same sign with respect to the substrate. The reason for the difference is that, in Figure 2.6, the plasma distribution is symmetric and therefore the plasma potential experienced by the edge all around the wafer is the same low value. In contrast, for Tsunokuni *et al.*'s case, the plasma potential at the lower edge of the wafer is higher than the plasma potential along the wafer edge away from the bottom. Since only one potential is supported in the substrate, the substrate potential that satisfies charge balance condition for the substrate must lay somewhere between the lowest and highest points. At such a potential, excess electrons are collected from the edge where the plasma potential is low to balance the excess ion current at the edge where the plasma potential is high.

Why did Tsunokuni *et al.* only observe increased breakdown frequency for capacitors that are located at the high negative ΔV_{FB} region but not the high positive ΔV_{FB} region? If Kawamoto's calibration curve for his MNOS device is any indication (not symmetric), the voltage developed across the gate-oxide during plasma exposure is probably higher for the high positive ΔV_{FB} region than the high negative ΔV_{FB} region. We can answer this question only after further details of charging damage mechanism are explained. We will return to this question at the end of Section 2.4.3.

2.2.2 Effect of Injecting and Removing Electrons on Charge Balance

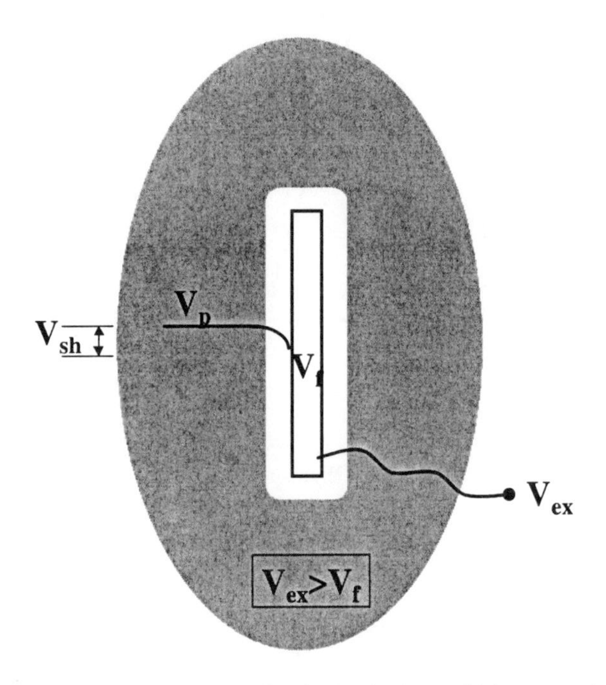

Figure 2.11a. An external contact forces the floating body to a higher potential by drawing electrons from the object.

We mentioned earlier that the sheath potential V_{sh} accelerates ions toward the wafer and that this ability to produce anisotropic ion flux is the main reason plasma is utilized for semiconductor processing. In many plasma-processing situations, this built-in self-bias of the plasma is not the optimal desired energy needed for the ion flux. In older diode type plasma reactors, a large average sheath potential (DC-bias) can be generated at the electrodes by the RF source. Usually, the same RF source is also relied upon to sustain the plasma. Thus one cannot control independently the ion flux and ion energy. In most modern plasma systems, the power source that sustains the plasma is separated from the source that generates the DC-bias to allow the control of the ion flux and ion energy to be decoupled. In our derivation of Equation 4, we have not taken the intentionally generated DC-bias into account. We must ask, does the sheath potential remain independent of the plasma potential in the presence of a strong DC-bias? In addition, does the large DC-bias itself cause charging damage?

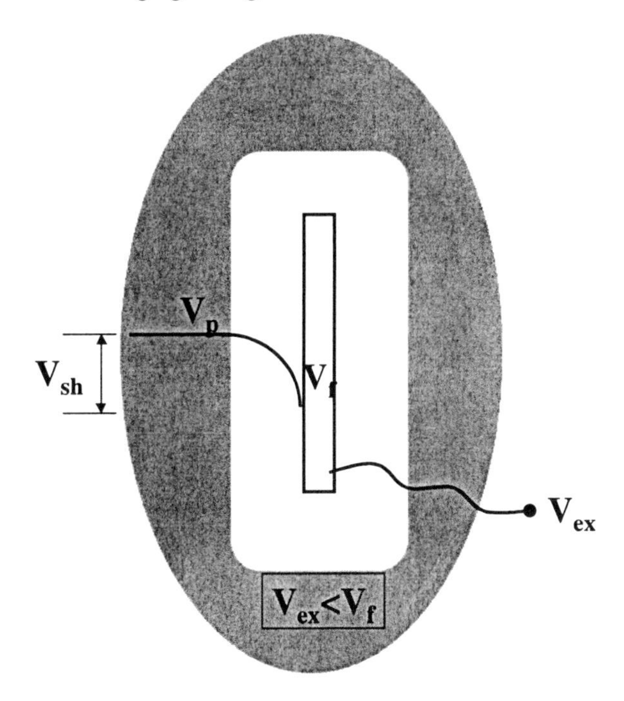

Figure 2.11b. An external contact forces the floating body to a lower potential by injecting electrons into the object.

Before we answer these questions, let us first take a look at how DC-bias is generated. We shall look at the general case of separate sources for plasma generation and DC-bias generation. The result should apply equally well to the special case where a single RF source is responsible for both plasma generation and DC-bias generation (a diode reactor). Returning to Figure 2.3, let us modify the situation a little by adding an external contact to the isolated object. With an external contact, the potential on the object can be controlled. There are two situations to consider. One, the external contact drives the object's potential to a higher value (Figure 2.11a). The other is to drive it to a lower value (Figure 2.11b).

In the case of Figure 2.11a, if we assume that the plasma potential does not change (anchored by the processing chamber walls), a higher potential on the object means a smaller sheath potential. The sheath width would reduce correspondingly. With a smaller retarding potential, Equation 2.2 says that the number of electrons reaching the object from the plasma increases drastically. Note that even under this condition, charge remains balanced. Since ion current remains unchanged, all the excess electrons must be drawn away by the external contact (that is how the potential of the object is raised in the first place). Note that the assumption that the chamber walls anchor the plasma potential means that the electron loss is replenished to the plasma by the chamber wall.

In the case of Figure 2.11b, a lower potential on the object means a larger sheath potential and a wider sheath. As a result, the number of electrons reaching the object from the plasma reduces. To maintain charge balance, electron must be supplied by the external contact since ion current does not change.

2.2.3 Asymmetric Response of Electron Current

Note that for the same magnitude of potential change, the amount of electrons flowing to or from the external contact in the two cases is quite different. If we plot the left hand side of Equation 2.2, that is electron current, versus V_{sh} changes in multiples of kT_e/e (Figure 2.12), the asymmetry of the two cases is quite clear.

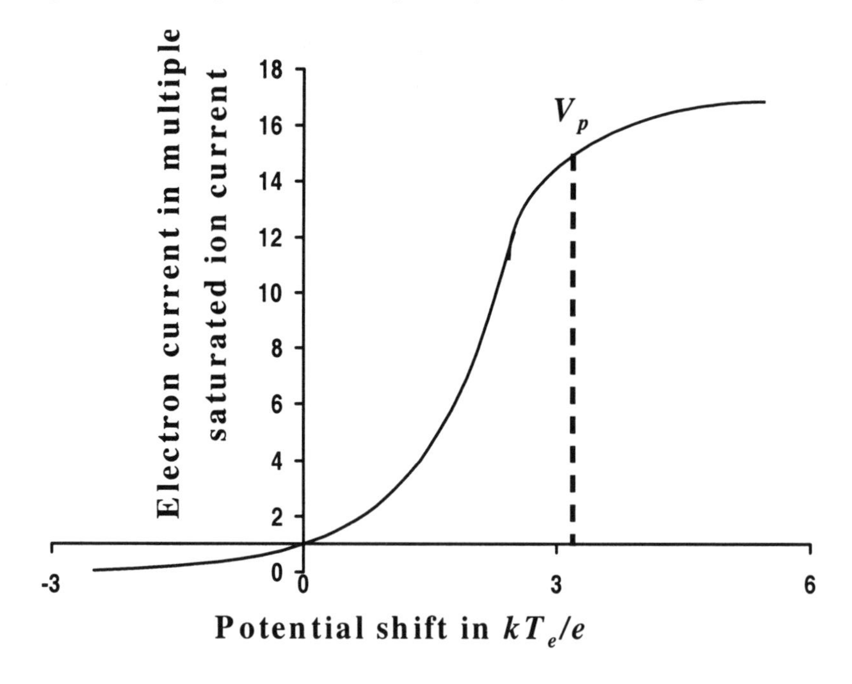

Figure 2.12. The asymmetric response of electron current from the plasma to the changes in floating potential of the object.

2.3 Charging in the Presence of an Applied Bias

2.3.1 DC-bias from Applied AC Voltage

Now consider the case where the external contact is coupled to an AC source through a capacitor (Figure 2.13). The capacitor isolates the object, which in this case is assumed to be a bare silicon wafer. The plasma is supported independently by an unspecified source. Before the AC voltage is applied, the wafer floats to a negative potential with respect to the plasma. When we start to apply to AC voltage, the positive-going half of the cycle raises the potential of the wafer. As a result, a large quantity of electrons flows to the wafer from the plasma. As we enter into the negative-going half of the AC cycle, the wafer potential follows the driving voltage down. Electron flux from the plasma decreases rapidly and eventually stops when the potential becomes lower than the original floating potential. The downward swing of the applied AC continues to drive the wafer potential further down. When the applied AC bottoms out, the wafer is at a potential more negative than the sum of the original floating potential and half of the peak to peak AC voltage. This is because the large amount of electrons collected by the wafer during the positive going cycle cannot be neutralized by the ion current that is unaffected by the AC voltage.

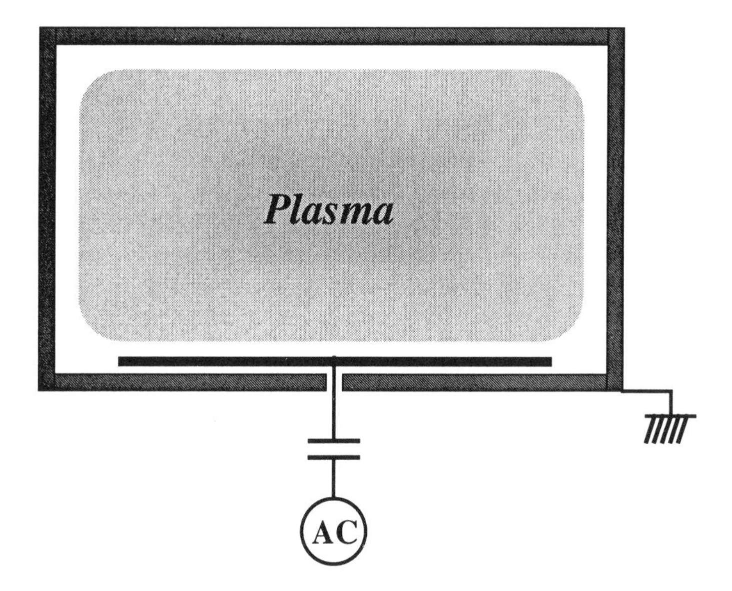

Figure 2.13. The external contact is through an AC coupling capacitor. The wafer remains floating from the DC point of view.

In the next positive-going half cycle, the wafer potential again rises toward the plasma potential. This time it does not quite reach to as high a potential as the last cycle because it starts from a lower potential point. As the wafer potential approaches the plasma potential, large electron flux flows to the wafer again. The negative-going half cycle again pushes the wafer potential down. This time it goes a little further because it starts from a lower potential. Note that during all this time, the positive ion current remains constant. The ion energy does vary since the accelerating potential that is provided by the sheath is now a time varying function. After a few AC cycles, the downward shift of the AC potential swing of the wafer stabilizes. At this point, the integrated electron flux during the positive peak of the wafer potential swing exactly equals the integrated ion flux during the entire AC cycle. The average wafer potential is now significantly more negative than the original floating potential (Figure 2.14). The average potential is the so-called "DC-bias". The magnitude of this DC-bias is obviously related to the amplitude of the applied AC voltage. If the applied AC voltage is large, the DC-bias is about half the peak to peak value.

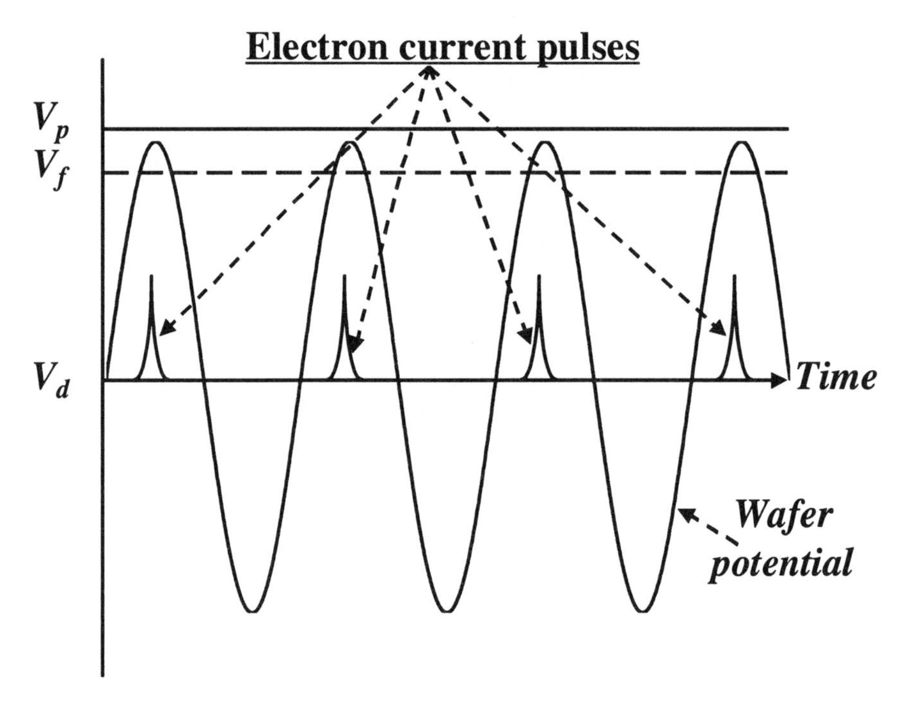

Figure 2.14. The steady state behavior of wafer potential, plasma potential, (V_p), DC-bias, (V_d), and electron current.

2.3.2 Relationship Between DC-bias and Charging Damage

In many etching processes, the DC-bias can be quite large. As much as 1000 volt is used in some cases. With such a large voltage, would the thin oxide be stressed or broken by it? The answer is obviously no or the etching process will never be adopted in production. Let us see why it is not a problem. Returning to Figure 2.13, let us assume that on top of the silicon wafer, there is a top electrode separated from the bulk of the wafer by a thin oxide layer. The top electrode and the silicon wafer thus form a capacitor. The top electrode is separated from the plasma (a conductor) by the sheath. One can model the sheath as a parallel plate capacitor as well. Then a capacitor network as shown in Figure 2.15 can represent the whole system.

When capacitors are connected in series, the AC voltage applied across the whole series is divided among the capacitors according to the inverse of the capacitance. In most processing plasma, the sheath thickness ranges from 100 microns (very high density plasma) to 10 mm whereas the typical thickness of the oxide layer ranges from 50A to 1 micron. The size of the external coupling capacitor (also called blocking capacitor) is normally chosen to be larger than the sheath capacitance. As a result, the applied AC voltage is largely dropped across the sheath, which is where it is most useful. The amount that dropped across the oxide is usually thousands times smaller and therefore cannot do any damage.

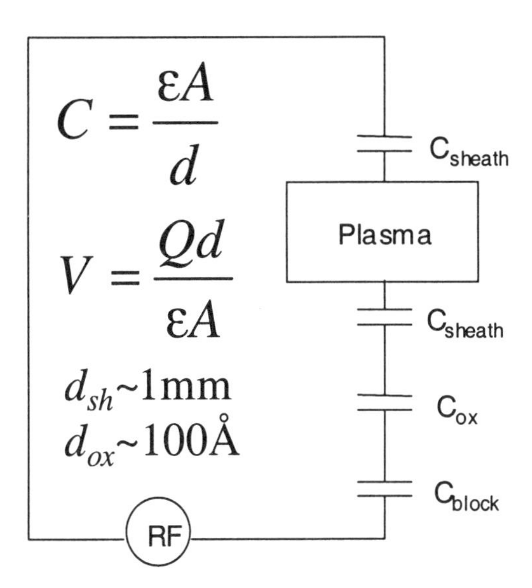

Figure 2.15. The equivalent circuit of the plasma system with the wafer is a network of parallel-plate capacitors connected in series. The equation for a parallel-plate capacitor is given in the figure. The applied RF voltage is shared by all capacitors. The voltage across each capacitor is inversely proportional to its capacitance. The voltage across C_{ox} is therefore extremely small.

2.3.3 Charge Balance with Large DC-bias

An important consequence of the DC-bias is that electrons flow from the plasma to the wafer in short bursts - one burst per cycle (see Figure 2.14). It is obvious that charge cannot be balanced at all times. It can only be balanced when the entire AC cycle is integrated. Since the plasma potential remains fixed, the oscillating floating potential implies that the sheath potential is also oscillating. We need therefore to modify the fixed retarding sheath potential in Equation 2.2 with an oscillating potential. Since most of the applied AC voltage is dropped across the sheath, we can simply use the AC voltage function as the retarding potential. In addition, we need to integrate the left-hand-side of Equation 2.2 over the entire RF cycle:

$$\frac{1}{4}n\sqrt{\frac{8kT_e}{\pi m}}\frac{1}{2\pi}\int_0^{2\pi}\exp\left(\frac{e\left(V_d+V_{rf}\cos\omega\tau\right)}{kT_e}\right)d\omega\tau = 0.6n\sqrt{\frac{kT_e}{M}} \tag{2.5}$$

where V_d is the DC-bias and V_{rf} is the amplitude of the applied AC voltage. After rearranging, we have:

$$\exp\left(\frac{eV_d}{kT_e}\right)\frac{1}{2\pi}\int_0^{2\pi}\exp\left(\frac{eV_{rf}\cos\omega\tau}{kT_e}\right)d\omega\tau = \sqrt{\frac{M}{2.3m}}$$

or, using Equation 2.2, we have:

$$\exp\left(\frac{eV_d}{kT_e}\right)\frac{1}{2\pi}\int\exp\left(\frac{eV_{rf}\cos\omega\tau}{kT_e}\right)d\omega\tau = \exp\left(\frac{eV_{sh}}{kT_e}\right).$$

Which can be written as:

$$\exp\left(\frac{eV_d}{kT_e}\right)I_0\left(\frac{eV_{rf}}{kT_e}\right) = \exp\left(\frac{eV_{sh}}{kT_e}\right),$$

where $I_0(eV_{rf}/kT_e)$ is a zeroth order modified Bessel function of the second kind. Taking the natural log and rearranging, we get:

$$V_d = V_{sh} - \frac{kT_e}{e}\ln\left(I_0\left(\frac{eV_{rf}}{kT_e}\right)\right) \tag{2.6}$$

Equation 2.6 is less well known than Equation 2.4. The validity of it was demonstrated experimentally by Mantei [5]. It worked remarkably well for an eV_{rf}/kT_e value up to at least 150 (Figure 2.16).

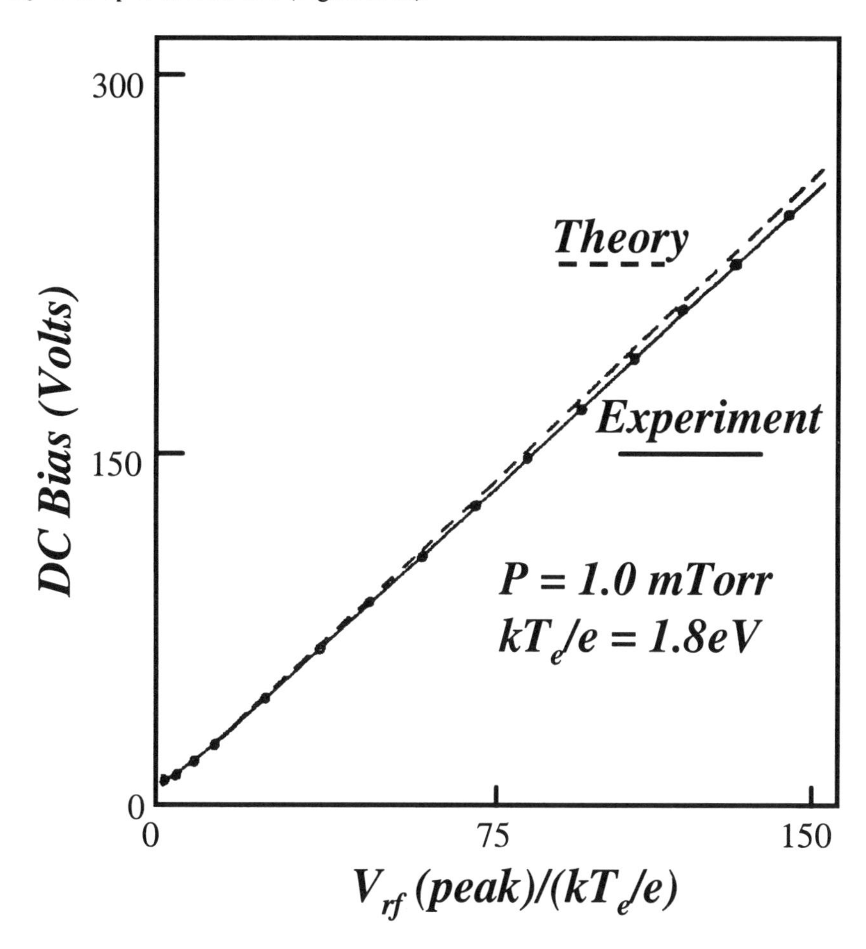

Figure 2.16. Comparing the prediction of Equation 2.6 with measured DC self-bias, excellent agreement was established. Taken from [5].

Equation 2.6 says that even with a large applied RF to generate DC-bias, the sheath potential still does not depend on the plasma potential. Even under a large DC-bias, the time-averaged floating potential of an isolated object still tracks the plasma potential. The argument for plasma charging to occur when the plasma potential is non-uniform remains valid even when a large DC-bias is present.

2.4 Fowler-Nordheim (FN) Tunneling and Charge Balance

2.4.1 Impact of FN Tunneling on Floating Potential Distribution

Let us now return to Figure 2.6 where the non-uniform plasma has set up a potentially damaging situation for the gate-oxide and take a closer look at what can happen. Suppose the plasma potential varies from center to edge by 30 volts and that the gate-oxide is 100Å thick. Assume also that the plasma density is such that the average ion current density is about 1mA/cm^2 (representing a moderate density plasma). At time zero (first moment of exposure to plasma), the top and bottom electrodes are not at the proper floating potential dictated by the plasma. Once exposure starts, the top and bottom electrodes are immediately driven toward the appropriate potential so that charges can be balanced.

Since the top electrode (gate) potential at the center is as much as 30 volts higher than the substrate potential, the electric field across the 100Å thick gate-oxide is as high as 30MV/cm - way above the breakdown field. One therefore expects the gate-oxide to be broken. However, this is not the case. As the gate potential rises above the substrate potential to satisfy charge balance, the field increases across the gate-oxide. When the oxide field reaches beyond 8MV/cm, a significant level of current starts to flow across the insulating gate-oxide by way of Fowler-Nordheim (FN) tunneling. Since the substrate is at a lower potential, the tunneling mode is substrate injection. That is, electrons are flowing from the substrate into the gate, similar to the case of Figure 2.11b.

The charge balance at the gate must now include the additional source of current. The sum of electron flux from the plasma and from FN tunneling must equal the ion flux. Since the ion flux is fixed, so should the sum of electron flux. When the FN current increases, the electron flux from the plasma must decrease. This is achieved by lowering the gate potential with respect to the plasma potential (increase the sheath potential). At some point, the electron flux from the plasma will be reduced to near zero. However, the gate potential lowering does not stop there. The FN current, which is an exponential function of oxide field, may still be significantly higher than the ion current. The excess electrons flowing into the gate will continue to lower the gate potential until FN current is reduced to the point that it is exactly the same as the ion current. The description we just made is, of course, not what actually happens. It is only a way to help understand the various factors at work. What really happens is that as the gate potential rises beyond some point, FN tunneling prevents it from going any higher. The gate potential never actually reaches 30V.

We can use the same thought process to understand what happen in the substrate. From the point of view of the substrate, it losses electrons to the gate via FN tunneling. Since the substrate is a conductor and is exposed to the plasma at the edge, it must maintain charge balance by collecting more electrons from the plasma at the edge. By losing electrons to the gate through FN tunneling, the substrate potential will rise, similar to the case in Figure 2.11a and a large amount of electrons will flow to the substrate at the edge from the plasma. The movement of gate and substrate potential is graphically shown in Figure 2.17.

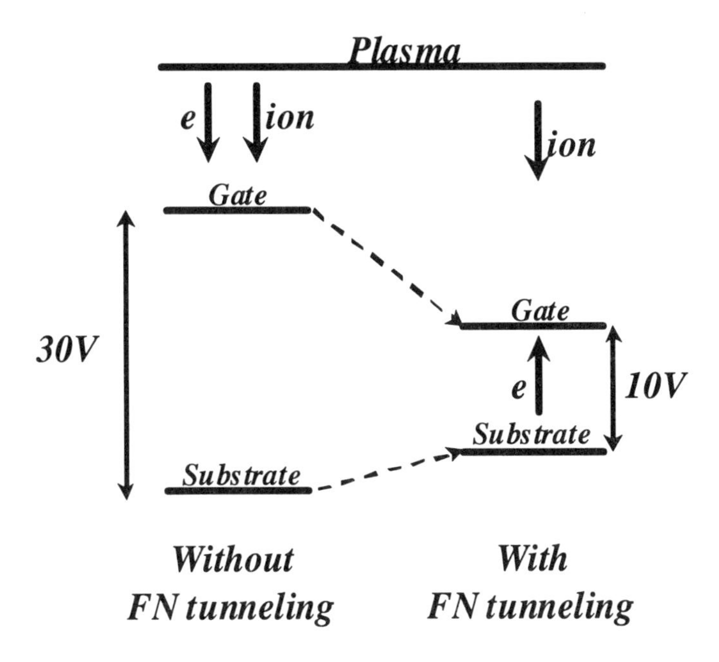

Figure 2.17. The potential of the gate and the substrate both adjust in response to the presence of FN tunneling to maintain charge balance.

Notice that in Figure 2.17 the increase in substrate potential is far less than the decrease in gate potential. This asymmetry is partly the result of the asymmetric response of electron flux to the floating potential change described earlier (Figure 2.12). When the substrate potential moves up, the electron flux from the plasma increases rapidly. In addition, the integrated area of the wafer edge where the substrate can collect electrons from the plasma is quite large. The overall effect is that a small increase in substrate potential will collect enough additional electrons from the plasma to maintain charge balance for the substrate. The final potential difference between the gate and substrate at the center of the wafer, as depicted in Figure 2.17, is 10 volt. This voltage is determined by the oxide field supporting enough FN current to balance the ion current. In this case, the FN current required is $1mA/cm^2$ (same as ion current density). The oxide field needed to support such tunneling current is around 10MV/cm. For a 100Å thick gate-oxide, that translates to 10 volts.

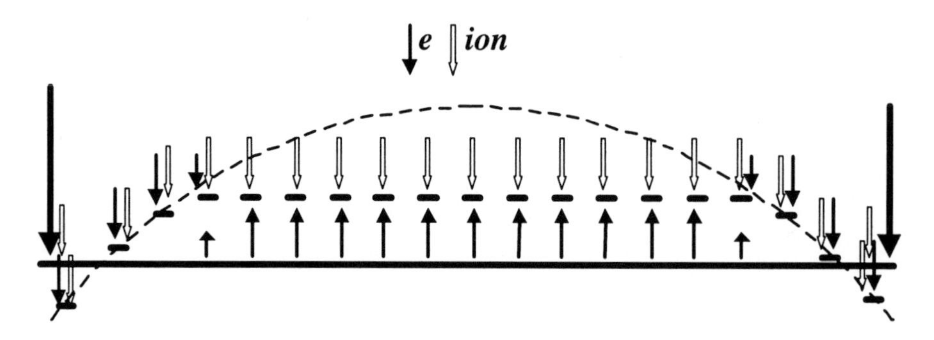

Figure 2.18. The potential distribution and the current components for the gates and the substrate in the presence of FN tunneling are shown. The broken curve indicates where the original floating potential distribution dictated by the plasma was.

By applying the same argument for each gate electrode, we can arrive at the final distribution of gate and substrate potential as depicted in Figure 2.18. To see how such a potential distribution is arrived at, let us imagine that FN tunneling was turned off at the beginning. The potential distribution in the absence of FN tunneling is shown in Figure 2.6. Next we turn on FN tunneling. Majority of the capacitors has a very high oxide field across it except those at the very edge of the wafer. All of these capacitors will have FN tunneling. All of them will lower their potential due to electron injection from the substrate. At the same time, the substrate will raise its potential due to loss of electrons. As the substrate potential rises, more electrons will flow from the plasma to the substrate (represented by the large arrow of electron flow at the edge). The potential difference between the substrate and various gates changes. As a result, the region of low oxide field starts to move away from the edge of the wafer. The gate potentials in the center region rise with the substrate potential to maintain the fixed oxide field to satisfy charge balance. The oxide-field for the capacitors at the edge starts to rise as well, but with the gate at lower potential than the substrate. If the oxide field is high enough, FN current will flow from the gates to the substrate. In this case, we choose to show that the field is not so high and no FN tunneling occurs at the edge of the wafer.

It must be emphasized again that the thought processes described above don't really happen. The final potential distribution is reached directly without the voltages going up and down as described. Nevertheless, the thought process helps to understand the many driving forces involved, and illustrates that plasma charging damage is governed by the close relationship between plasma and the detailed structure of the wafer.

As indicated in Figure 2.18, most of the gates at the center of the wafer are not receiving any electron from the plasma. For these capacitors, the charge balance condition is FN tunneling current equals the ion current. Let I_{is} denote the FN current of the ith capacitor in substrate injection mode and I_{jg} denote the FN current of jth capacitor in gate injection mode. Note that I_{is} and I_{jg} are of opposite sign, just like I_e and I_{ion}. Then for the substrate, charge balance means:

$$\sum_i I_{is} + \sum_j I_{jg} + I_e + I_{ion} = 0 \tag{2.7}$$

Since the substrate potential has risen above the original (without FN tunneling) floating potential, the I_e component is the largest of all the current components. For the gate of each capacitor, charge balance means

$$I_{FN} + I_e + I_{ion} = 0$$

Here I_{FN} denotes FN current regardless of the type of injection. For the capacitors at the center of the wafer, $I_e \sim 0$ and $I_{FN} = I_{ion}$. At the low oxide field region, $I_e = I_{ion}$ and $I_{FN} = 0$. At the edge of the wafer, I_{FN} and I_{ion} have the same sign and I_e equals the sum of the two.

2.4.2 Bipolar Damage by Tunneling

From a damage point of view, we are most interested in the magnitude of I_{FN}. We know its magnitude at the center and at the low oxide field region from the above discussion, but how do we know what its magnitude is at the edge of the wafer? For the example at hand, the gate oxide is 100Å thick. The potential difference between the substrate and the gate needs to be at least 8 volts before appreciable FN current will flow. This means that the substrate potential must rise at least 8 volt in its effort to maintain charge balance with the FN components of the rest of the wafer. From Figure 2.12, we know that I_e increases exponentially with the potential rise. However, this potential rise is measured in multiples of kT_e/e. That is, the magnitude of I_e depends on the electron temperature of the plasma. For a given magnitude of total FN current from the rest of the wafer, the amount of substrate potential rise necessary to produce an equal increase in I_e also depends on the electron temperature.

Let us ignore the FN current in the transition region between $I_e \sim 0$ and $I_{FN} = 0$ for now and assume the total gate area in the $I_e \sim 0$ region is A_{FN}. Assume the total exposed wafer edge area is A_{ed}, then:

$$\frac{I_e}{I_{ion}} = \frac{A_{FN}}{A_{ed}}.$$

From Figure 2.12, we see that for each kT_e/e of substrate potential rise, I_e rises 3 to 4 fold. Unless the area of exposed edge is very small, the substrate potential is not going to rise much more than one kT_e/e. For most plasmas, T_e is less than 8eV and therefore the oxide field is less than 8MV/cm at the edge of the wafer. As a result, the I_{FN} is negligible. However, if the gate oxide is 50Å thick, the oxide field can be very high for high T_e plasmas. For such cases, the I_{FN} will be substantial.

2.4.3 Floating Substrate

If A_{ed} =0, that is, no edge exposure, the substrate is initially completely isolated from the plasma. This is a common situation in wafer processing. The initial substrate potential (without FN tunneling) would be somewhere between the maximum and minimum of the gate potential distribution, determined by capacitive coupling. (Recall that the voltage-drop across the gate-oxide is very small on average (Figure 2.15) by capacitive coupling). Once FN tunneling started, the substrate charge balance dictates that the sum of all substrate to gate current (I_s) equals the sum of all gate to substrate current (I_g). Since the substrate is isolated, I_e and I_{ion} are both equal to zero. Equation 2.7 becomes:

$$\sum_i I_{is} + \sum_j I_{jg} = 0 .$$

Here

$$\sum_j I_{jg} = I_{FN}$$

at the edge of the wafer for the case at hand because gates are more negative than substrate. Since gate injection will pull the gate potential up, which leads to rapid increase in electron flux from plasma (Figure 2.12), the gate injection FN current components can be much higher than the saturated ion current. The only constraint is that for each gate, charge must be balanced. In other words,

$$I_{jg} + I_{ion} = I_e$$

The possibility of $I_e >> I_{ion}$ means I_{jq} can be quite large. A large I_{jg} requires a higher oxide field to support. The substrate potential is thus expected to positioned at a higher level when the wafer is isolated than when it is not (see Figure 2.19).

From the above discussion, it is clear that when charging damage is by gate injection, the damage is likely to be more severe. This is exactly the same conclusion Kubota *et al.* had come to after a combination of experimental and simulation effort [6]. If the substrate is truly floating, any charging damage by substrate injection at some part of the wafer must accompanied by gate injection damage at some other part of the wafer. Otherwise, the circuit is not complete and damage cannot occur. Returning to the experiment of Tsunokuni *et al.* that was discussed earlier (Section 2.2.1), we are now equipped to understand a few more details of their data. The ΔV_{FB} profile in the perpendicular case (left-hand side of Figure 2.7) is negative at the center and positive at the edge, exactly the opposite of Figure 2.19. One thing to keep in mind is that the total dielectric layer thickness (the nitride and oxide stack) in a MNOS device is quite thick. Little current can tunnel directly across the stack. The voltage it experiences or indicates is the full voltage the plasma can support. Even though the positive ΔV_{FB} at the edge indicates a higher potential can develop in the absence of tunneling, the actual voltage developed across the capacitor during plasma exposure is limited by the ion current density. On the other hand, the negative ΔV_{FB} at the center indicates the plasma can support a high voltage across the capacitor (although not as high as at the edge) as well. For these capacitors, the FN tunneling is by gate injection. As discussed above, the tunneling current is supported by the electron flux from the plasma, which can be 2 to 3 orders of magnitude higher than the ion current density. Thus we can understand why Tsunokuni *et al.* observed an increase in breakdown frequency only at the negative ΔV_{FB} locations.

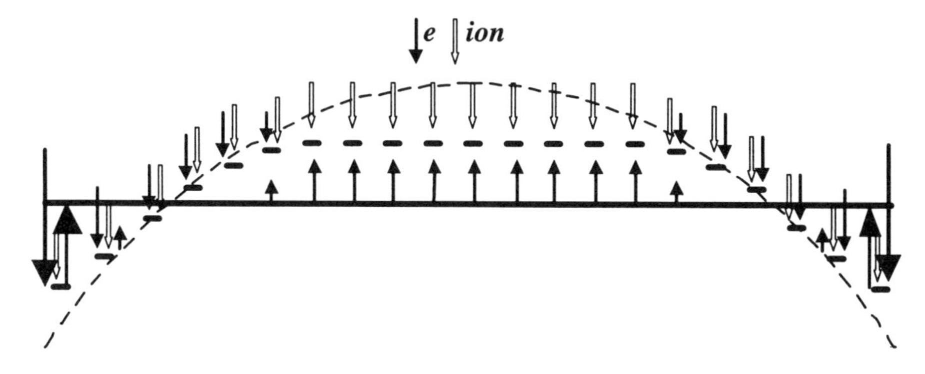

Figure 2.19. The potential distribution and current components for the gates and the substrate in the presence of FN tunneling are shown for the floating substrate case.

Obviously, damage had happened to the positive ΔV_{FB} region as well. However, Tsunokuni *et al.* was measuring the dead-on-arrival frequencies. Since gate-oxide requires a certain total injected charge to break, those capacitors at the positive ΔV_{FB} location did not receive enough FN current to break. If Tsunokuni *et al.* were using other more sensitive method to measure damage, they would have detected damage there as well.

Another point worth mentioning is that, if the substrate in Tsunokuni *et al.*'s experimental were floating (edge not exposed), even the capacitors in the negative ΔV_{FB} region might not have been broken. This is because all current flow must be balanced (the circuit must complete). The current injected from the gate at the center to the substrate must flow out to the plasma from somewhere. When the substrate is floating, it cannot flow directly back out to the plasma. It must then flow out to the plasma through FN tunneling into the gate at the edge. However, the FN current there is limited by the ion current and cannot increase. A fixed FN current means that the potential drop across the gate and the substrate is fixed. They can remain fixed if they are shifted by the same amount. Recall from Section 2.2.3 that shifting a floating potential downward is far easier than shifting it upward. We can expect then both the gate potential at the edge and the substrate potential will shift downward to reduce the potential difference between the gate and the substrate at the center. More accurately, the electrons injected into the substrate will pull down the substrate potential, which will drag the gate potential at the edge of the wafer down as well. However one like to think about it, the net result is that the FN tunneling current at the center will be reduced substantially. Note that the FN current at the center will still be higher than the edge. This is due to the total area difference between the center negative ΔV_{FB} region and the edge positive ΔV_{FB} region.

2.4.4 Charge Balance versus the Charge Imbalance Model

We have seen how the concept of charge balance leads to charging damage and how it can be used to explain complex charging situations. In the plasma charging damage literature, it is far more popular to describe plasma charging damage as the result of local charge imbalance.

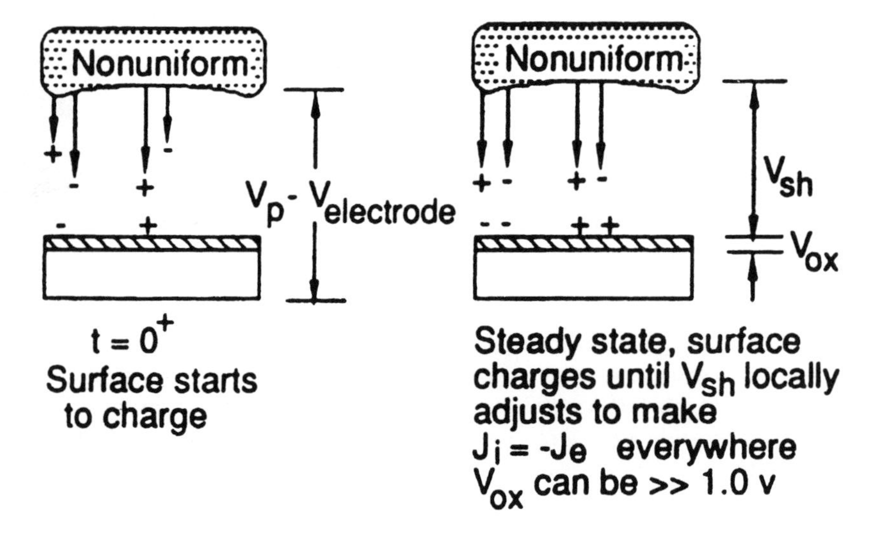

Figure 2.20. The charge imbalance model assumes that the sheath thickness is non-uniform due to the non-uniformity of the plasma. The non-uniform sheath thickness leads to imbalance in electron and ion flux from the plasma. Where the sheath is thin, the electron flux is larger than the ion flux and vice versa where the sheath is thick. At steady state, the surface will charge up enough so that the electron and ion flux are the same every where. Taken from [9].

The charge imbalance model was introduced by Fang *et al.* [7-10]. Figure 2.20 is taken directly from Fang *et al.*'s paper [9]. In their model, wafer coming into contact with non-uniform plasma will have a non-uniform sheath thickness. Because electron flux is an exponential function of sheath thickness, it should vary drastically across the wafer initially. The left-hand-side of Figure 2.20 shows that at the point where the sheath is thin, more electrons arrive to the wafer than ions, and at the point where the sheath is thick, more ions arrive to the wafer than electrons. This initial imbalance will cause local potential to build up, which in turn causes the electron flux to change until, at steady state, the ion flux and electron flux is the same every where (right-hand-side of Figure 2.20). On the other hand, if the oxide separating the gate from the substrate is thin, tunneling will occur at some point. When the local potential is high enough that the tunneling current equals the local current imbalance, potential build up will stop.

In Section 2.1.3, we discussed how a plasma sheath is formed from basic plasma physics. When an object comes in contact with plasma, both electrons and ions are bombarding it at a high rate. Electrons, being high in energy and lighter in mass, always come in at a rate 3 to 4 orders of magnitude higher than ions. So, at the initial stage, unless there exists some other means to inhibit electron flow, no where on the wafer would ion current be higher than electron current. The scenario in the left-hand-side of Figure 2.20 is, in general, not correct. In Figure 2.4, we have seen that in the absence of external connection, the floating potential tracks the plasma potential. This is the essence of Equations 2.4 and 2.6. The sheath potential and therefore the sheath thickness are independent of the plasma potential. Therefore, in the absence of external contact, non-uniform plasma does not lead to a non-uniform sheath thickness, nor does it lead to a non-uniform electron current. The non-uniform sheath thickness arises from the influence of an external contact. When the external contact supplies or takes away electrons from the floating object, it will change the sheath thickness so that electron flux from the plasma will be modulated to achieve charge balance. In other words, the non-uniform sheath is formed as a consequence of FN tunneling to or from the substrate, not the cause.

In the charge balance model, The key is to keep all current components sum to zero at every point (similar to Kirchhoff's law). The local difference in electron and ion current from the plasma is the consequence of floating potential variation. In the charge imbalance model, the local difference between electron and ion current from the plasma is taken as the cause of gate potential variation. In doing so it has created an impression that the current imbalance is the characteristic of the plasma under study. As a consequence of that impression, the electron and ion current imbalance at any location on the wafer is incorrectly expected to be independent of the substrate's potential. In other words, whether the substrate is floating or exposed to plasma should be completely irrelevant according to the charge imbalance model. Therefore to use the charge imbalance model to think about charging damage, one should always keep in mind that the description depicted in Figure 20 is not the real physics. It is simply a convenient way of describing the end results. If one can do that, then the charge imbalance model does not necessary lead to the conclusion of substrate potential being immaterial. However, few have been able to keep track of the real physics and the charge imbalance model has created a tremendous amount of confusion in the literature. In our discussion at the end of the last section, the electron current flowing from the plasma to the gates at the center of the wafer in Tsunokuni *et al.*'s experiment is highly dependent on whether the substrate is exposed to the plasma at the wafer edge or not. Such a result is intuitively inconsistent with the idea that the current imbalance is an inherent property of the plasma.

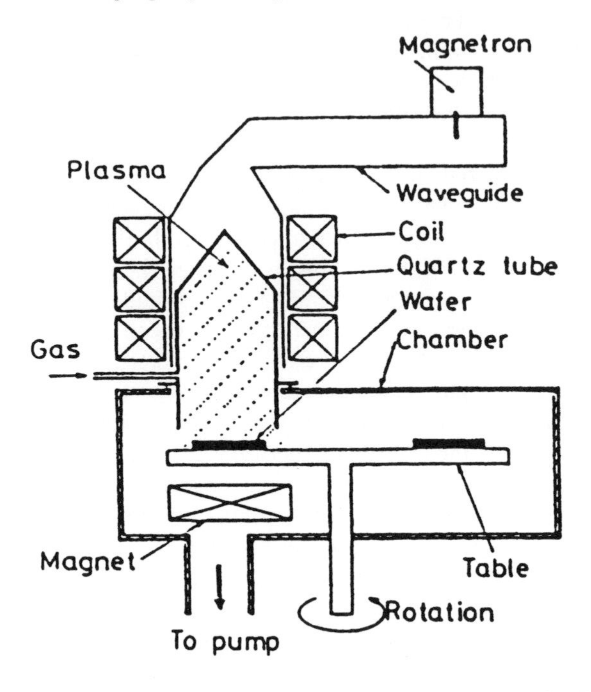

Figure 2.21. The ECR etcher used by Kawamoto [4] in his damage study. Wafers were laid out on a platen which rotates to bring the wafers in and out of the plasma. To understand the cause of damage, Kawamoto purposely stopped a wafer half way between in and out of the plasma.

The non-uniform floating potential being the cause of charging damage was first demonstrated by Kawamoto [4] a few years ahead of the charge imbalance model. Kawamoto studied the damage to gate-oxide in a rather peculiar plasma etcher (Figure 2.21). The plasma was provided by an electron-cyclotron-resonance (ECR) source. The peculiar part of the system was that wafers were laid out on a platen in a mini-batch format. As the platen rotated, wafers were brought under the plasma for processing one at a time. Since the plasma was always on, the wafers that were moving in and out of the plasma would at time be partly in the plasma and partly out of the plasma. Kawamoto notice that when he put a wafer half way in the plasma, the part that was not exposed was where a large flat-band shift was observed in MNOS devices (Figure 2.22). The calibration curve of his MNOS device has been shown before (Figure 2.9). The 3V value of ΔV_{FB} observed corresponds to >15V voltage across the nitride-oxide dielectric stack. When he replaced the MNOS devices with capacitors, sure enough, the breakdown frequency increase was observed only in unexposed the region (Figure 2.22). To explain his result, Kawamoto reasoned that for the part of the wafer in the plasma, both the gate and substrate were driven to a similar floating potential. The part of the wafer that was out side of the plasma, however, the gate was driven to a much higher floating potential than the substrate, which was pull down by the part that was exposed to the plasma. The reason for the gates out side of the plasma to have a much higher floating potential was that the ECR plasma was magnetically confined. Electrons would have to cross magnetic field lines to reach the gates that were out side the plasma. Crossing magnetic field lines was not allowed for electrons. It could only do so when a collision event occurs. Electrons thus had a much lower probability of reaching the part that was out side the plasma than ions. When ion flux is higher than electron flux, the floating potential becomes higher to attract more electrons from the plasma to achieve charge balance. From these results, Kawamoto correctly concluded that the cause for charging damage was floating potential variation across the wafer.

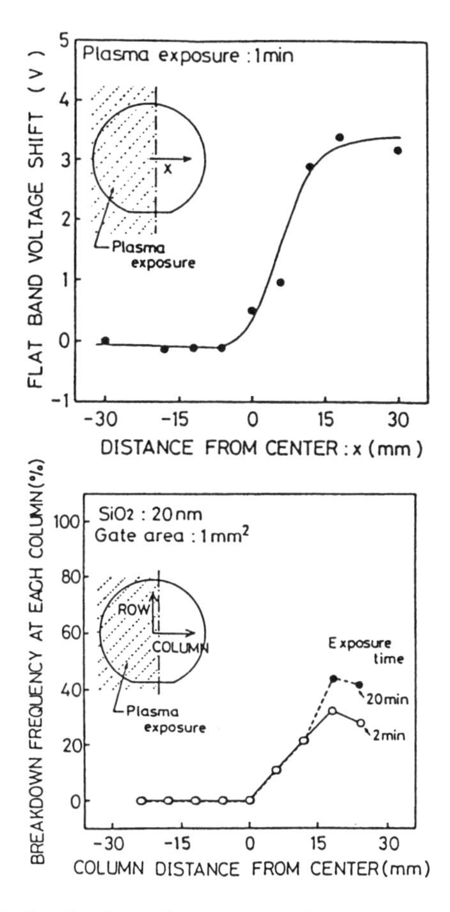

Figure 2.22. A large flat-band voltage shift is observed only at the locations that are outside of the plasma, indicating >15V of potential difference exist across the dielectric stack of the MNOS device. Capacitor breakdown is also only observed in the region outside of the plasma. Taken from [4].

To understand charging damage quantitatively, just identifying floating potential non-uniformity as the cause is not sufficient. One must be able to show how much current actually flows across the oxide under the influence of the potential. After all, for gate-oxide to break, a substantial amount of charge must flow across it under stress. The charge imbalance model provided a way to account for the tunneling current, but failed to recognize the role of the substrate. This failure was probably the result of two sources. One, Fang *et al.*'s experiments were all done using grounded substrates. Two, mistaking the local current imbalance as a property of the plasma under study. Two years after the charge imbalance model was proposed, the basic plasma physics of charging damage mechanism was reexamined and the successful charge balance model discussed here was introduced [11].

2.5 Antenna Effect

2.5.1 Area Antenna Effect

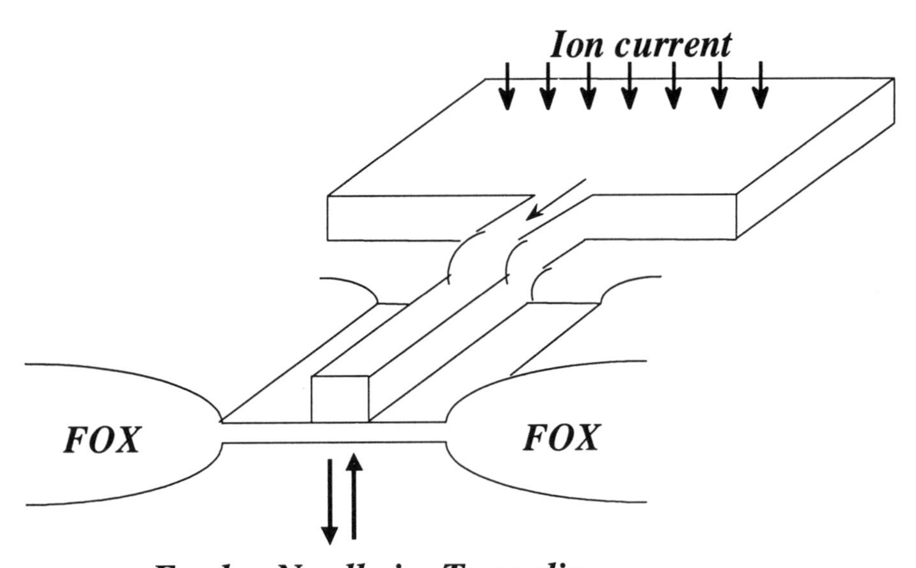

Figure 2.23. A large conductor plate on top of thick oxide (FOX) can collect the ion current and channel it to the thin gate-oxide region where the FN current must balance the total ion current.

We have seen in Section 2.4.1 that, for the case of damage by substrate injection, charge balance dictates that the magnitude of I_{FN} is the same as I_{ion}. That is, when the gate potential is lowered so much that no electron can overcome the barrier to arrive at the gate from the plasma, tunneling current is the only source of electrons that will balance the ion current. If we have a device that collects ion current from plasma over a large area and channels it to a small gate-oxide area where tunneling occurs, the tunneling current density will need to be substantially higher to maintain charge balance. This amplification of tunneling current density is the so called "antenna effect". The devices that help this happen are call antenna devices, and an example is shown in Figure 2.23. The large area that collects ion current must be a conductor and must be on top of a thicker layer of oxide so that tunneling in that region is negligible. This large conductor is called the antenna. The amplification factor for the tunneling current of the antenna device is simply the ratio of the total area of the conductor over thicker oxide to the area of conductor over thin oxide. This ratio is called the antenna ratio.

Using the same example in Section 2.4.1, if the antenna ratio is 1000:1, the total ion current collected by the antenna is now $1mA/cm^2$ x 1000 = $1A/cm^2$. For charge balance to be maintained, I_{FN} at the wafer center also needs to be $1A/cm^2$ instead of the $1mA/cm^2$ experienced by the simple capacitor. Not only does the tunneling current increase drastically, the voltage across the oxide is higher as well. While it takes only 10MV/cm to support $1mA/cm^2$ of FN current, 13MV/cm is required to support a $1A/cm^2$ of FN current. The voltage across the gate and the substrate is not the 10 volts as depicted in Figure 2.17. It should be 13 volts instead. For gate-oxide breakdown, it is not just the total amount of charge that flow across that is important. The field strength plays a very important role as well (see Section 1.5.1). With an antenna structure, both the FN current density and the stress field increase drastically. As a result, damage due to antenna structures is very serious.

Tsunokuni *et al.* [3] was the first to observe the antenna effect experimentally, in the same plasma system shown in Figure 2.7. In addition to MNOS devices and large capacitors, they used a small capacitor with large a polysilicon pad connected to the gate (Figure 2.24) to detect charging damage. As can be seen from Figure 2.10, they found that for capacitors that have 120Å thick gate-oxide, no increase in breakdown frequency is observed in all cases for the simple, large capacitors. However, when they looked at the small capacitors with a large polysilicon-pad attached, they found the breakdown frequency for the parallel case, positive ΔV_{FB} location increases with the area ratio of the polysilicon pad to gate ratio (Figure 2.24). Recall that in Section 2.2.1, we discussed the reason that no breakdown frequency increase was observed (even though the voltage at the positive ΔV_{FB} location for the parallel case is quite high) is that the plasma can supply too little current. With an antenna attached to the gate, it amplified the current supplied by the plasma many times. With the plasma also supplying a high voltage to support the high FN current, breakdown is readily seen.

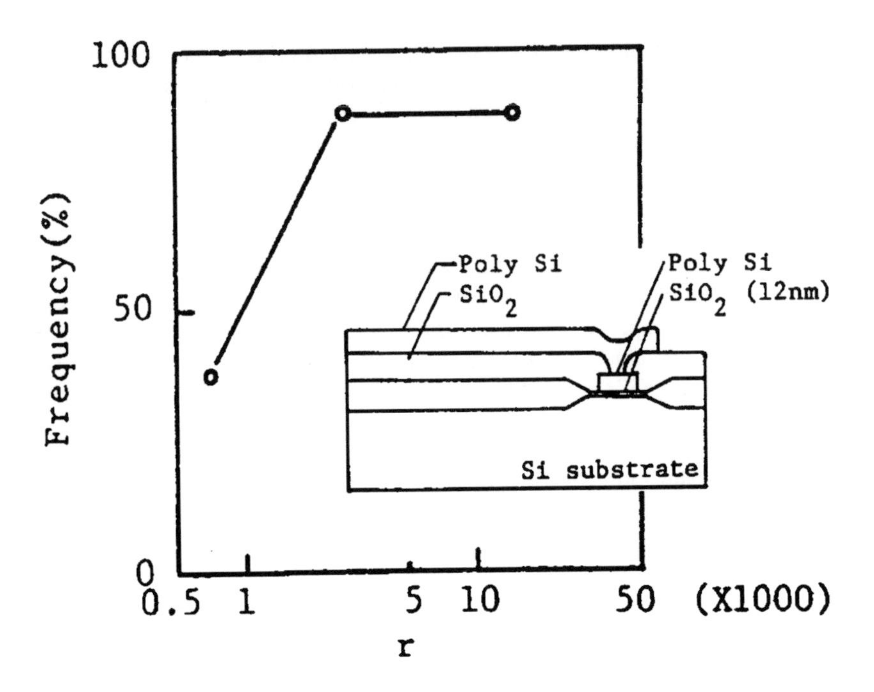

Figure 2.24. Effect of a large polysilicon pad connected to the polysilicon gate on breakdown frequency after plasma exposure. The breakdown frequency increases with the pad to gate area ratio. This is the first reported antenna effect. Taken from [3].

They observed also that the negative ΔV_{FB} location for the perpendicular case did not have an increase in breakdown frequency even for the capacitors with large polysilicon-pad attached. They reasoned that for breakdown to occur, first the plasma has to be able to support the necessary voltage that drives the FN current. Even though the antenna ratio is large, the voltage at the negative ΔV_{FB} location for the perpendicular case is not high enough. In the example we used earlier (Figure 2.17), we started with a potential of 30 volts and ended up with only 10 volts, the voltage needed to support the FN current, across the oxide. If we start out with only 9 volts, then the field across the 100Å oxide can only be 9MV/cm maximum. The FN current supported by such a field will only be ~0.1mA/cm^2, very much smaller than the 1mA/cm^2 ion current density. The gate potential will hardly change and the electron current from the plasma balances most of the ion current.

Tsunokuni *et al.*'s work clearly demonstrated another very important point in plasma damage. Namely both voltage and current have to be sufficient to cause damage. Missing either one, no damage can occur. One may therefore view the plasma as a current limited voltage source. First and foremost, it has to be able to support a large field across the gate-oxide before damage can happen. The antenna serves only as an amplifier to increase the available current for damage.

The area antenna effect has been widely reported. Whenever the full area of the antenna is exposed to the plasma, damage dependent on antenna area ratio is observed. The most common occasion for that to happen is in resist ashing operation after the regular etching is done. Shin *et al.* [12] studied the resist ashing damage after aluminum etch and found the actual charging current that flowed across the gate-oxide during ashing is almost linearly proportional to the aluminum pad area (Figure 2.25).

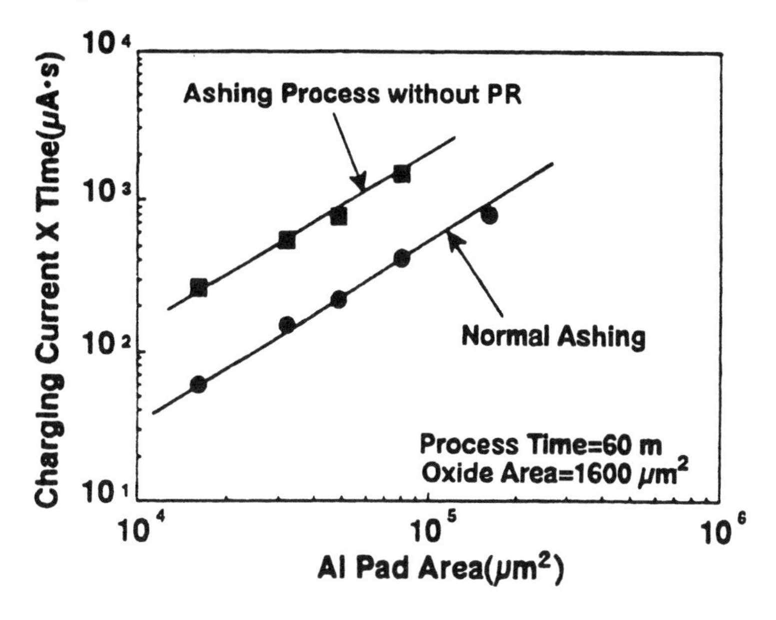

Figure 2.25. The measured charging current through the oxide during resist ashing is almost linearly proportional to pad area showing area antenna effect. The ashing process without the resist is shown here as a reference. Taken from [12].

2.5.2 Area Antenna Effect - Gate Injection Case

The above discussion of the antenna effect applies only to the substrate injection case, where the charge balance condition for charging damage is the tunneling current balancing the ion current. For the gate injection case, it is not as simple. The charge balance condition for damage in the gate injection case is the sum of FN current and ion current equals the electron current from the plasma. Since the electron current can be very large, gate injection damage can be serious without the help of an antenna. However, one thing is for certain, it is worse with the help of an antenna. For the gate injection case, the main determining factor is the ability of the plasma to support a high field across the oxide. Once that is attained, serious damage can occur. Attaching an antenna to the gate can worsen the damage, but going from very bad to worse may not produce a clear antenna dependent signal.

2.5.3 Perimeter Antenna Effect

To turn a large collection of individual transistors into functioning circuits, one must use conductors to connect them together. A MOSFET is a four-terminal device. All four terminals must be connected, sometimes by rather long conductor lines. These connections are the source of antennae. Antennae connected to anyone of the four terminals will impact charging damage to the transistor during fabrication by either affecting the gate or the substrate. However, during plasma etching of these antennae, an insulating mask layer keeps them covered. The covering mask layer can be photoresist or various kinds of hard masks. With the insulating layer on top, the antenna's collection of ions or electrons obviously will be inhibited. One might expect that damage would not be a problem under these conditions. Experience, however, shows otherwise.

The first clearly indicated damage case with a photoresist masked etching process was reported by Sekine *et al.* [13]. Figure 2.26 shows the structure they were etching with a magnetized plasma system. Sekine *et al.* did not discuss how damage could happen with the insulating mask. Although not specifically shown, many damage studies [14-19] before or around the same time of Sekine *et al.*'s work must have involved masked antennae as well.

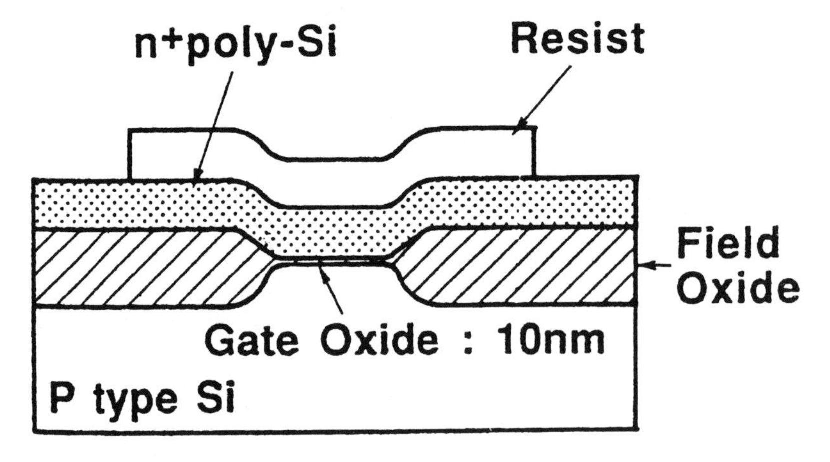

Figure 2.26. Defining polysilicon capacitor in a magnetized plasma using a photoresist mask. Damage was observed even though the polysilicon is covered by the insulating photoresist. Taken from [13].

Before we discuss how resist-covered antennae amplify the charging current to damage the gate-oxide, we need to back track a little bit. We started out to discuss how non-uniform plasma drives isolated objects to various floating potentials and as a result, how damage could occur. In wafer processing, most of the time the conductor on a wafer enters the plasma processing system as a solid sheet. It is the plasma process that will divide the conductor into millions of isolated islands that can be driven to various floating potentials. Figure 2.27 is a representation of the etching process to define a conductor layer into millions of isolated islands.

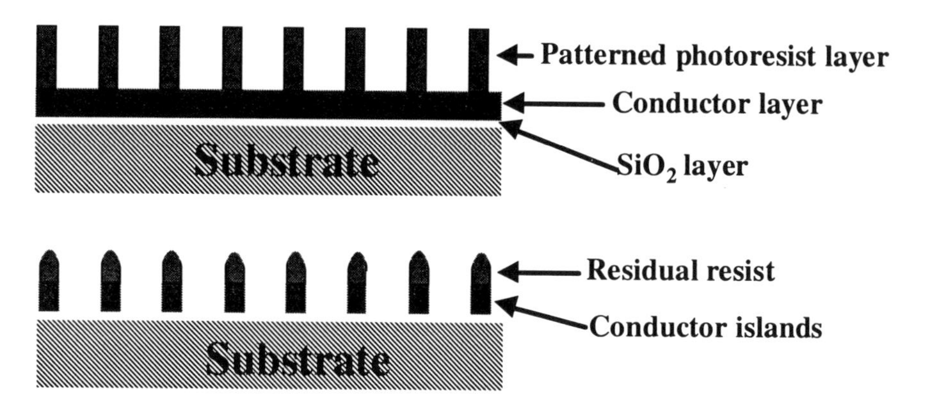

Figure 2.27. An illustration of an etching process that defines a conductor layer into millions of isolated islands. The top half is the condition before the etching starts condition and the bottom half is the finished condition.

Before the conductor is divided by etching, a good part of it is exposed to the plasma where charging can certainly happen. However, since the conductor is a solid sheet, it, like the substrate, does not support more than one potential. So even if the plasma is very non-uniform, there will only be one potential that the conductor layer will float to. Figure 2.28 shows one example. Here the plasma is very non-uniform. The dotted line represents the plasma potential offset by the sheath potential (floating potential curve for isolated objects). Line A is where the floating potential of the conductor should be. It is located near the bottom of the dotted line because of the highly asymmetric response of electron current from the plasma to floating potential shift. Where line A is above the dotted line, the electron flux from the plasma is much larger than the ion flux. Where line A is below the dotted line, electron flux from the plasma diminishes. Charge balance is maintained everywhere by allowing electrons to flow within the conductor. Line B is the floating potential of the substrate if it were floating (not exposed to plasma at any place). It is determined by the capacitive coupling as described in Section 2.3.2 and shown in Figure 2.15. Line C is the floating potential of the substrate if it is exposed to the plasma at the edge of the wafer.

It is often stated [9] that when the conductor is a solid sheet so that current can easily flow within the conductor to eliminate the current imbalance from the plasma, charging damage cannot occur. The difference in floating potential between the conductor layer and the substrate shown in Figure 2.28 for the case of exposed substrate can in fact cause charging damage. It is not too difficult to see that the potential difference between line A and line C can be high enough to cause gate-oxide to breakdown. Note that current will not be a limiting factor because there are bound to be weak points in the gate-oxide when the same field is applied across the wafer. Where there is a weak point, all current will be channeled there and the current density will be very high. It is lucky for us that most of the time, when conductors are being etched, the substrate is either tied directly to the conductor layer (when it is a solid sheet) or floating. As a result, it is most likely that before the conductor is etched into many small islands damage would not occur.

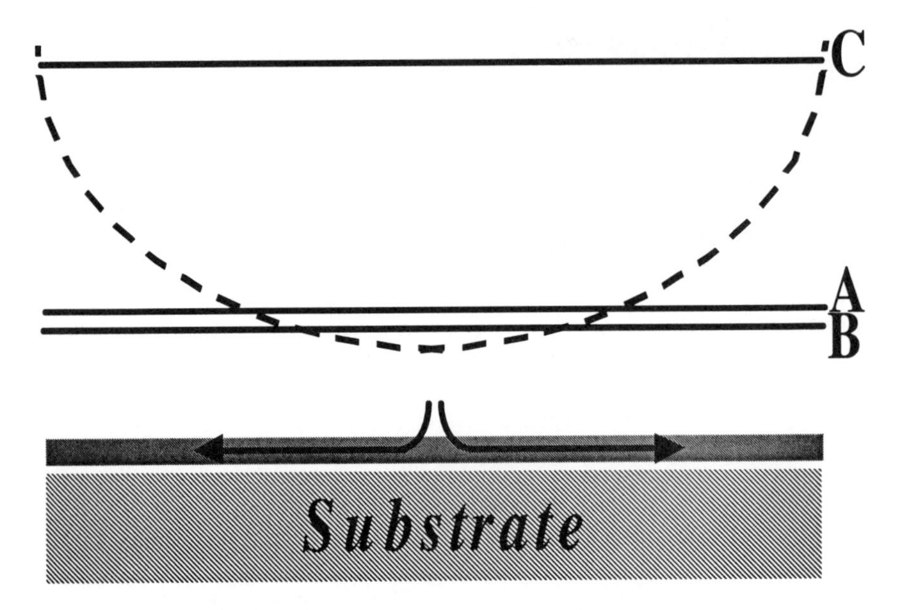

Figure 2.28. The potential diagram at the beginning of etching a conductor. Dotted line represent a very non-uniform plasma potential offset by the sheath potential. Line A is the conductor's floating potential. It is pinned at the center where it is collecting large amount electrons from the plasma. Line B is the substrate potential if it is floating. Line C is the substrate potential if it is exposed to plasma at the edge.

If damage does not occur before the conductor is broken up (top half of Figure 2.27), does it occur after conductor etching is done (bottom half of Figure 2.27)? It does, but most of the damage occurs before that. Before we continue, let's make clear what we are talking about. When we talk about the charging damage to the conducting islands, we assume that these conducting islands are each connected to a gate that is on top of thin gate oxide. Similar to the area antenna structure of Figure 2.23, the current collected by the conducting island will be channeled to the thin oxide area to balance to FN current flow across there. Figure 2.29 shows a representation of the structure.

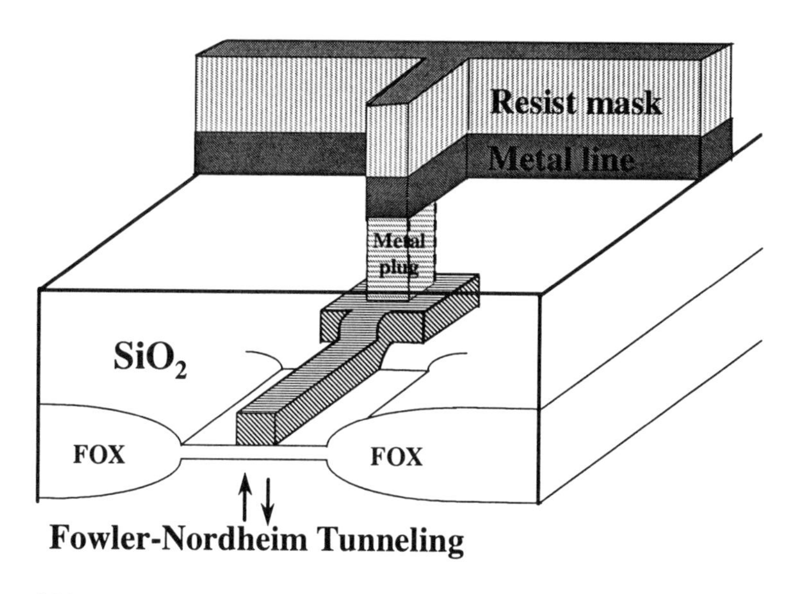

Figure 2.29. An antenna structure with insulating mask on top can only collect current at sidewalls of the conducting pad. The collected current is channeled to the thin oxide area to balance the tunneling current.

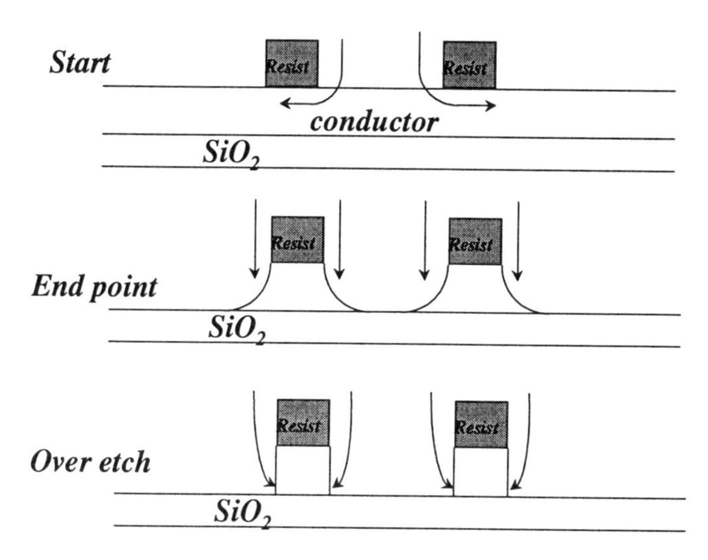

Figure 2.30. Between the initial condition and the final condition, there is an "end point" condition. At "end point", all the little conductor islands have "feet" to collect current from the plasma. After the "end point" is the over etch regime. Even though the "feet" are gone, charges can be bend by the field toward the conductor islands and damage can continue.

For a masked antenna (perimeter antenna), most of the damage occurs when "end point" is reached. "End point" is when the conductor first breaks into many small islands. At that point, the shape of the conductor islands is not quite like those shown at the bottom half of Figure 2.27. At "end point", each of the conductor island has "feet" at the bottom as shown in Figure 2.30. These "feet" obviously can collect current from the plasma and, at this point, current flow within the conductor to maintain charge balance is no longer possible. The total "feet" area along the perimeter of the masked antenna structure governs the efficiency of the antenna [20]. This "feet" at "end point" was later called the "halo effect" by Fang *et al.* [10]. As etching continues, the "halo" becomes smaller and smaller. Eventually, it disappears. How large and how long do these "halos" last are highly dependent on etching conditions as well as the density of the conductor pattern. Obviously, the "halo effect" should produce damage that is dependent upon the antenna perimeter length rather than the area. This is indeed the case. Shin *et al.* [21] studied the actual charging current that flowed across the gate oxide of antenna capacitors during aluminum etching and found that it is almost linearly proportional to the perimeter length of the antenna (Figure 2.31).

Fang *et al.* [10] demonstrated the "end point" effect through both experiment and simulation (Figure 2.32). The measured capacitor flat-band shift, which is an indication of charge trapping in the gate-oxide due to FN stress, rises sharply at "end point" and levels off. The simulated FN current also peaks at "end point". Damage by the perimeter antenna effect happening mostly at "end point" has been widely observed. For example, in an aluminum etch, Lin *et al.* observed that most of the damage happened during aluminum over etch and the clearing of barrier metal [22]. Note that in this case, it is during aluminum over etch and clearing of barrier metal that the "feet" exists.

Would damage continue after the "feet" are gone? The answer is that it does, but not as bad as during the conductor clearing in most cases. In Figure 2.32, I_{FN} does not go to zero after "end point". Instead, it remains at a low value. As indicated in Figure 2.30, charged particles can be bent toward the conducting island by the electric field. Figure 2.33 illustrates how the electric field develops and bends the trajectory of charged particles. The left-hand-side of Figure 2.33 is very similar to Figure 2.17 that illustrates the effect of FN current injecting into the gate on the potential of both gate and substrate. The starting gate and substrate potentials are what they should have been if no FN tunneling occurs. Damage is assumed to be significant during "end point". When the "feet" disappears, the amount of current intercepted by the conducting island is reduced. Charge balance dictates that I_{FN} also reduces. For the substrate, fewer electrons are lost to the gate by FN tunneling, meaning that its potential will drop back down somewhat. For the gate, it will drop further down so that the oxide field and therefore I_{FN} are reduced to keep track with the smaller I_{ion}. At the open area, conductor has been cleared away. It must rise up to the potential of no FN tunneling level (these are not thin gate-oxides). So in between the pattern conductors, the potential is significantly higher than the potential of the conductors. A lateral field thus exists to deflect ions toward the conductor islands.

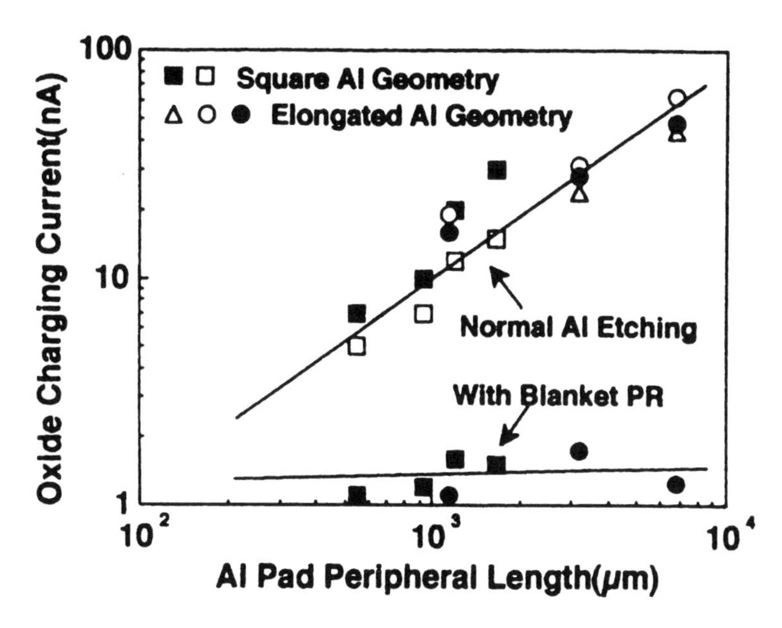

Figure 2.31. The measured charging current through the oxide is found to be linearly proportional to the perimeter length of antennae. When the antennae are completely covered by resist (no exposed edges), there is no measurable charging current. Taken from [21].

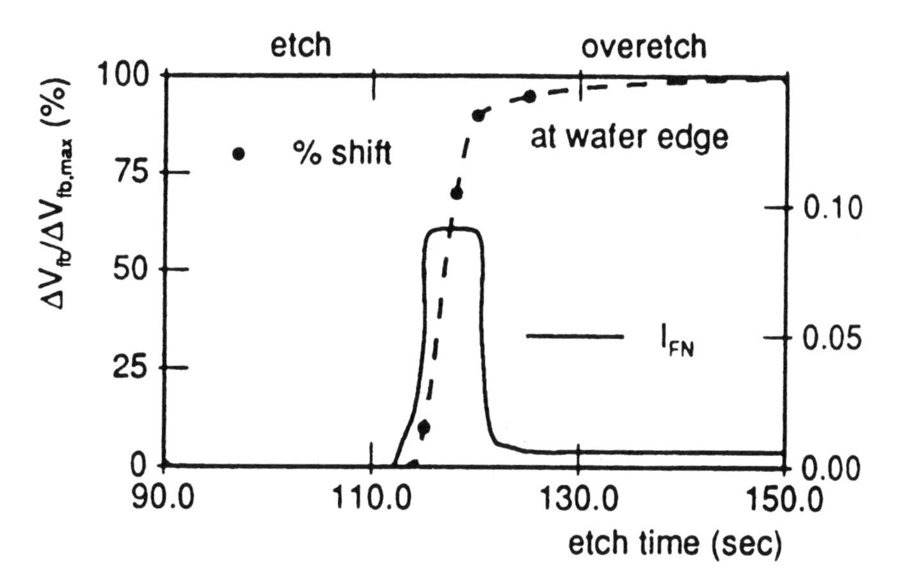

Figure 2.32. The measured flat-band shift of a capacitor etched in plasma is produced mainly at the end point. Simulated FN current also peak at the same time. Taken from [10].

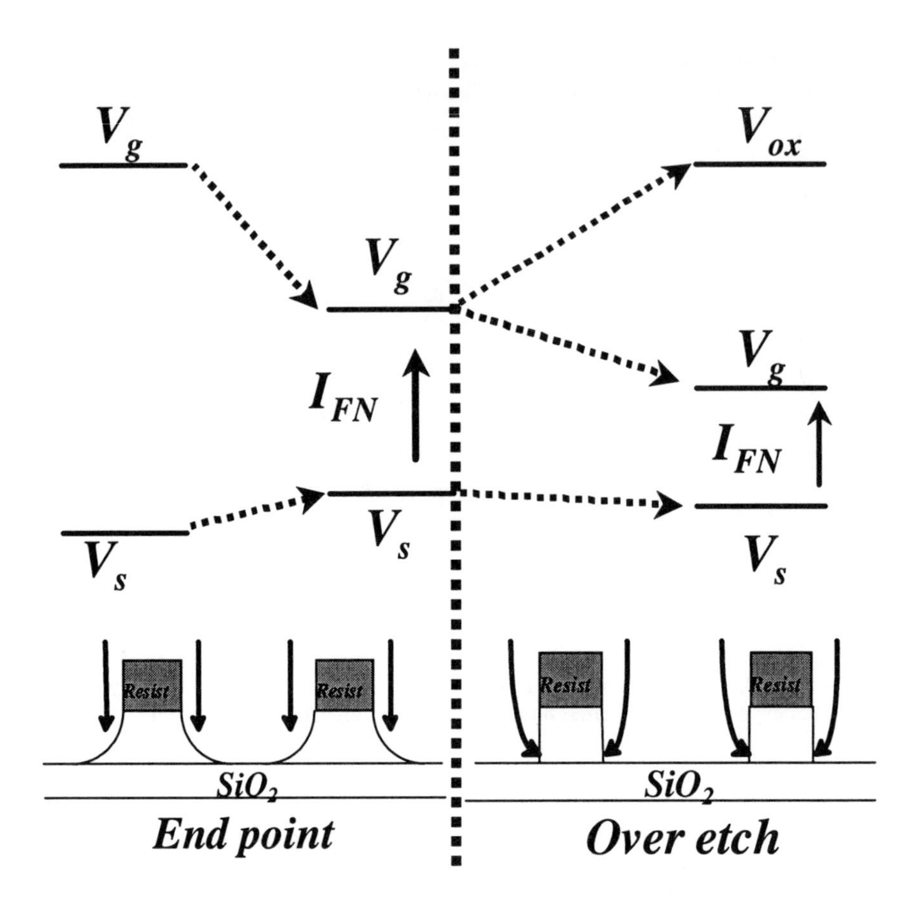

Figure 2.33. Starting from the no FN tunneling at the "end point" potentials, the gate potential shifts downward and the substrate potential shifts upward when FN tunneling is included. As the "feet" disappear, the collected ion current and therefore the FN current reduce. The substrate potential drops back down because less charge is lost. The gate potential drops further to keep charge balanced. The potential at the open oxide region floats up to the starting no FN tunneling level. An electric field thus exists horizontally to bend ions toward the conductor islands.

The case illustrated in Figure 2.33 is for substrate injection damage. Similar argument can be made for the gate injection case as well. All the signs will have to be reversed, and electrons rather than ions are deflected toward the conductor islands. Fang *et al.* [10] simulated the electron trajectory during over etch for the gate injection case (Figure 2.34). This is the source of the low-level I_{FN} shown in Figure 2.32 after "end point". In this simulation, a wide-open area is assumed. The electron trajectory is expected to be different for small spacing between conductors and therefore the collection efficiency will be different. In addition, the simulation shown in Figure 2.32 assumes that electrons are all initially traveling perpendicular to the wafer surface. In reality, electrons should be coming in from all directions.

Experimentally, clear demonstration of this post "end point" damage is seldom reported. The best example is probably the one reported by Shin *et al.* [23]. In Shin *et al.*'s aluminum etching experiment, the damage mode is substrate injection as shown in Figure 2.35. Positive charging current (I_{FN}) was measured all across the wafer. The "end point" is reached at 42s. Shin *et al.* noted that the interface state density increase (a measure of damage) at 45s is minimum. The continuous increase in interface state density with time (Figure 2.36) is therefore due to over etch. In Figure 2.36, the number of measurement points is limited. It is still clear that the rise after "end point" is faster. Note that "end point" marks the time at which the conductor is broken into islands. The "feet" will remain for a while longer. The faster rising early part may be due to the "halo effect". The slower but definitely rising part is clearly due to over etch.

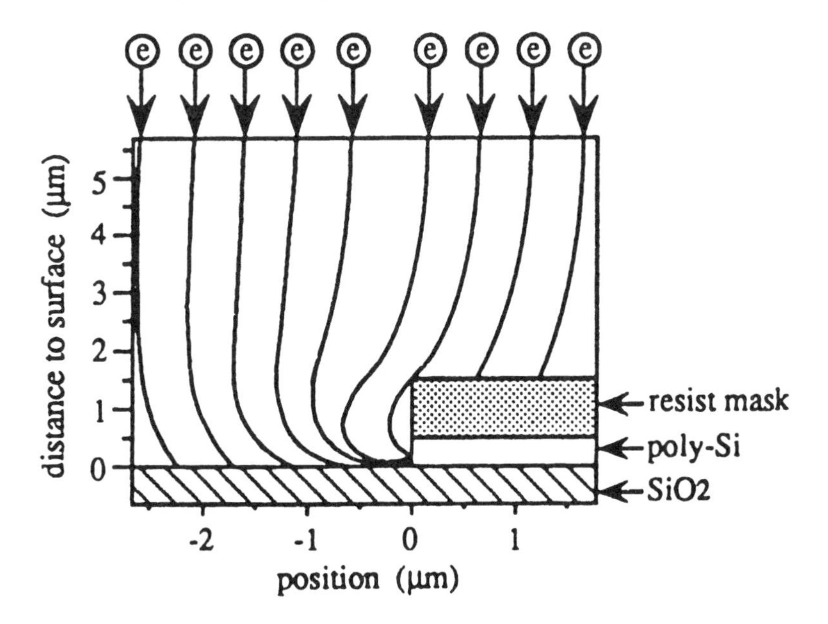

Figure 2.34. The simulated electron trajectory during over etch time for a gate injection case. Taken from [10].

Note that if charging was not severe enough to cause significant I_{FN} to flow during "end point", then nothing will happen during the over etch time. This is because the conducting island would have been at the same potential as any insulating surface near by. One cannot have a situation that charging only happens during over etch time but not during "end point" for etching a perimeter antenna structure. On the other hand, it is possible to design an etching recipe that the "end point" duration is very short so that the total damage during "end point" is negligible compared to over etch.

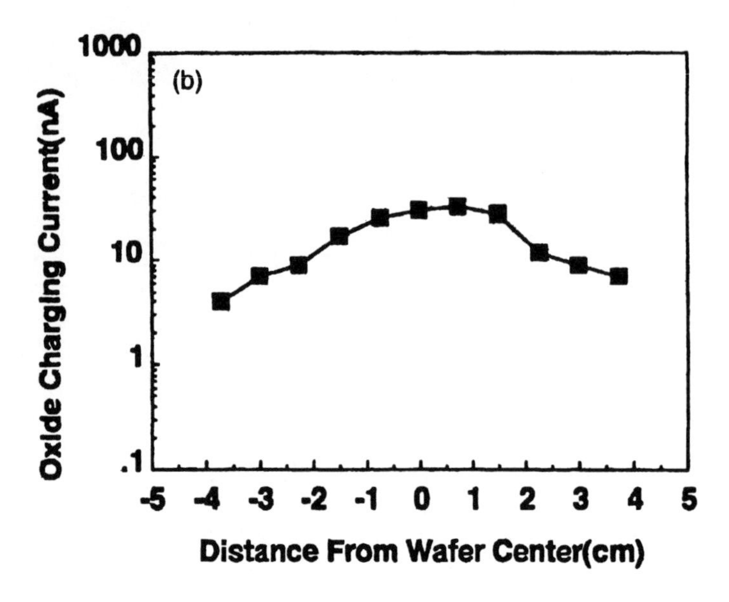

Figure 2.35. The positive charging current (substrate injection) distribution across the wafer in Shin *et al.*'s aluminum etch experiment. Taken from [23].

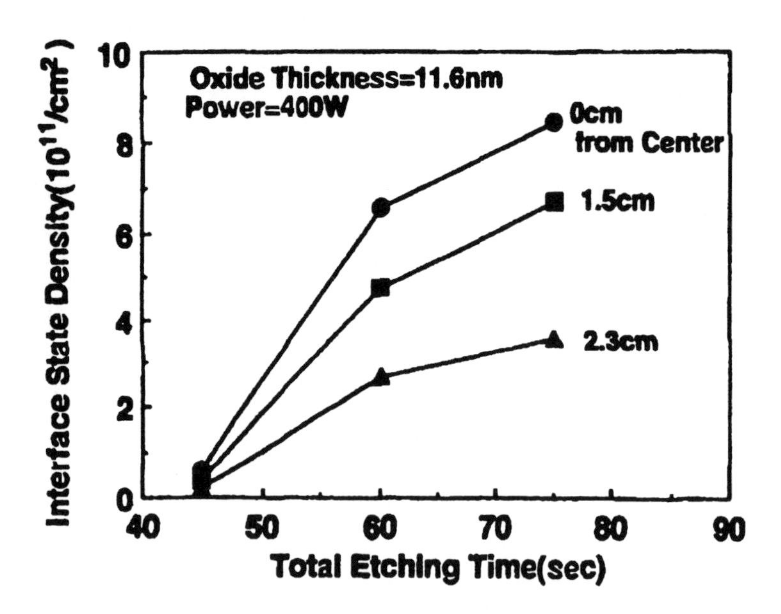

Figure 2.36. The measured interface state density increases (damage) as a function of etch time. The "End point" was at 42s. At 45s, little interface state density increase was observed. Damage is concluded to occur mainly during over etch. Taken from [23].

2.6 Uniformity of Electron Temperature

Since floating potential non-uniformity is the cause for charging to occur, anything that affects floating potential uniformity needs to be considered carefully. The charging damage discussion up to this point is based on Equations 2.4 and 2.6, which make floating potential track the plasma potential by showing that sheath potential is independent of plasma potential. However, Equation 2.4 and 2.6 do show that sheath potential is linearly dependent on electron temperature. If electron temperature is not uniform, it can cause the floating potential V_f to deviate from tracking the plasma potential V_P. We have so far ignored the effect of electron temperature. We have, in fact, implicitly assumed that electron temperature is uniform under most plasma processing conditions. Is this assumption reasonable?

Electrons travel at very high velocity. At low pressure where the mean free path is larger or comparable to the processing chamber dimension, it is not possible to have hot spots for electron temperature. Of course, as mentioned at the beginning of this chapter, power that sustains the plasma is coupled into the plasma through the acceleration of electrons by the externally applied electromagnetic field. Where the power is deposited, it is possible to have a higher population of energetic electrons. However, this higher population of energetic electrons cannot be too localized because the energetic electrons simply quickly traverse the small distance of a processing chamber. Instead, the entire plasma will have a high-energy tail for electron energy distribution that deviates from Maxwellian. This is rather common in low-pressure plasma. At pressures of a few Torr and above, where the mean free path starts to be significantly shorter than the chamber size, localized hot spots can exist. In most etching processes, the pressure is below 100mTorr. We are therefore safe to assume that electron temperature is uniform across the plasma for etching processes. On the other hand, plasma-enhanced chemical vapor deposition processes tend to have typical pressure of many Torr. For these processes, the assumption is not as valid. Even here, however, the impact is not expected to be too large.

If there is a magnetic field existing within the plasma, however, things can be quite different. Magnetic field limits the movement of electrons to orbital motions along the field line. Crossing the magnetic field line is possible only when a collision event occurs. We have encountered this effect in Section 2.4.4 when we discussed how Kawamoto explained the high floating potential outside the plasma. The electron mobility is thus very anisotropic and can cause significant non-uniformity in electron temperature even at low pressure. Based on Equation 2.4, even if the plasma is uniform, a non-uniform electron temperature can cause the floating potential to be non-uniform, which can in turn lead to charging. Recently, Tokashiki *et al.* [24] reported the observation of plasma charging damage due to non-uniform electron temperature distribution. In Tokashiki *et al.*'s experiment, a very high density plasma is generated by a Helicon source, which relies on the magnetic field to support the Helicon mode. The plasma is allowed to expand downward toward the wafer. The expansion is guided by another magnetic field so that the plasma density becomes uniform at the wafer surface (Figure 2.37a). The magnetic field that guides the plasma to the wafer surface prevents electrons from traveling sideways. This inability to intermix horizontally allows a non-uniform electron temperature distribution to be supported across the wafer (Figure 2.37b). When they measured charging damage using various antenna ratio devices, they found that the charge to breakdown (Q_{BD}) decreases with increasing antenna ratio (Figure 2.37c) indicating increasing damage. Tokashiki *et al.* examined three different plasma conditions. All produce fairly uniform plasma potential distributions but different degrees of electron temperature non-uniformity. They found that the more non-uniform the electron temperature distribution, the more severe the damage (Figure 2.37a, b, and c).

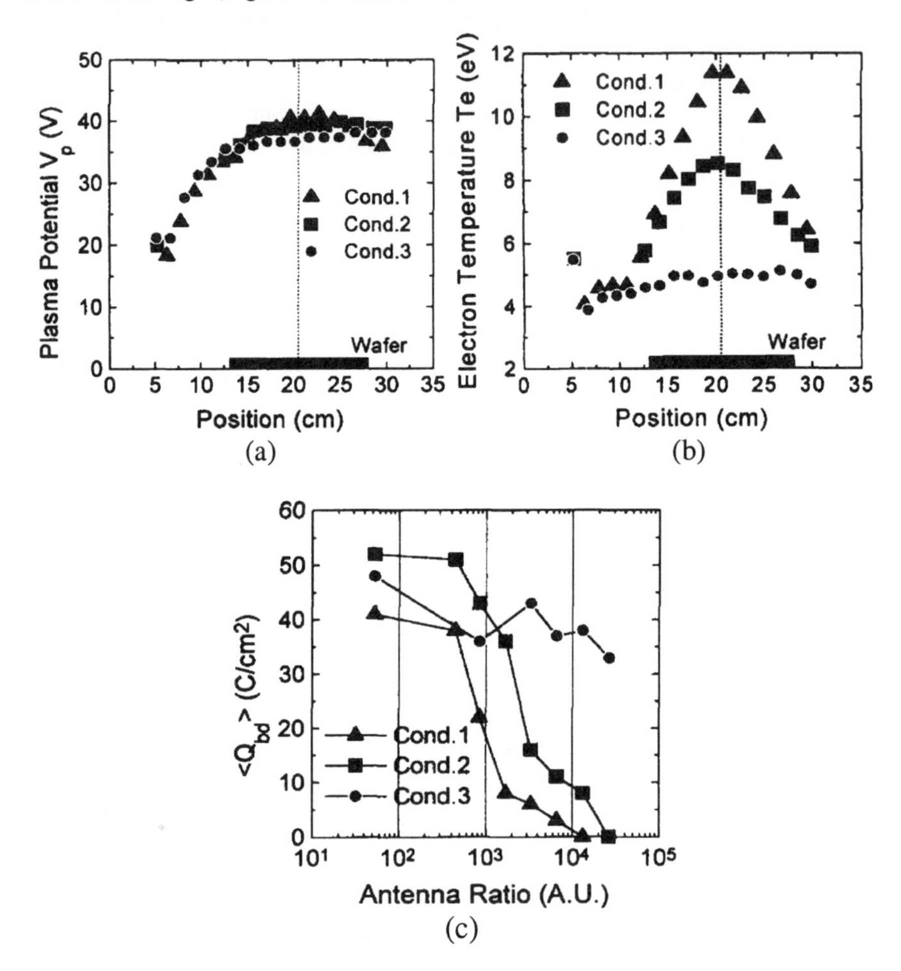

Figure 2.37. The plasma potential non-uniformity across the wafer surface was quite small (a). However, the electron temperature non-uniformity across the wafer was quite large for conditions 1 and 2 (b). When the antenna devices were tested for charge to breakdown, those that were from condition 1 and 2 had much lower Q_{BD} values (c). Taken from [24].

While Tokashiki *et al.*'s result is interesting and quite believable, one must be cautioned from taking it as conclusive. Electron temperature measurement using an electrical probe, particularly in the presence of a magnetic field, is highly difficult. There is a strong tendency to error on the high side. A more detailed discussion on the pitfalls involved in electron temperature measurement will be given in next chapter.

2.7 Charging Damage by High-density Plasma

We have been using a hypothetical plasma potential distribution through out most of the discussions in this chapter. The shape of the non-uniform plasma potential shown in Figures 2.18 and 2.19 does exist in many older plasma systems. The magnitude of the plasma potential variation assumed also is realistic only in some old systems. In modern IC manufacturing, most plasma systems use highly uniform plasma that is generated by sources that support a plasma potential distribution rather different from the shape shown in Figure 2.18 and 2.19. While the general discussion using the simple shape of plasma potential distribution can be applied to other shapes as long as we follow the physical argument rather than the specifics, there is an important difference in advanced plasma systems that warrants further discussion.

In the discussions up to this point, we considered only the limiting case where the plasma potential variation is large (many multiples of kT_e/e). It is under this assumption that we concluded $I_{FN} \sim I_{ion}$ when $I_e \sim 0$ and that the stress current is directly proportional to antenna ratio in the substrate injection case. Figure 2.38 shows where the limiting case would apply. Note that the horizontal axis of Figure 2.38 is floating potential deviation from charge balance, not the potential difference between gate and substrate. In other words, Figure 2.38 plots the electron flux change as the floating potential shifts to compensate tunneling. The magnitude of plasma potential variation is, at the minimum, equal to the sum of the shift and the potential across the oxide. As seen from the figure, only when the shift is more than twice the value of kT_e/e toward lower potential would we be in the regime where stress current is linearly proportional to the antenna ratio. Conservatively, the estimated minimum plasma potential variation would be at least 3 times the value of kT_e/e. The value of kT_e/e for most modern plasma is in the range of 2eV to 4eV. We need a plasma potential variation ranging from 6V to 12V. Most modern plasma processing systems have uniformity much better than that.

The limiting case of stress current linearly proportional to antenna ratio in highly non-uniform plasma has been reported in the literature [21], [25]. For example, Shin *et al.* reported that the measured stress current in aluminum etching is linearly proportional to the perimeter length of the aluminum pads (Figure 2.31). In this experiment, stress current during plasma charging was determined from the change in capacitance-voltage characteristics of capacitors after exposed to plasma. Since the aluminum pads are defined by photoresist, only the edges of the aluminum pads are exposed to the plasma. The linear relationship between perimeter length of the pad and the stress current is a clear demonstration of stress current linearly dependent on antenna ratio. While reports like these lend support to the simplified model discussed up to this section, one should be careful not to generalize such results, particularly when dealing with modern production plasma systems.

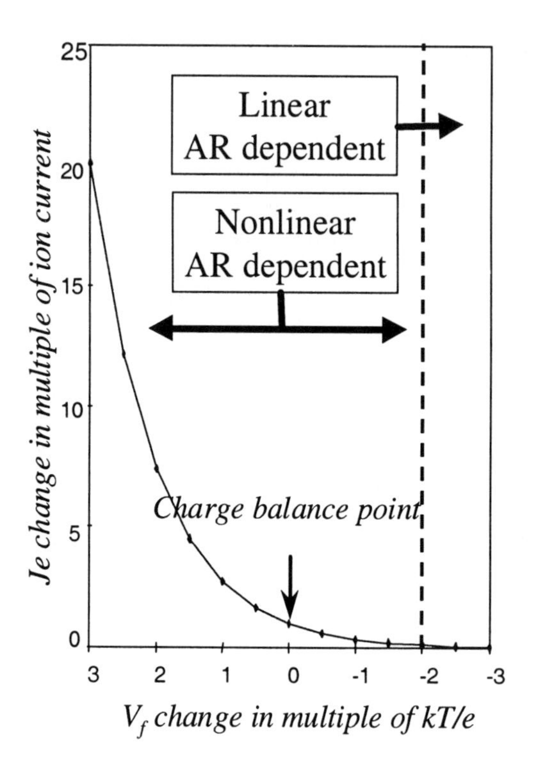

Figure 2.38. The electron flux reaching the wafer from the plasma as a function of the deviation amount from the floating potential in the absence of tunneling (charge balance point).

To treat the general problem of antenna ratio dependent stress current, we must return to the basic principle of charging damage, namely charge balance. Other than a very short initial transient, the following always hold:

$$I_{tun} = L \cdot AR \cdot (I_{ion} - I_e)$$

for every antenna device with antenna ratio AR and an antenna efficiency L. The antenna efficiency term is introduced to account for the fact that in perimeter intense antenna the effective exposed area per unit perimeter length is not easily defined. If we express I_e in multiples of I_{ion} as is done in Figure 2.38, we can write:

$$I_{tun} = L \cdot AR \cdot I_{ion}\left(1 - \exp\left(-e\left(V_0 - V\right)/kT_e\right)\right) \tag{2.8}$$

where V_0 is the floating potential at which the ion flux is exactly balanced by the electron flux and V is the floating potential of the antenna. We have made use of the fact that electron flux arriving to the wafer is exponentially dependent on the barrier potential. Since I_{FN} is a function of gate to substrate potential difference only, it is convenient to interpret V_0 and V as the gate to substrate potential difference. Thus plasma charging would have driven the gate and substrate to a potential difference of V_0 had there been no tunneling current across the gate-oxide. Due to the existence of tunneling current, the actual voltage difference developed is only V. Such interpretation of V_0 and V is entirely valid and does not involve any approximation. It is merely a change of the point of reference from the plasma potential to the substrate potential.

If we know V_0 and if we can express I_{tun} as a function of V, we could in principal solve Equation 2.8 to obtain stress voltage V as a function of the product of AR for a given value of LI_{ion}. Once we know the stress voltage, the stress current can be found directly from the current-voltage curve of the oxide in question. V_0 and LI_{ion} can be measured directly using methods to be described in Chapter 5, or can be obtained from yield data of more than one AR [26]. This later method will be discussed in more detail in Chapter 7.

When gate-oxide is thicker than 40Å, the tunneling mode encountered in charging damage will be exclusively due to Fowler-Nordheim (FN) tunneling. The expression for FN tunneling is well known (Equation 1.1). The only problem is that it is dependent on oxide field rather than gate to substrate voltage. While it is in principal possible to calculate the oxide field from the gate to substrate voltage, to do so accurately requires the knowledge of doping level in both the gate and the substrate and can be done only by computer simulation (see Chapter 1). If we know the flat band voltage V_{FB} of the device in question, it is possible to approximate the oxide field as:

$$E_{OX} \approx \frac{V - V_{FB}}{T_{OX} + 8\text{Å}} \tag{2.9}$$

Here we approximate the combined effect of polysilicon depletion and quantum confinement as 8Å additional oxide thickness. Unless the transistor in question is extremely poorly made, this approximation should be sufficient. Combining Equation 2.9 with Equation 1.1 of Chapter 1, we have an approximate expression for I_{tun} as a function of V.

When gate-oxide is thinner than 40Å, we must also consider direct tunneling. As discussed in Section 1.2.3, there is no simple analytical expression for direct tunneling that works well for the full range of oxide field. Fortunately in plasma charging damage, we are always dealing with high field stress and Equation

1.2 is adequate for this purpose. Since Equation 1.2 includes Equation 1.1 as a limiting case, we can use Equation 1.2 for all oxide thickness.

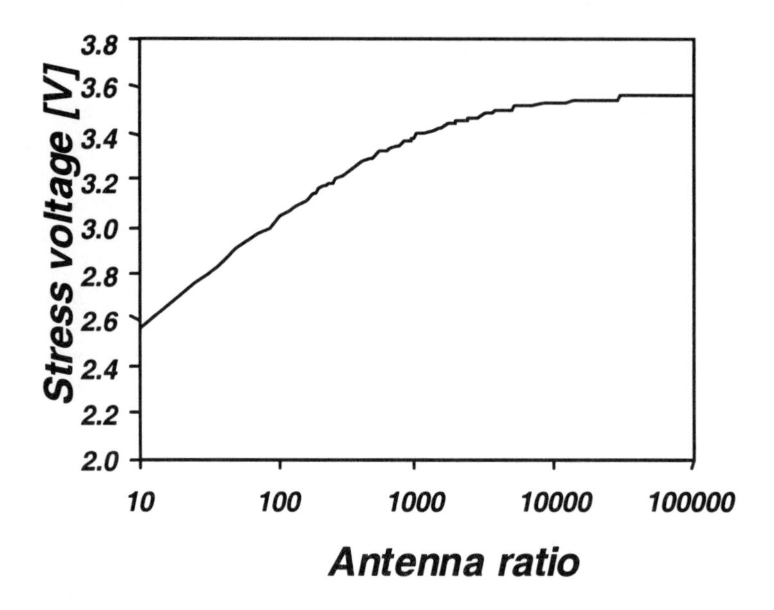

Figure 2.39. The calculated stress voltage as a function of AR during a dielectric deposition process using high-density plasma.

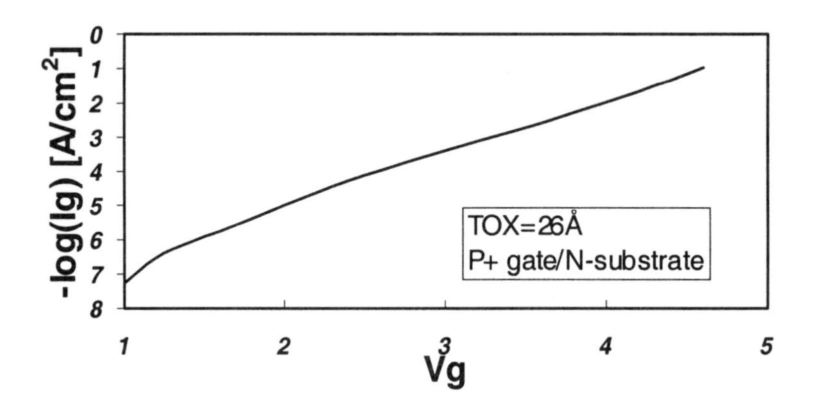

Figure 2.40. The current-Voltage (I-V) curve for the 26Å gate-oxide used the experiment of Figure 2.39.

Figure 2.39 shows the stress voltage calculated using the method described above for plasma charging damage during dielectric deposition using a current state of the art production tool. The gate-oxide was only 26Å thick and its I-V characteristic is shown in Figure 2.40. Combining data from Figure 2.39 and 2.40, the stress current during charging damage as a function of AR is obtained (Figure 2.41). Also shown in Figure 2.41 is the line representing stress current linearly proportional to AR (slope =1). For this particular example, a few important

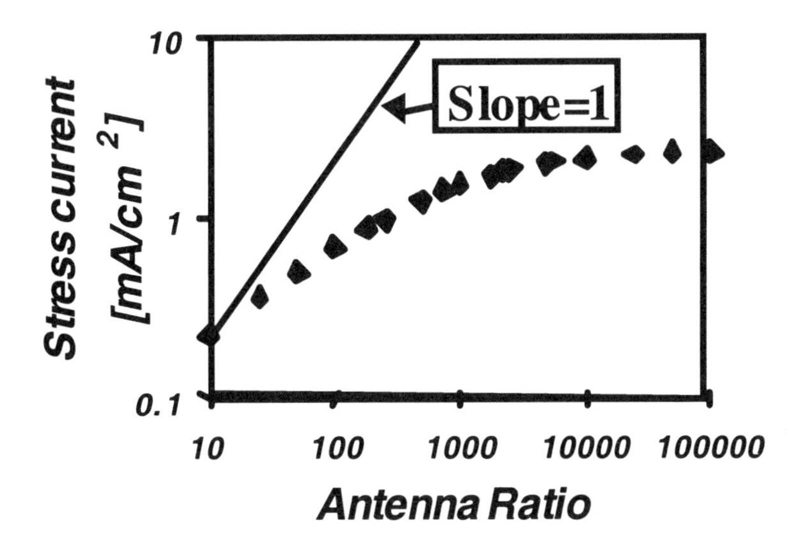

Figure 2.41. Calculated stress current as a function of AR from Figures 2.39 and 2.40.

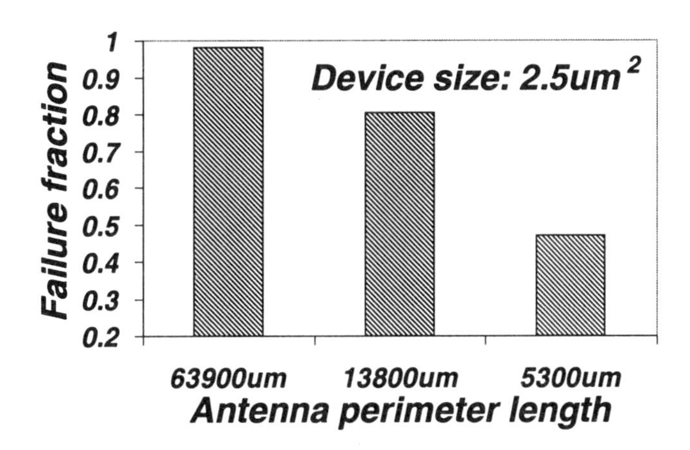

Figure 2.42. Failure fractions (F) for antenna devices with different antenna ratios.

characteristics of plasma charging stress in modern, highly uniform plasma system are demonstrated. First is the much lower than linear AR dependent stress current at low to medium AR. Second is the stress current's AR dependence weakens as AR increases and eventually becomes saturated. Third is that even though the stress voltage is quite low, the damage to ultra-thin oxide can still be significant. Figure 2.42 shows the antenna device failure rate after being exposed to the plasma in question.

References

1. Lieberman, M.A. and A.J. Lichtenberg, *Principles of Plasma Discharges and Materials Processing*. 1994, New York: John Wiley & Son.
2. Chapman, B., *Glow Discharge Processes*. 1980, New York: John Wiley & Son.
3. Tsunokuni, K., *et al.*, *The effect of charge build-up on gate oxide breakdown during dry etching*. in *19th Conf. Solid State Devices & Materials*. 1987. p. 195.
4. Kawamoto, Y., *MOS Gate Insulator Breakdown Caused by Exposure to Plasma*. in *Dry Process Symp*. 1985. p. 132.
5. Mantei, T.D., *Substrate Biasing for Plasma Etching*. J. Electrochem. Soc, 1983. **130**: p. 1958.
6. Kubota, M., *et al.*, *Simulational study for gate oxide breakdown mechanism due to non-uniform electron current flow*. in *IEEE International Electron Device Meeting (IEDM)*. 1991. p. 891.
7. Fang, S. and J.P. McVittie, *A model and experiments for thin oxide damage from wafer charging in magnetron plasmas*. IEEE Electron Dev. Lett., 1992. **13**(6): p. 347.
8. Fang, S. and J.P. McVittie, *The role and "antenna" structure on thin oxide damage from plasma induced wafer charging*. in *Mat. Res. Soc. Symp.* 1992. p. 231.
9. Fang, S. and J.P. McVittie, *Charging damage to gate oxides in an O2 magnetron plasma*. J. Appl. Phys., 1992. **72**(10): p. 4865.
10. Fang, S., S. Murakawa, and J.P. McVittie, *A new model for thin oxide degradation from wafer charging in plasma etching*. in *IEEE International Electron Device Meeting (IEDM)*. 1992. p. 61.
11. Cheung, K.P. and C.P. Chang, *Plasma-charging damage: A physical model.* J. Appl. Phys., 1994. **75**(9): p. 4415.
12. Shin, H. and C. Hu, *Monitoring plasma-process induced damage in thin oxide.* IEEE Trans. Semicond. Manufacturing, 1993. **6**(2): p. 96.
13. Sekine, M., *et al.*, *Gate Oxide breakdown Phenomena in Magnetized Plasma*. in *Dry Process Symp*. 1991. p. 99.
14. Wu, I.-W., *et al.*, *Breakdown yield and lifetime of thin gate oxides in CMOS processing*. J. Electrochemic. Soc., 1989. **136**(6): p. 1638.

15. Wu, I.-W., *et al.*, *Damage to gate oxides in reactive ion etching*. in *SPIE conf. Dry Processing for Submicrometer Lithography*. 1989. p. 284.
16. Shone, F., *et al.*, *Gate oxide charging and its elimination for metal antenna capacitor and transistor in VLSI CMOS double layer metal technology*. in *VLSI Technology Symp.* 1989. p. 73.
17. Greene, W.M., J.B. Kruger, and G. Kooi, *Magnetron etching of polysilicon: Electrical damage.* J. Vac. Sci. Technol. B, 1991. **9**(2): p. 366.
18. Gabriel, C.T., *Gate oxide damage from polysilicon etching.* J. Vac. Sci. Technol. B, 1991. **9**(2): p. 370.
19. Hoga, H., *et al.*, *Charge build-up in magnetron-enhanced reactive ion etching.* Jap. J. Appl. Phys., 1991. **30**(11B): p. 3169.
20. Gabriel, C.T. and J.P. McVittie, *How plasma etching damages thin gate oxides.* Solid State Technolog., 1992. **June-92**: p. 81.
21. Shin, H. and C. Hu, *Dependence of Plasma-Induced Oxide Charging Current in Al Antenna Geometry.* IEEE Electron Dev. Lett., 1992. **13**(12): p. 600.
22. Lin, M.-R., *et al.*, *Characterization and optimization of metal etch processes to minimize charging damage to submicron transistor gate oxide.* IEEE Electron Dev. Lett., 1994. **15**(1): p. 25.
23. Shin, H., G. Park, and C. Hu, *Plasma charging damage during over-etch time of aluminum.* Solid-State Electronics, 1998. **42**(6): p. 911.
24. Tokashiki, K., *et al.*, *Correlation Between Electron Temperature Uniformity and Charging Damage in High Density Plasma Etching Tool.* in *International Symposium on Plasma Process-Induced Damage (P2ID).* 1997. p. 207-210.
25. Ma, S., J.P. McVittie, and K.C. Saraswat, *Prediction of Plasma Charging Induced Gate Oxide Damage by Plasma Charging Probe.* IEEE Electron Dev. Lett., 1997. **18**(10): p. 468.
26. Mason, P., *et al.*, *Quantitative Yield and Reliability Projection from Antenna Test Results - A Case Study.* in *Symp. VLSI Technology.* 2000. p. 96.

Chapter 3

Mechanism of Plasma Charging Damage II

In modern VLSI manufacturing, extremely tight control on every production step is required. Plasma processing steps are no exceptions. Plasma processing tool has been developed at great expense to achieve excellent uniformity across the wafer. Gone are old tools such as barrel etcher with poor plasma uniformity. Yet, in modern IC manufacturing, plasma charging damage is more prevailing. The reason for this increase in incidences of damage is due mostly to the continued scaling down of gate-oxide thickness. Thinner gate-oxide requires a lower voltage to support Fowler-Nordheim (FN) tunneling and therefore can be damaged with a less severe plasma non-uniformity. On the other hand, if one were to calculate the plasma potential variation across the wafer based on the measured ion current or etch-rate in most modern plasma processing tools, one would have found that the voltage variations are often lower than what is needed to cause damage. Clearly, the actual floating potential variation across the wafer must be higher than what is suggested from these measurements. It is, of course, quite possible that the measurement method is inadequate and that the real variation in plasma potential is larger. This possibility will be explored in the next chapter. In this chapter, the focus is on some of the mechanisms with which charging damage can arise even with perfectly uniform plasma.

3.1 Electron-shading Effect

Electron-shading effect is an important charging damage mechanism. Unfortunately, there are still many unanswered questions about this effect. The discussion below therefore follows the format of a review rather than a presentation of well formulated subject. The details of various experimental and theoretical works are presented and critiqued to give a general feel of the current state of understanding of the subject.

3.1.1 Basics of Electron-shading Effect

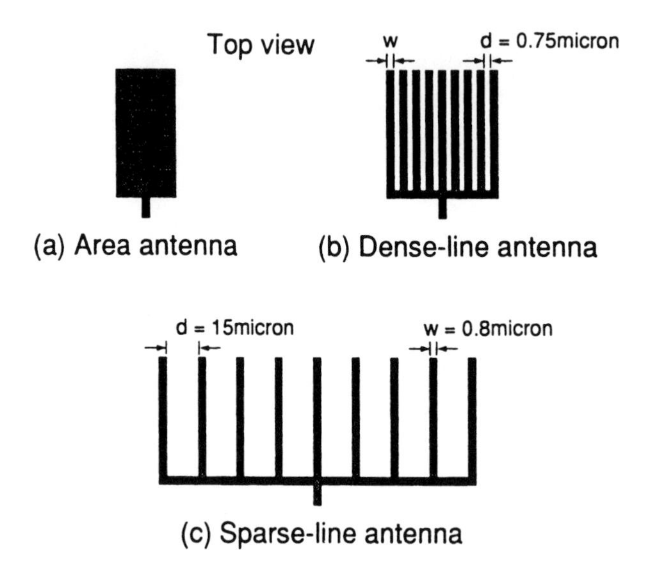

Figure 3.1. Three types of antenna with the same area ratio are shown. The two perimeter antennae have the same perimeter length. The only difference between them is the spacing between lines. Taken from [1].

In the course of studying plasma charging damage with antenna testers, Hashimoto [1] came across a peculiar result that did not fit the prevailing theory of charging damage. In his experiment, he had three types of antenna structures. All three of them had the same antenna area ratio. Two of the antennae were perimeter type that shares the same perimeter to gate area ratio (Figure 3.1). One had sparse-line (large spacing) while the other had dense-line. After plasma etching, the yield of sparse-line antenna capacitors was the same as the area antenna capacitors while the yield for the dense-line antenna capacitors was significantly lower (Figure 3.2). Since the etching was using a photoresist mask, a more accurate description of the area antenna is a perimeter antenna with very small perimeter length. The fact that the yield for sparse-line antenna and area antenna was the same suggests that capacitor yield was antenna perimeter ratio independent. Normally, one would conclude from this result that there was no charging damage or charging was too minor to impact breakdown behavior of capacitors. To make sure that this was indeed the case, Hashimoto exposed large area antenna capacitors without any resist on top to the same plasma. When no impact to breakdown yield was found, he concluded there was indeed no charging from the plasma of the conventional type. In other words, dense-line antenna capacitors were heavily damaged in the absence of non-uniform plasma induced charging damage. This was a puzzling result.

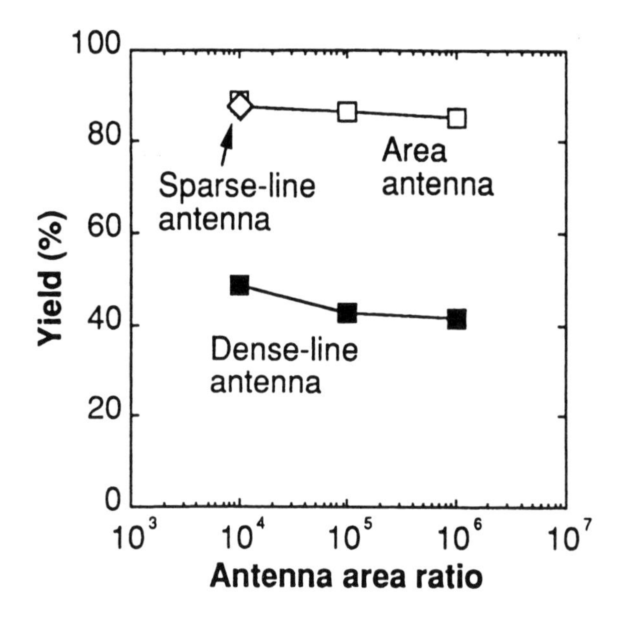

Figure 3.2. The yield for capacitors (MOS) with sparse-line antenna was the same as the yield for capacitors with area antenna. Only the dense-line capacitors suffer higher loss, indicating more damage. Taken from [1].

To explain the observed damage to dense-line antenna capacitors, Hashimoto considered the effect of insulating photoresist being charged up by the difference in directionality of electrons and ions. Such a charging effect had been considered in the past [2, 3] as the cause for etch profile distortion in polysilicon etching. This particular type of charging effect can be explained, after Hashimoto [1], with the help of Figure 3.3. At "end point", the sparse-line antennae have "foot" at the bottom while the dense-line antennae are still connected. The slower etch rate in narrow space is a common phenomenon in etching. It is called Aspect Ratio dependent Etch-rate (ARDE). If there is conventional charging damage due to non-uniform plasma, ARDE would have caused the dense-line antennae to be damaged more because the "foot" will be larger and last longer. However, this was not the case since Hashimoto already established that there was no non-uniform plasma related charging damage. The real reason, he argued, was that, similar to the etch profile distortion problem, fewer electrons can reach the bottom of narrow spacing than ions due to the isotropic characteristic of electron flux from the plasma. The narrow spacing of the dense-line will act as a spatial filter that allows only a small fraction of electrons to reach the conductor underneath the resist. On the other hand, ions are accelerated by the sheath and therefore highly directional toward the wafer. They have no problem reaching the bottom of dense-line antennae.

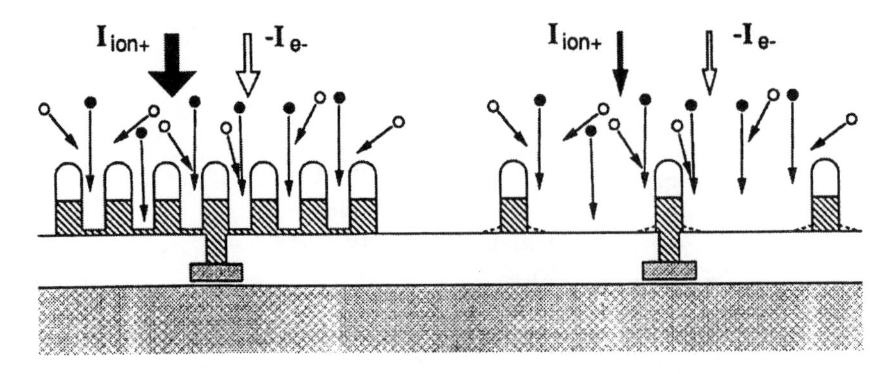

Figure 3.3. At "end point", both dense-line and sparse-line antennae have "feet" at the bottom. Both can collect charged particles. However, electrons, being more isotropic than ions, have difficulty reaching the bottom of the dense-line antenna. The antenna charges up to attract more electrons. Taken from [1].

Using the idea from charge imbalance model, Hashimoto reasoned that since the plasma was uniform, the electron flux and ion flux must balance everywhere in the absence of the shading effect of narrow spacing. In the space between narrow lines, fewer electrons than ions can reach the bottom means that there is charge imbalance there. One therefore expects charging damage to the dense-line antenna capacitors. As a result, even though there is no charging damage due to non-uniform plasma, charging does exist in the dense-line antenna just as he has observed experimentally. This explanation, however, was proven not adequate in a subsequent experiment.

In order to eliminate plasma non-uniformity as potential charging damage source, Hashimoto [4] made antenna structures out of small pieces (5mm) of wafer and put them into plasma that was designed to be uniform across the whole wafer. He used three types of structures as shown in Figure 3.4. One type had dense-line antenna capacitor with an open area to expose the substrate. Another type had the same dense-line antenna but no exposed substrate and the last type had flat area antenna with exposed substrate. As shown in Figure 3.4, of the three types only the one with dense-line antenna *and* exposed substrate showed antenna size (perimeter length) dependent damage. Clearly, dense-line antenna structure by itself is not the sufficient condition for electron-shading damage to occur.

Once again, we see that one must always think in terms of floating potential difference between the gate and the substrate, not the local current imbalance. In Section 2.1, we showed that floating potential was the mean to maintain charge balance on an isolated object. It is affected by the flux of both electrons and ions. Any change in the ratio of electron and ion flux must be reflected in floating potential change. Compared to the wide open space that exposed the substrate to plasma, the electron flux reaching to bottom of dense-line antenna with insulating mask is much smaller. The floating potential at the bottom of the dense-line antenna must be less negative relative to the substrate *with respect to* the plasma. In other words, the gate that is connected to the dense-line antenna must be at a higher potential than the substrate. If this potential difference is large enough, damage can occur provided enough current is available.

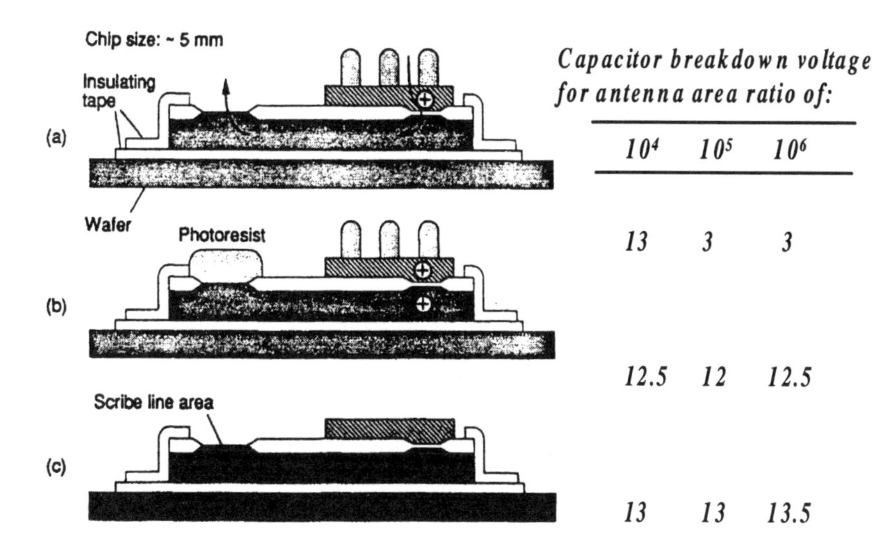

Figure 3.4. Use small chips supported on a wafer to eliminate charging due to non-uniform plasma to prove the presence of electron-shading damage. Capacitor breakdown shows that damage occurred only on dense-line antenna with exposed substrate. Taken from [4].

A more graphical representation of electron-shading damage is shown in Figure 3.5. Here the distribution of electron velocity vectors are shown flatter than isotropic to account for the fact that it must overcome a retarding potential traveling across the sheath. Ions are accelerated and are highly directional. Negative charges are accumulated at the top as well as in between narrow spacing of the resist. The accumulation at the top is normal for any floating object and is the reason for the formation of the sheath potential. The accumulation of negative charge on the resist sidewalls in between narrow line is higher than on the top. This is because the ion-flux is much more directional and does not neutralize the negative charge build-up on the sidewalls. The resulting floating potentials in the absence of FN tunneling is as follow. The top of the resist has the same potential as the wide open exposed substrate; the resist side wall is more negative than the top; and the conductor underneath the dense resist lines is more positive than the top. When FN current is flowing, the substrate potential will rise while the potential of the conductor underneath the dense resist lines will drop until the potential across the gate-oxide is what it needs to support the FN current. The key questions are how high will the gate potential be above the substrate in the absence of FN tunneling and how large is the FN current when it flows?

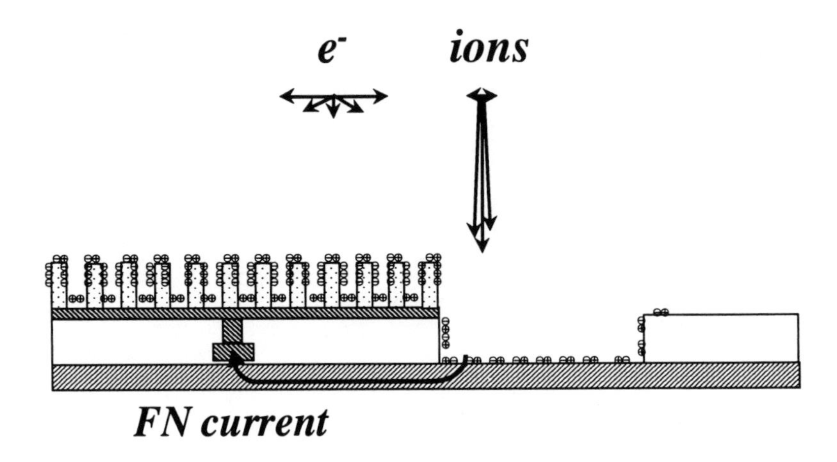

Figure 3.5. A graphical illustration of how electrons are shaded while ions are not.

3.1.2 Analytical Model

Vahedi *et al.* [5] attempted to model the electron-shading effect along with the conventional charging effect due to non-uniform plasma in one unified equation. Their approach is based on a structure similar to Figure 3.5. For the non-uniform plasma part, since floating potential track plasma potential, which in turn is a function of plasma density, one can write:

$$V_{f1} - V_{f0} = V_{p1} - V_{p0} = T_e \, ln\left(\frac{n_1}{n_0}\right) \tag{3.1}$$

Here V_{f1} and V_{f0} are floating potentials at the dense-line antenna and the exposed substrate respectively. Equation 3.1 is simply the well-known exponential dependent of plasma density on plasma potential [6]. For electron-shading effect, Vahedi *et al.* introduced shading factors k_i and k_e to represent the reduction factors for ion and electron fluxes arriving the conducting antenna underneath the insulating resist. Combining the electron-shading effect and plasma non-uniformity effect, they wrote:

$$\Delta V_f = T_e\left(ln\left(\frac{k_i}{k_e}\right) + ln\left(\frac{n_I}{n_0}\right)\right) \tag{3.2}$$

Here ΔV_f is the total floating potential difference between the antenna and the substrate. By expressing the shading effect in the same form as the electron density in non-uniform plasma, they have made it easy to solve the problem analytically. Equation 3.2 is the maximum voltage developed across the gate-oxide without FN tunneling. They call this the open circuit voltage. By solving Equation 3.2 simultaneously with Fowler-Nordheim equation, they can obtain the actual voltage across the gate-oxide for a given set of plasma parameter and shading factors. Figure 3.6 shows an example of the calculated result.

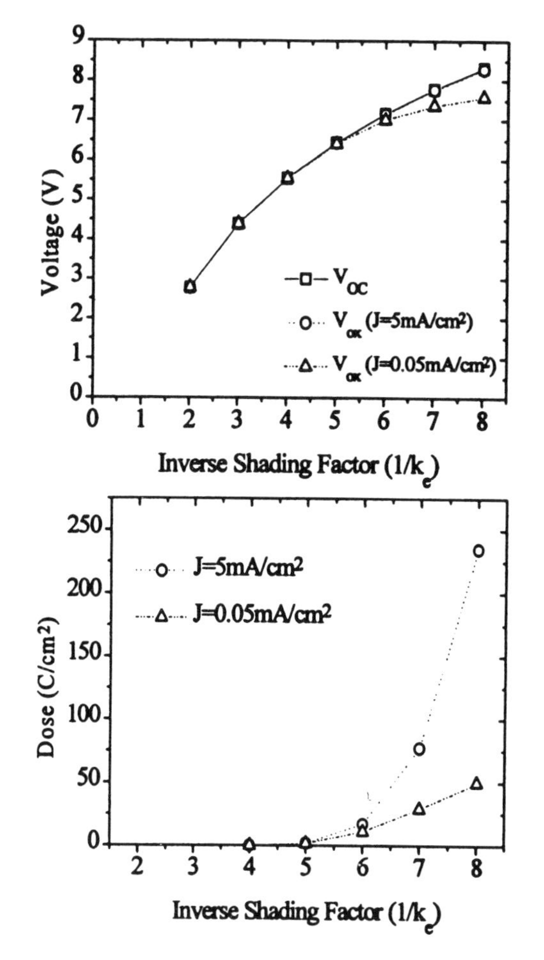

Figure 3.6. The stress voltage and current-density across the gate-oxide due to pure electron-shading effect according to Equation 3.2 and FN equation. Parameters are: Antenna ratio: 100000:1, open area to antenna ratio:100:1, $T_e = 4\text{eV}$, $T_{OX} = 5\text{nm}$, $k_i = 1$. Taken from [5].

By expressing the shading effect in the same form as the electron density in non-uniform plasma, an implicit assumption about the physics involved has also been made. This assumption can be understood by referring to Equation 2.4, which says the electron flux increases exponentially with decreases in barrier height. This exponential relationship between barrier and flux is a direct consequence of the Maxwellian distribution of electron velocity. Intuitively, electron-shading is a geometric effect. In other words, the dense resist line serve as a spatial filter. Without considering any distortion of trajectory due to charging, the filtering function is such that only those electrons that has a forward to sideways momentum ratio greater than certain number will make it to the bottom. The way Equation 3.2 is written, the potential rise serves the purpose of increasing the forward momentum of the electrons arriving to the wafer surface without changing the sideways momentum. As a result, the amount of electrons that satisfy the spatial filtering function increases. Simply put, Equation 3.2 implies that the bottom potential rises up to the point that it can bend the trajectory of electrons reaching the top opening of the narrow spacing so that more of them can reach the bottom to neutralize the ion flux.

The electron temperature T_e used in the calculation that produced the result shown in Figure 3.6 was 4eV, which is a reasonable number for high-density plasma [7, 8]. Notice that the potential rise at the bottom of the dense line is very low until the electron-shading is very severe. For example, the potential rise is less than 3 volts when 50% of the electrons cannot reach the bottom. Only when electron flux reduction reach 85% does the potential rise reach 7.5 volts. In Hashimoto's original experiment [1], his gate-oxide was 80Å thick. To breakdown the capacitors during the short plasma exposure, the field required is about 15MV/cm, which is 12 volts. To reach 12 volt using Equation 3.2 assuming the same electron temperature, the flux reduction is 95% (k_e =20). The aspect ratio in Hashimoto's experiment was about 2.5. It seems unlikely that shading can reach 95% with an aspect ratio of only 2.5.

Equation 3.2 is purely electronic; the role of ion is absent other than that the ion-shading factor k_i is not exactly equal to 1 due to the finite spread of momentum (non-zero ion temperature). While RF bias can increase the ion energy and thereby push the ion-shading factor closer to 1, it does not have an explicit role in Equation 3.2. Note that ions are accelerated toward the wafer and that ion temperature is normally much lower than electron temperature, we expect that the ion-shading factor k_i should be quite close to 1 even in the absence of a RF bias. The effect of RF bias is to make k_i go from very close to 1 to extremely close to 1. In other words, RF bias should have minimum impact on electron-shading damage according to Vahedi *et al.*'s model. A study of the effect of RF bias on electron-shading damage would be a good test of the model.

3.1.3 Impact of RF Bias on Electron-shading Damage

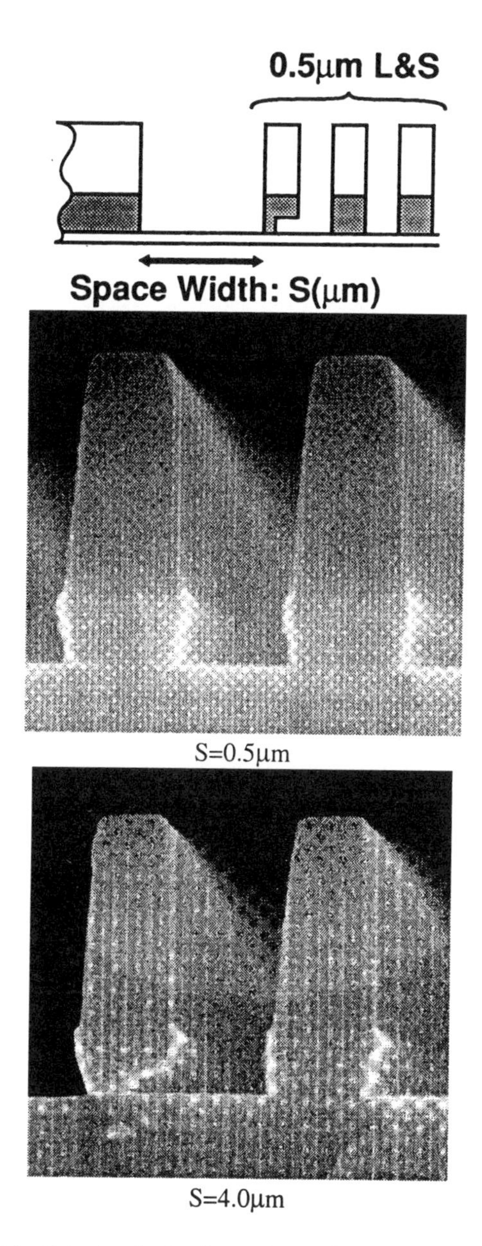

Figure 3.7. Etch profile distortion in the form of a wedge at the inner bottom of the outer most line. Taken from [9].

The impact of RF bias on electron-shading damage has been studied by more than one group. However, early studies are inconclusive. For example, Fujiwara *et al.* [9] studied the impact of RF bias on the notch depth in polysilicon etching. Notch formation in polysilicon etching is the most widely studied electron-shading damage. In this case, not the gate-oxide is damaged, but rather a wedge shape notch is formed at the bottom of the polysilicon line. A typical notch formation problem in polysilicon etching is shown in Figure 3.7 [9]. The plasma used was a high-density electron cyclotron resonance (ECR) source. Notch is observed only at the inner-side bottom of the outer most line and only when the open space is large enough. Typically, the dense-line are on top of thick oxides so that FN tunneling is not possible. The lines are not connected together and therefore each is free to float to any potential that satisfies charge balance. The mechanism for the notch formation is that electron-shading causes the poly lines underneath the photoresist to rise in potential. The outermost line rises the least amount because the open side of the outermost line is not shaded. The larger the open space, the more this outermost line can collect electrons. Note that the raised potential on this outermost line has the effect of attracting more electrons than the geometry may allow. In-between the outermost line and the second line, because the second polysilicon line is at a higher potential than the outermost polysilicon line, a lateral electric field exist to deflected ions toward the outermost polysilicon line during over etch to create the notch. As the space between polysilicon lines clears, the exposed oxide will charge up quickly and add to the deflection field for notch formation. Since the outermost line's ability to collect electron depends on the size of the open space, so should its potential. The more electrons it collects, the lower the potential rises and, as a result, the larger the lateral-field in the space between the outermost line and second line. The notch depth is therefore a function of the width of the open space (Figure 3.8).

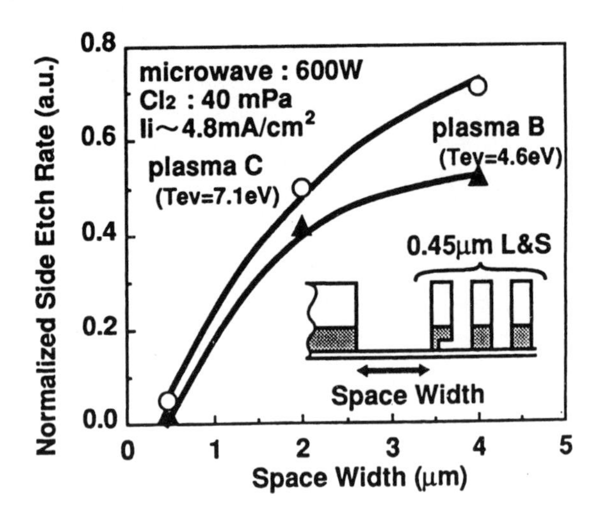

Figure 3.8. Side-etch rate as a function of space width for two different plasma conditions that have different "vertical electron temperature". Taken from [9].

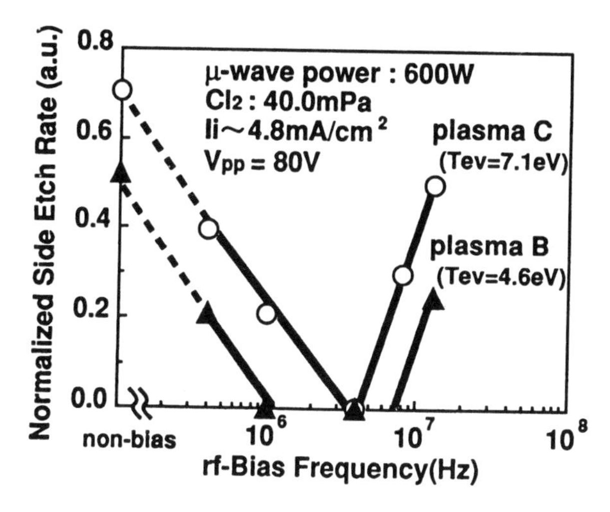

Figure 3.9. Side-etch rate as a function of bias frequency for two different plasma conditions. Taken from [9].

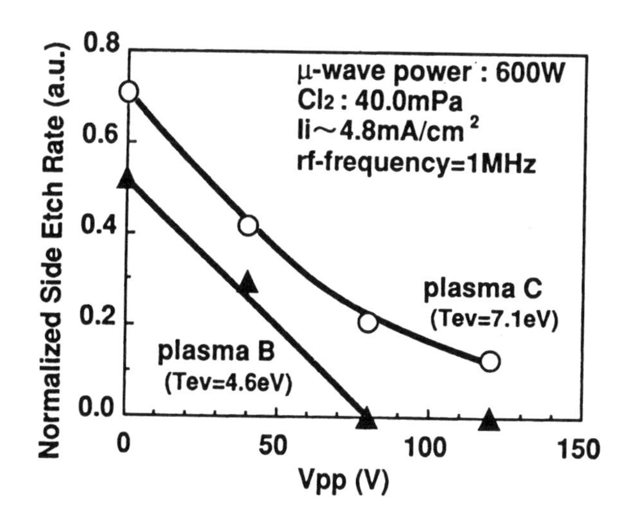

Figure 3.10. Side-etch rate as a function of bias voltage for two different plasma conditions. Taken from [9].

Fujiwara *et al.* found that the side-etch rate decreases at first with increasing bias frequency, reaches a minimum at some frequency and then increases sharply (Figure 3.9). They also found that the side-etch rate decreases with increasing bias (Figure 3.10). These results suggest that shading effect is a strong function of RF bias. The interpretation of both observations is, however, complicated by the change in photoresist erosion rate with RF bias. The bias dependent etching selectivity of polysilicon over resist is also a function of bias frequency and the trend is very similar to the side-etch rate for the etching condition studied (Figure 3.11). As selectivity decreases, more photoresist erodes during the polysilicon etching process. As a result, lower selectivity leads to a lower aspect ratio during over etch when notches are formed. Thus the bias frequency dependence of side-etch rate may be simply a direct result of lower shading effect when the aspect ratio is lower, not a real ion energy effect. A further complication is that higher resist erosion rate may help in generating polymers to protect the sidewalls and reduce notching. Similarly, the bias voltage dependent can also be explained by selectivity decreases with increasing bias voltage.

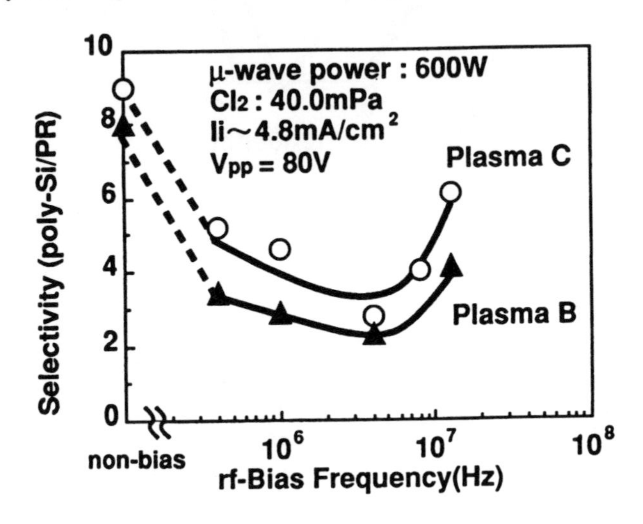

Figure 3.11. Selectivity as a function of bias frequency for the same two plasma conditions in Figure 3.9. Taken from [9].

Other studies suffer from a similar difficulty. For example, Hashimoto *et al.* [10] reported that electron-shading damage (to gate-oxide) in a metal etch process increases with bias power at first, and then decreases with further bias power increase. The very next year, they reported [11] that the observed effect was due to aspect ratio change caused by resist erosion. In this new report, Hashimoto *et al.* changed the shading insulator from photoresist to SiO_2, which maintained a high selectivity over metal in the etching process. Thus the aspect ratio remained fairly constant for all the bias condition studied. The result was that electron-shading damage increases with bias voltage and independent of bias frequency (Figure 3.12). In this study, a structure similar to the one shown in Figure 3.4a was used. The gate-oxide was 4nm thick and the aspect ratio of the shading SiO_2 was 2. The high-density inductive coupled plasma (ICP) was on for only 8 seconds and the bias for 5 seconds. As the applied bias increased from 0 volt to 50 volts and to 250 volts peak-to-peak RF amplitude, the antenna ratio at which significant yield loss occurred decreases, indicating more serious charging damage. This bias dependent electron-shading damage clearly cannot be explained using the model of Vahedi *et al.* Equation 3.2 clearly is not describing the real physics involved, or at least not all the physics involved.

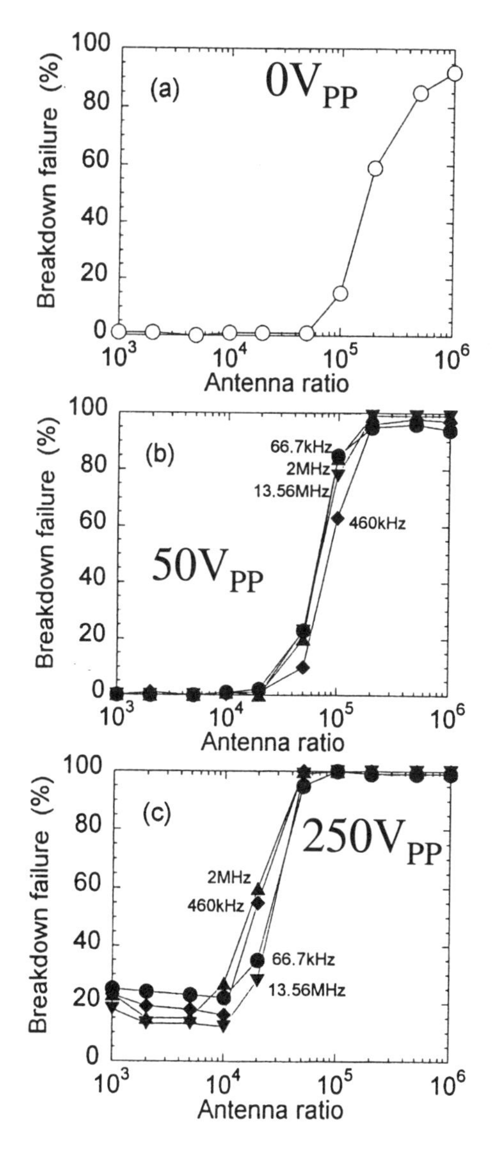

Figure 3.12. The antenna ratio at which breakdown failure increases significantly decreases with the bias voltage in an electron-shading effect experiment using structures similar to those in Figure 3.4. The bias frequency had no impact on damage. Taken from [11].

3.1.4 Ion Repulsion Model of Electron-shading Damage

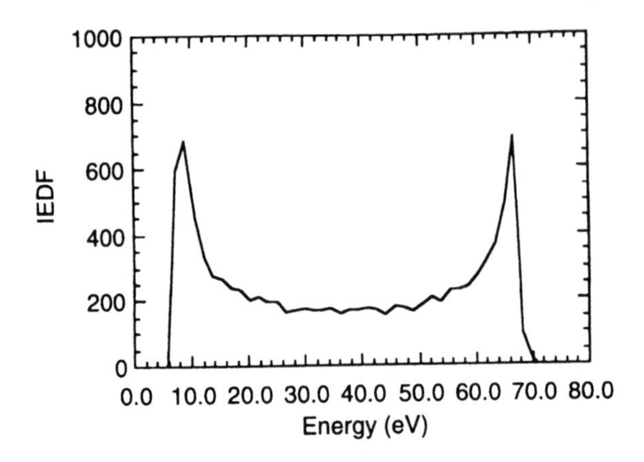

Figure 3.13. Ion energy distribution function at the wafer surface. The bimodal distribution is due to the oscillating sheath that is driven by the RF bias. Taken from [12].

The role of ions, particular low energy ions in electron-shading damage was recognized by Kinoshita *et al.* [12]. They numerically simulated the effect of rising bottom potential on the trajectory of incoming ions (assuming electrons arrive to the wafer surface isotropically). The plasma was assumed to have a T_e of 4eV, $n = 1\text{x}10^{12}/\text{cm}^3$ (high density) and a mean sheath voltage of 37 V. The ion energy distribution function at the wafer surface was calculated to have the form shown in Figure 3.13, a bimodal distribution. This bimodal distribution is the consequence of an oscillating sheath (RF driven). The resulting potential distribution at steady state is shown in Figure 3.14. At the top end of the resist sidewalls, negative potential develops (with respect to floating potential at open space) while at the bottom of the narrow space the potential is positive. The top negative potential tends to bend the ion trajectory toward the sidewalls. The bottom positive potential tends to repel low energy ions toward the sidewalls (Figure 3.15) while higher energy ions will reach the bottom (Figure 3.16). Charge balance is maintained by deflecting and repelling low energy ions toward the sidewalls to reduce the ion flux arriving to the bottom. The steady state bottom potential must therefore rise high enough to reduce the ion flux reaching the bottom to the same level of electron flux at the bottom.

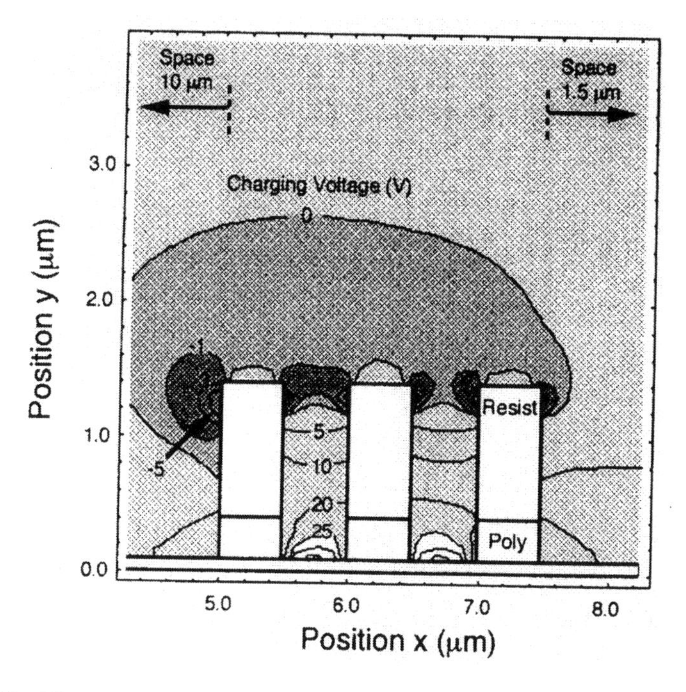

Figure 3.14. The equal potential contours in and around the dense lines due to electron shading. Potentials are shown with respect to the floating potential at open space. Isotropic electron trajectory was assumed. The negative potential developed at the top of resist sidewalls while the positive potential developed at the bottom. Taken from [12].

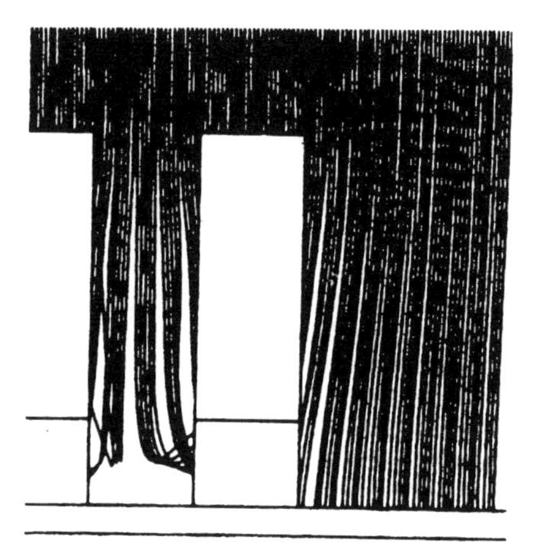

Figure 3.15. Trajectory of 20eV monoenergetic ions arriving to the wafer. The bottom potential is high enough that the ions are repelled toward the sidewalls. The negative potential near the top also attracts them toward the sidewalls. Taken from [12].

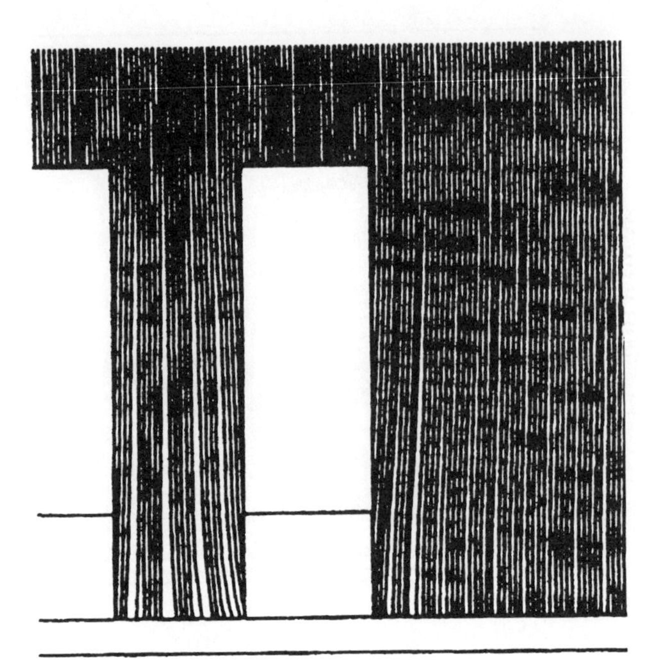

Figure 3.16. Trajectory of 60eV monoenergetic ions arriving to the wafer. Most can reach the bottom without problem. Taken from [12].

Kinoshita *et al.*'s model is exactly the opposite of Vahedi *et al.*'s. Instead of the bottom potential rising up to attract electrons so that enough electron will reach the bottom to balance the ion flux, the bottom potential rises up to repel enough ions so that the ions flux reaching the bottom equals the shaded electron flux. Note that Kinoshita *et al.*'s model does not preclude electron from being attracted toward the bottom of the narrow space. In their simulation, each charged particle's trajectory is calculated based on electrostatics. The bending of electron trajectories is already included. Even with the attraction force from the positive bottom potential, the net electron flux reaching the bottom is still far from matching that of the full ion flux without deflection.

The ion repulsion model of Kinoshita *et al.* has an obvious advantage over the electron attraction model of Vahedi *et al.* That is, bias voltage dependent is a natural consequence of the model. All else being equal, when bias voltage increases, the mean ion energy increases. The ion energy distribution width becomes wider. The population of low energy ion decreases. To repel the same fraction of ions from the bottom, the potential must rise up higher.

Kinoshita *et al.*'s model was successful in explaining the open-space width-dependent notch depth in polysilicon etching using high density plasma with property similar to the one they simulated [12]. Figure 3.17 shows the calculated charging potential (potential rise at the bottom of narrow spacing) inside and outside the outermost polysilicon line under the photoresist. One can see that the potential minimum in the polysilicon line is higher for the second line than the outermost line. The potential is even higher at the oxide surface in between the

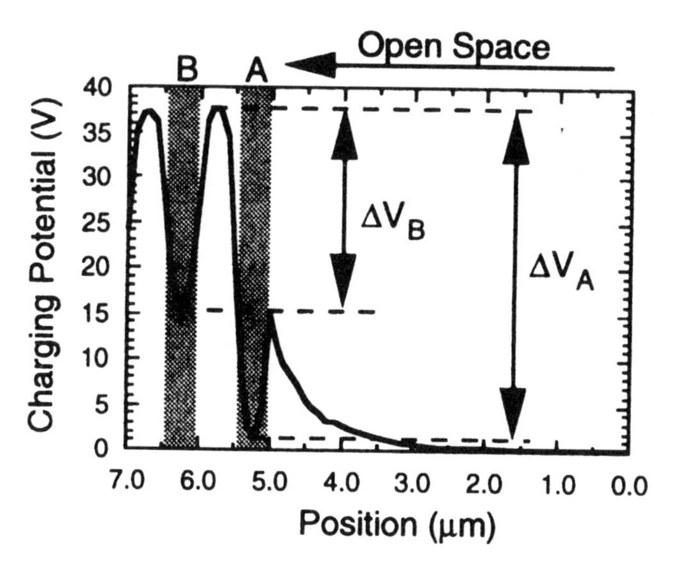

Figure 3.17. The calculated charging potential distribution for the dense-lines ending with an open space. The potential minimum in the poly line is lower for the outermost line. The potential is highest on the oxide surface in between lines. Taken from [12].

outermost line and the second line. When the ion energy is too high to be repelled and too low to reach the bottom, it can be deflected toward the polysilicon. The deflection is stronger toward the inner side of the outermost line. Figure 3.18 shows the trajectory of medium energy ions (monoenergetic in the simulation). More ions are bent toward the inner bottom of the outermost line. These ions can etch a notch into the polysilicon. Kinoshita *et al.* even showed the simulated potential difference between the maximum, in the narrow space, and minimum, in the outermost polysilicon line, follows a similar trend as the notch depth (Figure 3.19).

When doing simulation, some simplifications are unavoidable. Kinoshita *et al.*'s work was no exception. Hwang *et al.* [13] refined Kinoshita *et al.*'s simulation by allowing the charge distribution in the polysilicon line to be affected by the positive potential on the oxide in the narrow space, and by the negative charges on the resist. These subtle electrostatic effects tend to exacerbate the notch formation problem. The result is that the potential in the narrow space rises higher and becomes very asymmetric (Figure 3.20). The electric field that deflects the ions to the inner side bottom of the outmost line is very much stronger. Furthermore, positive charging can build up in the notch to further deflect ions deeper into the notch (Figure 3.21).

Although much work on electron-shading effect has been devoted to the study of notch formation problem in polysilicon etching, that problem itself is really not very important. There are many ways to reduce the notching problem. In fact, one has to work hard to create the problem so that it can be studied. These works, however, do produce a lot of useful information on the electron-shading effect, which can help understand and control the other more serious electron-shading damage, namely the damage to thin gate-oxides.

Figure 3.18. Medium (40eV) monoenergetic ion trajectories. More are bent toward the inner bottom of the outer most line. These ions can lead to notch formation. Taken from [12].

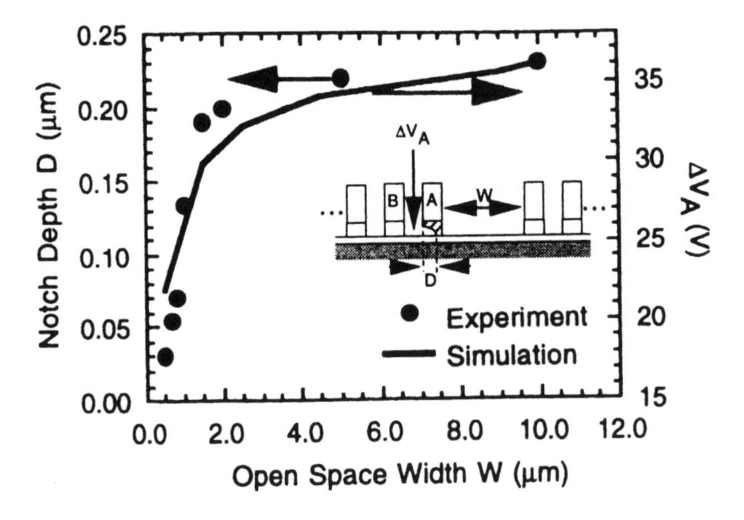

Figure 3.19. The potential difference between the maximum in between line and the minimum in the outer most poly line tracks the notch depth as a function of open space width. Taken from [12].

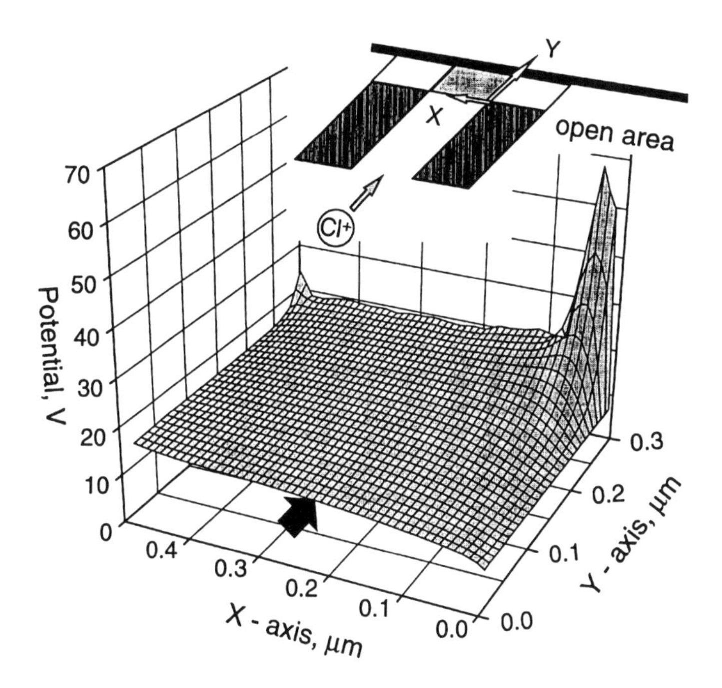

Figure 3.20. When charge distribution in the polysilicon is allowed to be influenced by the potential in the narrow space and by the negative charges on the photoresist, the potential on the oxide in the narrow spacing becomes higher and asymmetric. It peaks very near the inner bottom of the outermost line. The resulting lateral electric field is very strong and thus exacerbates the notching problem. Taken from [13].

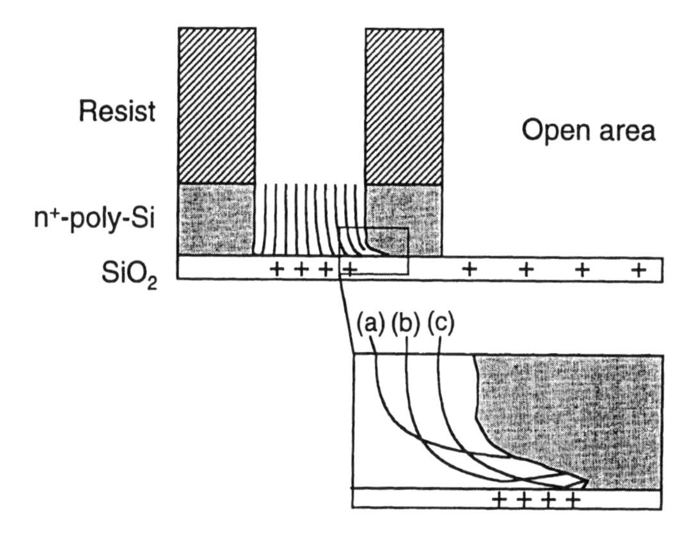

Figure 3.21. Charging can take place even inside the notch so that ions can be further deflected toward the deep end of the notch, allowing the formation of notch to continue. Taken from [13].

3.1.5 Electron-shading Effect in the Presence of FN Tunneling

Let's now apply the ion repulsion model to explain Hashimoto *et al.*'s bias dependent breakdown experiment [11] mentioned in Section 3.1.2 (Figure 3.12). Since the bottom potential rises to repel low energy ions, we need to know how bias affects the ion energy distribution. Ion energy distribution is difficult to measure directly, particularly when a RF bias is applied [14]. Most attempts to measure ion energy distribution through the RF biased electrode suffered from artifacts arising from not driving the energy analyzer to follow the RF electrode [15]. Because of this, we can only rely on theoretical predictions on how the ion energy distribution will change as a function of bias amplitude and frequency.

Figure 3.22 shows measured ion energy distribution as a function of ion mass [14] through a hole in the grounded electrode of a diode type RF discharge. As can be seen, the heavy ion Eu^+ has only one narrow peak whereas lighter ions H_2O^+ and H_3^+ has double peaks. The peak separation is larger for the lighter H_3^+ ions. These results can be understood by recognizing that ions take time to traverse the sheath. Because kinetic energy of the ions is limited (determined by the sheath voltage and the initial velocity according to the Bohm criterion), heavier ions have lower velocity and take longer time to cross the sheath.

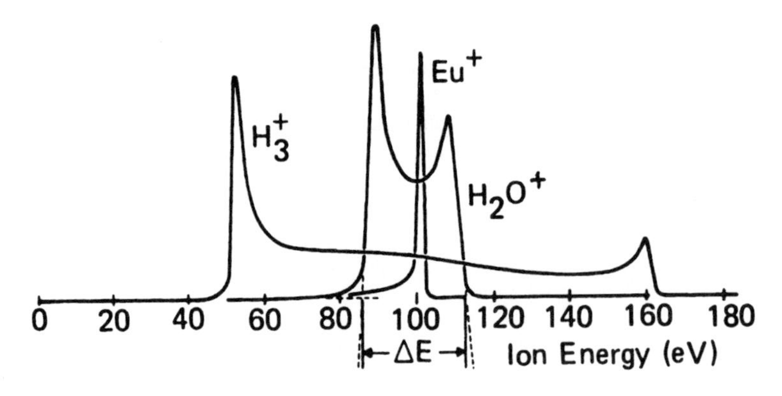

Figure 3.22. The ion energy distribution as a function of ion mass measured at the grounded electrode in a diode type RF discharge. Taken from [14].

The Eu^+ ions have a molecular weight of 151. At 100eV, their velocity is 3.0×10^6cm/sec. The average velocity is thus ~ 2.0×10^6cm/sec. The average sheath thickness for the discharge was ~1cm. The time for Eu^+ ions to traverse the sheath was therefore ~500ns, almost 7 RF periods (13.56MHz). The Eu^+ ions therefore acquired only the average sheath voltage and therefore had a single peak as terminal kinetic energy.

For the H_3O^+ ions, the transit time is somewhat more than 1 RF period. The energy for the H_3O^+ ions must then be a function of the sheath voltage when they started the sheath crossing. The spread in the energy distribution function is therefore wider. If the crossing starts at the high sheath voltage part of the cycle, the terminal kinetic energy will be higher than the Eu^+ ions. If the crossing starts at the low sheath voltage part of the cycle, the terminal kinetic energy will be lower than the Eu^+ ions. The median energy remains exactly the same as the Eu^+ ions.

For the H_3^+ ions, the transit time is a little less than half of a RF period. The energy is highly dependent on the sheath voltage when they started crossing the sheath. The energy distribution is therefore the widest. Those that traverse the sheath at the high voltage portion of the cycle will have a high terminal kinetic energy while those traverse the sheath at the low voltage portion will have a low terminal kinetic energy.

Note that even for the H_3^+ ions, the transit time is still a good fraction of a RF period. Their terminal kinetic energy still represents the average of a significant fraction of a RF cycle. Obviously, if we have something even lighter, the energy spread will be even wider. This broadening can continue until the transit time is only a very small fraction of a RF period so that the terminal kinetic energy is a good approximation of the instantaneous sheath voltage. This saturation of broadening becomes quite prominent when the transit time drops to about 1/20 of a RF period or less.

In the above discussion, we have neglected the sheath oscillation. The sheath thickness varies with the sheath voltage. According to the Child's law for space charge limited current, the sheath thickness is related to the sheath voltage by [16]:

$$s = \frac{\sqrt{2}}{3}\lambda_{De}\left(\frac{2eV_{sh}}{kT_e}\right)^{3/4} \tag{3.3}$$

where λ_{De} is the Debye length of the plasma and is given by:

$$\lambda_{De} = \left(\frac{kT_e\varepsilon_0}{n_e e^2}\right)^{1/2}.$$

Clearly, the sheath thickness is shrinking faster as a function of sheath voltage than ion velocity, which is proportional to the square root of sheath voltage. In addition, the initial velocity of ions due to the pre-sheath acceleration (Bohm criterion) does not change with the sheath voltage. Thus we have more ions crossing the sheath at low energy than at high energy, which resulted in a higher peak height at the low energy end.

When the bias frequency is lowered, the RF period becomes longer and, as a result, even the slower ions will traverse the thickness of the sheath at less than a RF period. All ion energy distributions will become broader. However, the median energy remains unchanged. So the low-energy peak become lower in energy and the high-energy peak becomes higher in energy. Note that in the example given above, if the RF frequency were lowered to 3MHz, the Eu^+ ions would have an energy distribution similar to the H_3O^+ ions at 13.56MHz, and the H_3^+ ions would have an energy distribution that is beginning to approach the saturation width. Similarly, if the average sheath thickness were lowered (shorter Debye length, which can be obtained by higher plasma density n_e or lower electron temperature T_e), the ion energy distribution would be broader. Again, for the example given above, if the plasma density were increased by a factor of 20, the sheath thickness would reduce to about 2mm, and the Eu^+ ion's energy distribution would resemble the H_3O^+ ions' distribution at the original plasma density. For modern high-density plasma commonly used in processing, average sheath thickness of the order of 100μm is not uncommon. With these plasmas, it is almost impossible not to get double peak type of, or saddle-shaped, ion energy distribution. In fact, most ions will have a rather wide energy distribution.

All the above discussions are about ion energy distribution at the grounded electrode. For the RF biased electrode, we can extrapolate the grounded electrode situation by modifying the sheath voltage accordingly. Combining Equation 2.4 and 2.6, we get for the average sheath voltage an expression like:

$$\overline{V}_{sh} = \frac{kT_e}{2e} ln\left(\frac{M}{2.3m}\right) + \frac{kT_e}{e} ln\left(I_0\left(\frac{eV_{rf}}{kT_e}\right)\right).$$

If we write [16]:

$$I_0\left(\frac{eV_{rf}}{kT_e}\right) \approx \left(\frac{kT_e}{2e\pi V_{rf}}\right)^{1/2} exp\left(\frac{eV_{rf}}{kT_e}\right),$$

We have:

$$\overline{V}_{sh} = V_{rf} + \frac{kT_e}{2e}\left(ln\left(\frac{M}{2.3m}\right) - ln\left(\frac{2e\pi V_{rf}}{kT_e}\right)\right).$$

Thus the instantaneous sheath voltage is:

$$V_{sh} = V_{rf}\left(1 + cos\,\omega\tau\right) + \frac{kT_e}{2e}\left(ln\left(\frac{M}{2.3m}\right) - ln\left(\frac{2e\pi V_{rf}}{kT_e}\right)\right) \tag{3.4}$$

The minimum sheath voltage (V_{min}) is:

$$V_{min} = \frac{kT_e}{2e}\left(ln\left(\frac{M}{2.3m}\right) - ln\left(\frac{2e\pi V_{rf}}{kT_e}\right)\right) \tag{3.5}$$

The maximum sheath voltage (V_{max}) is:

$$V_{max} = 2V_{rf} + \frac{kT_e}{2e}\left(ln\left(\frac{M}{2.3m} \right) - ln\left(\frac{2e\pi V_{rf}}{kT_e} \right) \right) \tag{3.6}$$

That is, V_{max} is V_{min} plus the peak-to-peak RF voltage.

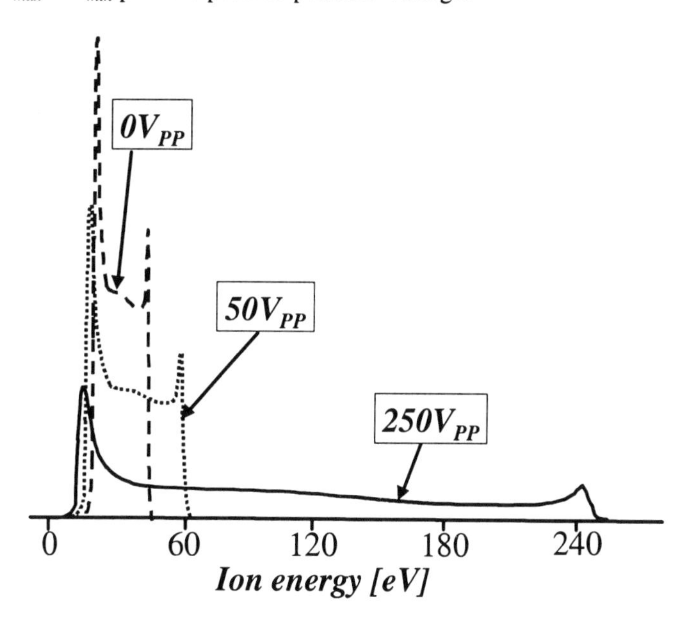

Figure 3.23. Simulated ion energy distribution for the 0V, 50V and 250V bias cases in Hashimoto *et al.*'s experiment.

Equation 3.5 and 3.6 indicate that when the RF bias is increased, while V_{min} becomes smaller, V_{max} becomes bigger. Figure 3.23 shows the simulated ion energy distribution for the plasma used in Hashimoto *et al.*'s experiment [11]. The 3 bias conditions of $0V_{PP}$, $50V_{PP}$ and $250V_{PP}$ are simulated. Note that even the zero bias case produces a double peak distribution. This is due to sheath potential oscillation driven by the source RF. The integrated ion current is the same for all three bias conditions. The most striking feature of the ion energy distribution in Figure 3.23 is that as bias goes up, the amount of low energy ions drops rapidly.

When we were discussing the mechanism of charging due to non-uniform plasma, we made use of some thought process to break down the complex interaction involved to help us understand how the final result is obtained. We can do the same here. Figure 3.24a shows the electron and ion trajectories "before" any charging occurs. This is purely geometric line-of-sight effect. Ions are so directional that none will hit the sidewalls on their way to the bottom. Electrons are so isotropic that most will hit the sidewalls and will stick.

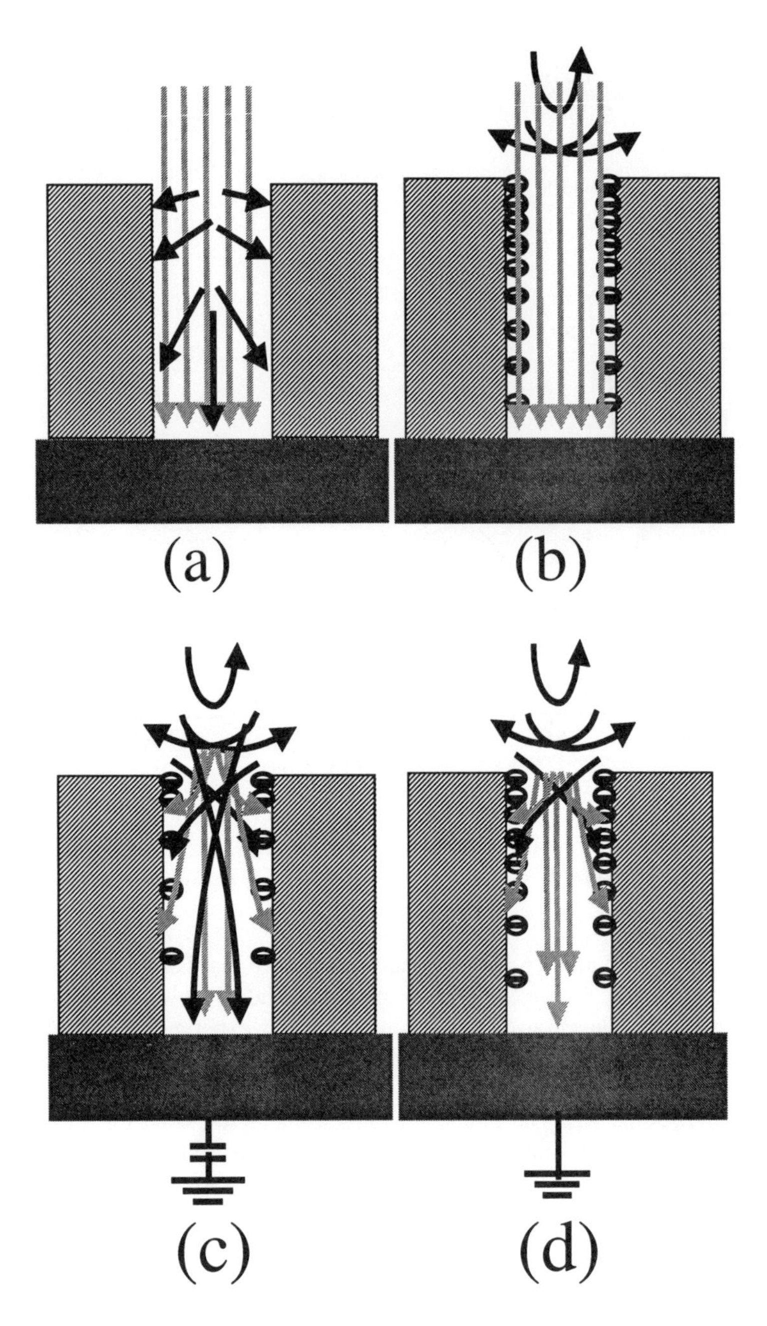

Figure 3.24. (a) Only pure geometrical effects are included without any charging. (b) Charging due to electrons is included. Ion is assumed to be completely stiff. Bottom potential is fixed at zero. (c) Bottom potential is allowed to float and ions are allowed to be deflected. (d) Bottom potential is fixed at ground, but ions are free to be deflected.

Figure 3.24b shows what happen when charging due to electron is taken into account and that the bottom potential is fixed at zero. Here we assume that ions are so energetic that the negative potential at the sidewalls does not affect them. Since the sidewalls are collecting only electrons, a negative potential will build up until all electrons reaching the opening of the narrow space are turned away. This potential can be quite large because it must repel even the few electrons at the high-energy tail of the Maxwellian distribution.

Figure 3.24c shows that if we allow ions to be attracted by the negative potential on the sidewalls and the bottom potential to float, significant amount of the lower energy ions will hit the sidewalls. This deflection of ions is due both to the negative potential at the sidewalls and the positive bottom potential, which rises up due to more ions than electron can reach the bottom without it. With more positive potential at the bottom, more electrons are reaching there as well. At steady state, charge is balanced everywhere. When the bottom potential is free to float, the ions flux and electron flux that reach the bottom are always lower and higher, respectively, than simple geometric consideration would predict.

The limiting case of a low aspect ratio is, of course, the wide-open space. Shading effect would not exist and both electron and ion fluxes are at 100%. For the limiting case of a very high aspect ratio insulating structure, the top of the opening is completely covered by a negative potential. The bottom positive potential cannot penetrate this negative field and therefore cannot attract any electrons that have not already overcome the repulsing field at the opening of the narrow space. However, with almost no ions hitting the top sidewalls, the negative potential will be so high that the fraction of electrons that penetrate the negative field is very small. The bottom potential must continue to rise until significant amount of ions can be deflected toward the top of the sidewalls. To a first approximation, the electron flux at the top of the sidewalls is 100% due to the isotropic nature of the electron velocity distribution. The only way charge can be balanced is that the 50% of the ions are deflected toward the top sidewalls so that the barrier is lowered enough for 50% of the electrons to enter the narrow space. Note that ions cannot be turned away before entering the space because all it sees is an attractive negative potential. Most of the ions and electrons that pass through the top region of the narrow space end up hitting the sidewalls below it. Very few or none will reach the bottom. That is to say, the bottom potential reaches V_{max} given by Equation 3.6. For aspect ratios anywhere in between these two extremes, the bottom potential will depend on the aspect ratio, the bias voltage, electron temperature and ion temperature [17].

Figure 3.24d shows the case where the bottom potential is fixed at zero. This case differs from the case of Figure 3.24b in that ions are attracted to the sidewalls by the negative potential. By not allowing the bottom potential to rise, far fewer ions are deflected toward the sidewalls. The sidewalls become more negative everywhere. At the top of the sidewalls, the negative potential must be high enough to attract some ions or charge will not be balances. It is therefore expected to be somewhat higher than V_{min} given by Equation 3.5. For Ar plasma with $V_{rf}=0$ (no bias), we have a V_{min} of $5.2T_e$. If the aspect ratio is high, this negative potential completely covers the entrance of the opening. Only electrons with energy higher than $5.2T_e$ will be able to enter the narrow space. Assuming a Maxwellian distribution, this means only 0.5% of the electrons can do that. The most important effect of pinning the bottom potential to zero is that it becomes very easy for the negative potential to completely cover the opening of the space between insulated lines, even when the aspect ratio is not very high. For charge to be balanced, almost 100% of the ion current must be neutralized by electrons supplied by the source that pins the bottom potential to zero.

If the bottom potential were pinned at some low positive value other than zero, more ions would be deflected toward the sidewalls. The potential at the sidewalls is less negative. It will also be less likely for the negative potential to completely cover the opening. More electrons will enter the narrow space. The additional amount of ions to be deflected depends strongly on the bias. If the bottom potential is V_B, the additional amount of ion deflected is roughly:

$$\frac{I}{I_{ion}} = \frac{V_B}{2V_{rf}}$$

Figure 3.25 illustrate how this relationship is derived. The maximum width of the ion energy distribution curve is $2V_{rf}$. The full distribution is above zero (or floating potential at open area) by V_{min}. At the bottom of the narrow spacing, the distribution

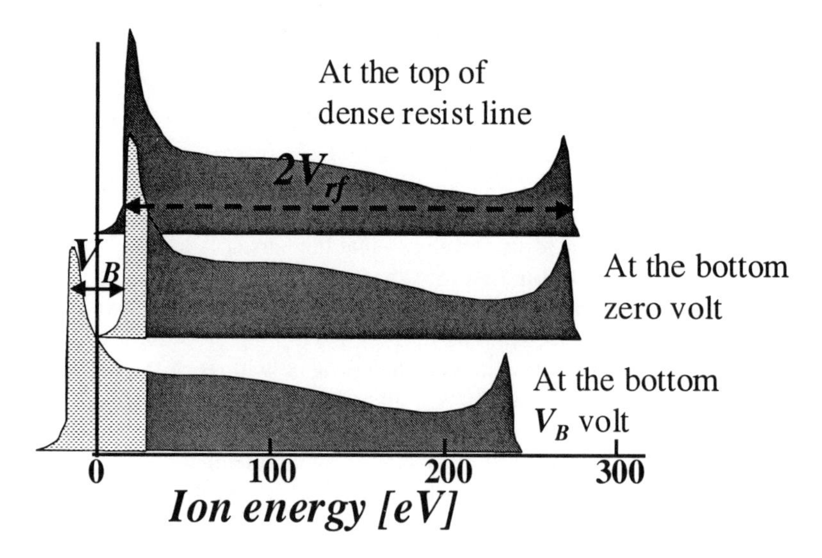

Figure 3.25. The ion energy distribution before entering the narrow space; at the bottom with zero bottom potential; and at the bottom with V_B bottom potential are illustrated.

lost the low energy ions (the dotted region) to the negative sidewalls. At zero bottom-potential, the distribution has no shift in energy. This is because while the negative potential at the top accelerates the ions before entering the narrow space, they are decelerated by the same amount before reaching the bottom (bottom potential at zero means no net changes in energy for the ions). At V_B bottom potential, more ions are lost to the sidewalls while the entire distribution is shifted toward lower energy by V_B amount. If we approximate the ion energy distribution after the first peak as a rectangle, then the additional amount of ion lost is simply the fraction $V_B/2V_{rf}$. Not a very good approximation, but it does give a good feel on how the bias voltage affects the amount of ion flux reduction. Clearly, the impact of bottom potential being pinned at a low positive value is larger for higher bias situation.

Assume that for the aspect ratio used in Hashimoto *et al.*'s experiment, a substantial amount of ions are deflected to the sidewalls at charge balance (without FN tunneling but including all the charging effects). The fractions of deflected ion are not the same for the 3 bias cases. The bottom potential required to do that is marked (roughly) on the energy distribution curves (Figure 3.26) as V_1, V_2 and V_3 for the case of $0V_{PP}$, $50V_{PP}$ and $250V_{PP}$ bias, respectively. As we discussed earlier, the higher the bias, the higher is the bottom potential. Note that V_1 and V_2 are quite low in value. This is because even if the bottom potential did not rise, a good fraction of the ions is deflected to the sidewalls already by the negative potential at the sidewalls. The rise in bottom potential will increase the fraction by deflecting additional ions to the sidewalls, as shown in Figure 3.25. For the $0V_{PP}$ and $50V_{PP}$ cases, the fraction of ions represented by a low voltage increase is very much larger than the $250V_{PP}$ case.

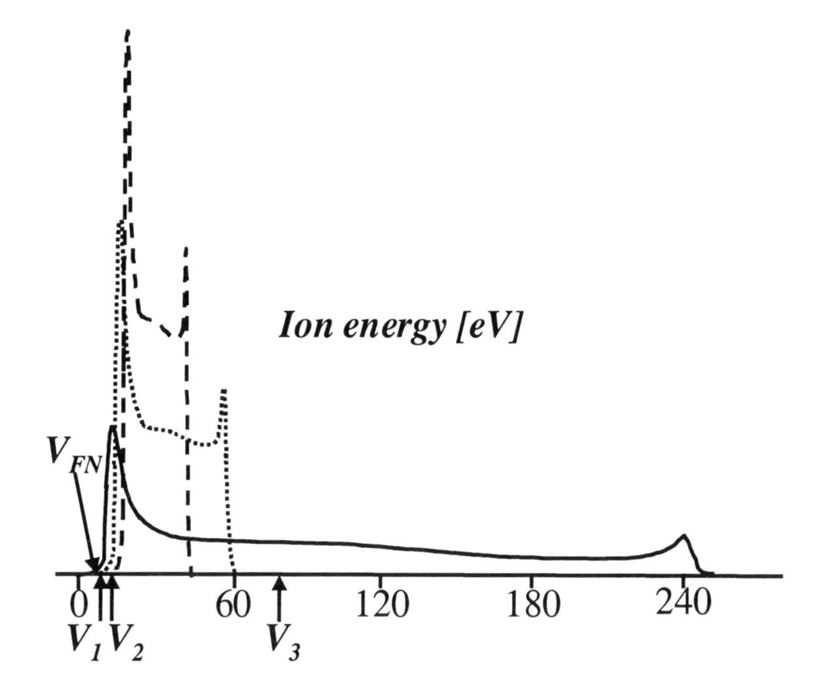

Figure 3.26. The antenna voltage V_1, V_2 and V_3 for the 3 bias cases in Hashimoto *et al.*'s experiment [11] when FN tunneling is neglected. The arrow indicates the antenna potential when FN tunneling is present.

In Hashimoto *et al.*'s experiment [11], the antenna was connected to the gate on top of a 4nm thick gate-oxide. With the substrate potential fixed by the open area (Figure 3.4a), the antenna potential rise will be limited by the FN tunneling. Regardless the bias voltage, the potential rise can only be about 6V or less (an oxide field of 15MV/cm will breakdown a good percentage of the oxide in a few seconds), which is lower than V_1, V_2 or V_3. This is the situation of pinning the bottom potential at a low positive value discussed above. We therefore expect an increase in ion flux and a decrease in electron flux reaching the bottom for all 3 bias cases. This simultaneous increase and decrease in ion flux and electron flux reaching the bottom need not be a symmetric change. The electron flux is expected to be more sensitive to potential changes.

In Hashimoto *et al.*'s experiment [11], the plasma exposure time is fixed. Their result was reported in terms of percent capacitor breakdown as a function of antenna ratio. If we fix the percent breakdown at 50 (median failure), then the antenna ratio at which it occurs for each bias can serve to estimate the difference in shading. From Figure 3.12, we can read out the median failure antenna ratios. They are 1.08×10^5, 7.5×10^4 and 2×10^4 for the $0V_{PP}$, $50V_{PP}$ and $250V_{PP}$ cases, respectively. The FN stress for all three cases are the same. The FN currents per unit antenna area are, taking the zero bias case as 1, 1, 1.44 and 5.4. This ratio agrees qualitatively with the FN values shown in Figure 3.26. Based on the discussion so far, one can expect that for the $250V_{PP}$ case, the electron flux reaching the bottom is essentially zero whiles >90% of the ion flux reach there. For the low and zero bias cases, the increase in ion flux at the bottom is very modest. As a result, the electron flux drop is not too drastic. Taking the charge flux mismatch in the $250V_{PP}$ case as 90% of ion current, the $0V_{PP}$ and $50V_{PP}$ cases are 16.7% and 24% respectively. These mismatches are the sum of electron flux decrease and ion flux increase.

The example of Hashimoto *et al.*'s experiment [11] shows that the impact of FN tunneling is to lower the bottom potential so that ion flux increases while electron flux decreases. The larger the change from the bottom floating potential to the potential pinned by FN tunneling, the bigger the resulting flux imbalance. This flux imbalance is the FN tunneling current that damages the gate-oxide. Eventually, the imbalance saturates. At that point, electron flux is zero, ion flux is close to 100% and the FN voltage is largely determined by the thickness of the gate-oxide. The bottom floating voltage is a function of aspect ratio and applied RF bias. The higher the aspect ratio, the lower the RF biases at which saturation occurs. For Hashimoto *et al.*'s experiment [11], the aspect ratio was 2 and the saturation occurs below $250V_{PP}$ bias.

3.1.6 The Effect of Electron Temperature on Electron-shading Damage

There is wide spread belief that lowering of the plasma's electron temperature is the main method to reduce or eliminate electron-shading damage. This belief originated from Fujiwara *et al.*'s observation that lower electron temperature plasma has lower side-etch rate in their polysilicon notching experiment (Figure 3.8) [9].

Subsequently, a number of modeling works [5, 18-20] supported this conclusion. Experimentally, other than Fujiwara *et al.*'s original experiment, the only other direct correlation between measured electron temperature and electron-shading damage was reported by Arimoto *et al.* [21].

The most widely studied approach to reduce or eliminate electron-shading damage has been the use of pulse modulated plasma. There have been quite a lot of reports in the literature [22-29]. Most of these reports attributed the reduction of electron-shading damage in pulse modulated plasma to the reduction of electron temperature during the off cycle. Thus created an impression that there is tremendous amount of evidences supporting the notion of electron temperature is the most important factor that influence electron-shading damage.

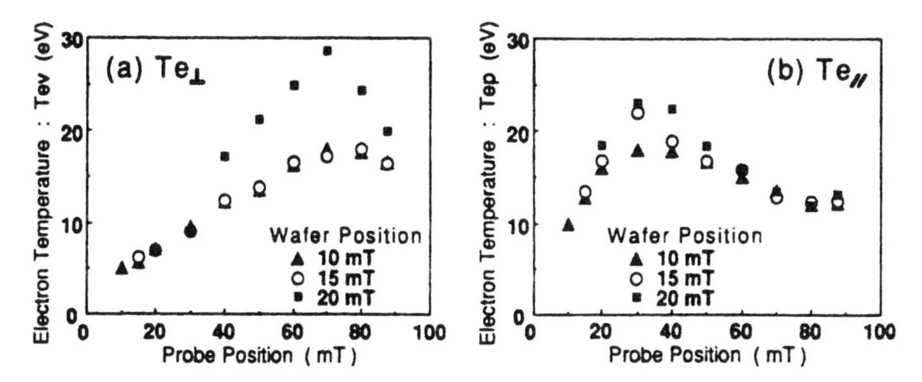

Figure 3.27. The measured electron temperature parallel with and vertical to the magnetic flux in an ECR plasma are shown. The solenoid coils were adjusted to produce 3 different magnetic field strengths, and thus electron temperatures, at the wafer surface. Taken from [30].

In a separate report, Fujiwara *et al.* [30] elaborated their method of measuring as well as changing the electron temperature in their experiment. They used ECR plasma in their experiment. Since there was a magnetic field, the electronic motion was anisotropic. They thus used two different types of probes to measure the electron temperature. One probe would measure the electron temperature (T_{ep}) along the magnetic flux line, which is perpendicular to the surface normal of the wafer. The other probe would measure the electron temperature (T_{ev}) vertical to the magnetic flux. To change the magnetic field gradient at various point, they adjusted the positions of the solenoid coils (there were four of them for their system) that generate the magnetic field. The measured result is shown in Figure 3.27. The plasma B and C of Figure 3.8 are the 10mT and 20mT conditions in Figure 3.27, respectively. Notice the measured electron temperatures are very high in value. As the probe move toward higher magnetic field region, the T_{ev} reaches as high as 30eV. Such high value for electron temperature is very unlikely to exist in "cold" plasma commonly used in plasma processing.

Measuring electron temperature using electrical probe is extremely difficult [31], particularly in the presence of a magnetic field [7]. Bowden *et al.* [8] studied the electron temperature in ECR plasma using Thomson scattering, which is an non-invasive method and found that the electron temperature ranges from 2 to 3.5eV (Figure 3.28). These low values agree well with theoretical expectation (Figure 3.29) for typical high-density plasma [16]. The very high T_e values found in Fujiwara *et al.*'s experiment are thus highly unusual.

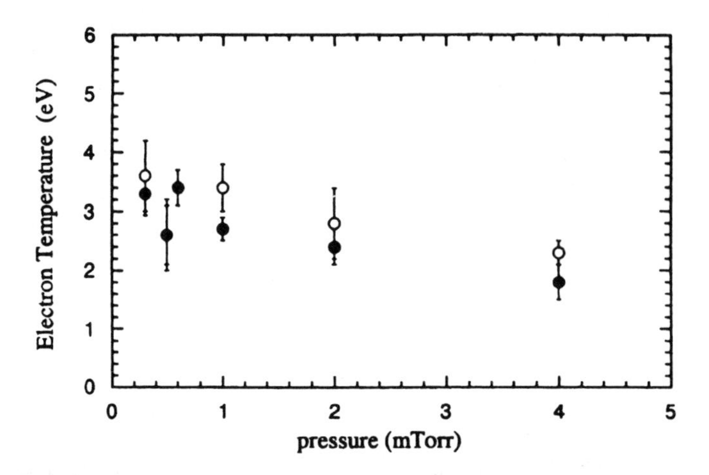

Figure 3.28. The electron temperature measured with a Thomson scattering method. The solid circles are electron temperatures without an electrical probe nearby. The open circles are electron temperatures perturbed by a nearby electrical probe. Taken from [8].

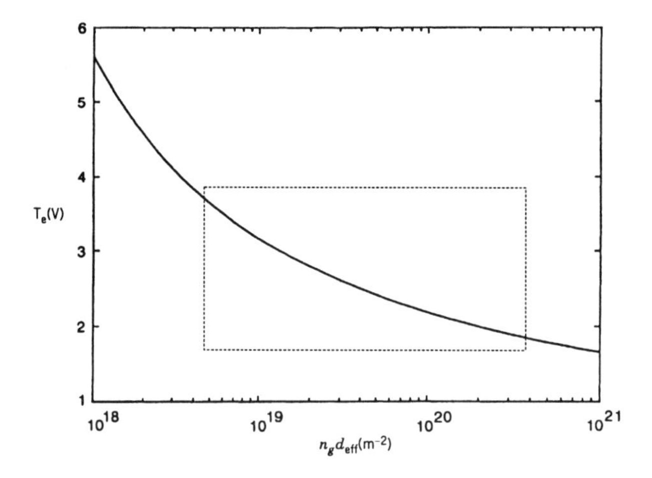

Figure 3.29. The theoretical electron temperature of an Ar plasma as a function of $n_g d_{eff}$. Where n_g is number density of molecules and d_{eff} is the discharge dimension. The broken rectangle indicates the typical processing condition. That is n_n from 1mTorr to 40mTorr and d_{eff} is 25cm. Taken from [16].

Arimoto *et al.* [21] also used an ECR plasma in their work. The breakdown of their dense-line antenna capacitor localized at one side of the wafer. When they used a single Langmuir probe to measure the electron temperature, they found that where the antenna capacitors are broken, the electron temperature is particularly high (Figure 3.30). As can be seen, the measured electron temperature is also very high, although not quite as high as those measured by Fujiwara *et al.*

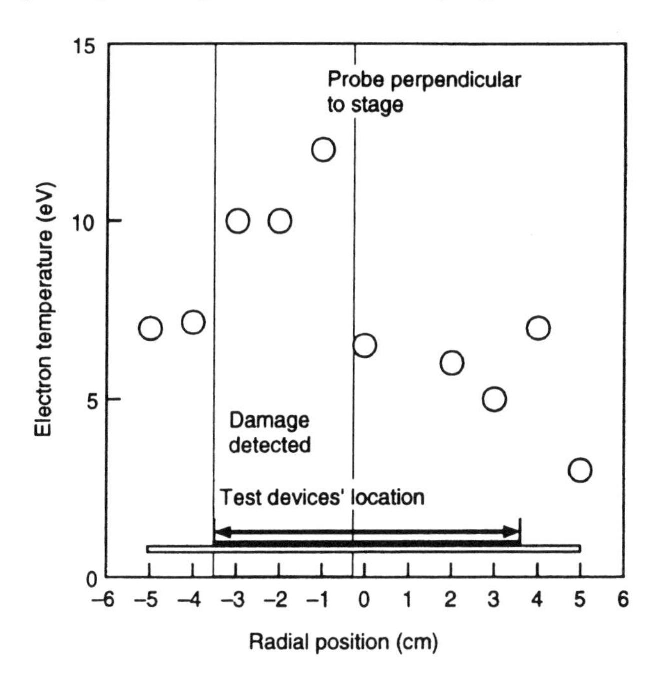

Figure 3.30. The correlation of electron temperature and dense-line antenna capacitor breakdown (damage) is plotted. Where damage was detected, electron temperature was particularly high. Taken from [21].

In electron temperature measurement using electrical probe, any error that occurs would almost always lead to a higher measured value. One of the most common problems of making electrical probe measurement in processing plasma is the lack of adequate reference electrode [32]. This is due to the use of insulating chamber wall. The result of insufficient reference is that the measured probe I-V characteristic tends to stretch out, which will give a much higher apparent electron temperature [31]. In the existence of a magnetic field, electrical probe measurement is even trickier. The local field is greatly distorted by obstacles that are much larger than the electron Larmor radius. Thus when using an electrical probe to measure magnetized plasma, only the I-V section that is very close to the floating potential can be used [7]. That is, only the high-energy tail of the electron energy distribution is measured. The high-energy tail is precisely where the distribution function most likely to deviate from Maxwellian, particularly when electronegative discharge such as those used in both Fujiwara *et al.* and Arimoto *et al.*'s experiments. This is because electron attachment is very efficient in these gases. This inelastic mode of scattering tends to deplete the high-energy tail's population. Figure 3.31 shows the measured electron energy distribution function of Cl_2 discharge in a transformer-coupled plasma (TCP) [33]. As pressure increases, the depletion of higher energy electron becomes more pronounced. Fujiwara *et al.*'s experiment was done at ~0.3mTorr so it is less affected by this problem. Arimoto *et al.*'s experiment, on the other hand, was done at 10mTorr and was definitely affected by this problem.

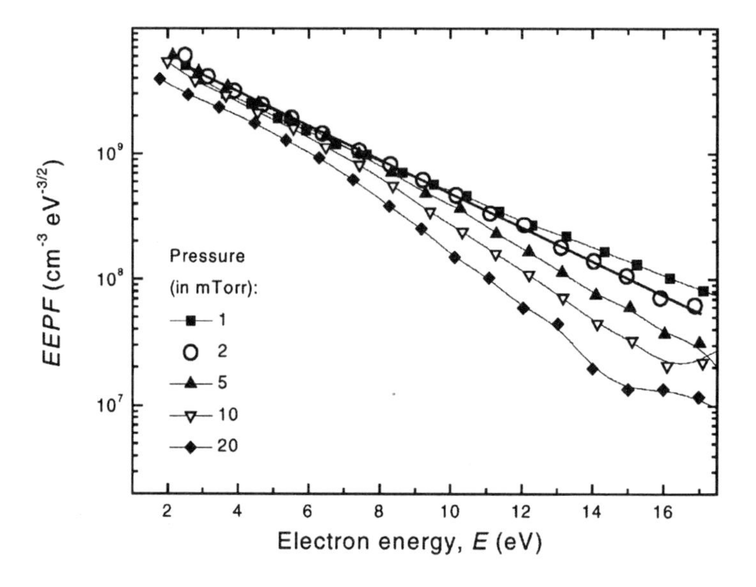

Figure 3.31. The electron energy distribution function of a TCP Cl_2 discharge. The depletion of high-energy electron is more severe at higher pressure. Taken from [33].

Careful measurements of high-density plasma [7, 8, 33, 34], particularly those that use non-intrusive methods, have confirmed the expected theoretical electron temperature range shown in Figure 3.29. There is, therefore, serious doubt on the reliability of the measured electron temperatures in both Fujiwara *et al.*'s and Arimoto *et al.*'s experiments. If we discount these experiments, we could find no direct evidence to support that electron-shading damage is strongly affected by electron temperature.

Theoretically, electron temperature has been shown to have major impact on electron-shading damage [5, 18-20]. We have already pointed out the shortcomings of Vahedi *et al.*'s model. The rest are numerical simulation and they share some common features such as relatively low bias voltage ($100V_{PP}$ or less), not connected to thin gate-oxide (no pinning of bottom potential) and comparing drastically different electron temperatures. Under these conditions, it makes good physical sense that electron temperature should have significant influence on shading damage. Lowering of electron temperature would help in two ways. Ion energy distribution would shift down proportionally with T_e. More low energy ion means more are deflected toward the sidewalls to neutralize the negative charges. When the bottom potential rises up, its influence on electron is larger when the negative potential at the sidewalls is lower. Low energy electron trajectories are also more readily bent toward the bottom [20].

When shading damage is referring to notch formation in polysilicon etching, these numerical simulations are valid. This is because polysilicon etching uses low bias voltage to achieve high selectivity over the gate-oxide and notch formation does not involve pinning of bottom potential. The only exception is that changing electron temperature drastically is more difficult to achieve in practice. Even in notch formation type of shading damage, when the aspect ratio is high, the influence of electron temperature will diminish. This is the limiting case discussed in last section in association with Figure 3.24c.

When shading damage is referring to thin gate-oxide degradation, the validity range of these simulations is very limited. As discussed in last section in association of Figure 3.24d, when the bottom potential is pinned at a low positive value, fewer ions are deflected toward the sidewalls and fewer electrons will be able to penetrate the negative barrier at the top of the narrow opening. This is the saturation of shading effect. Once saturated, few electrons can reach the bottom. The onset of saturation does depend on the electron temperature, but it depends much more on the aspect ratio and bias voltage. Electron temperature has an effect only when both the aspect ratio and the bias are low.

Very recently, Hwang *et al.* [17] did include the effect of FN tunneling through thin gate-oxide in their electron-shading damage simulation. However, they were still only simulating the low bias case ($60V_{PP}$). When T_e is 8eV, they found that the top of the narrow spacing with aspect ratio of 2 is fully blocked by a negative barrier. This blocking negative potential is not quite as negative as a fully saturated situation would need. In their simulation, they made the assumption that current will flow readily when the electric field on the photoresist exceeds 1MV/cm. This may help reduce the negative potential somewhat. Whether current is readily flowing at 1MV/cm for photoresist or not is unclear. At some point, the field will exceed the breakdown strength of photoresist. Picking the value of 1MV/cm may not be too far off. When T_e is only 2eV, the top negative potential is much lower and the net tunneling current dropped off significantly. Note that in Hwang *et al.*'s simulation, they used only 5 lines and 4 narrow spacing. The simulated current includes the two outer edges that are not shaded. The resulting current is thus a combination of effects.

Hwang *et al.*'s simulation [17] was to explain the electron-shading damage during metal overetch in inductively coupled plasma (ICP) [35, 36]. The damage has been identified to happen during overetch of barrier metal. It is assumed that the situation depicted in Figure 3.32 exist. In other words, barrier metal in between dense-line remains connected for a little longer than isolated line. During metal etching, some metal lines are inevitably connected to the substrate. So the substrate potential is not floating. Electron-shading effect was thought to be the source of damage to antenna test devices [35]. It was found that when the plasma volume was expanded by raising the gap between the power electrode and the wafer platen, damage reduces significantly [35], [36]. Colonell *et al.* speculated that the expansion of the plasma might have reduced the electron temperature, which in turn reduced the shading damage.

Even if the expansion of the plasma reduced T_e, it is not likely that T_e could have changed from 8eV to 2eV as Hwang *et al.* had simulated. Theoretically, we expect that the amount of T_e change by doubling the plasma size to be negligible. Careful measurement using non-intrusive method later confirmed that the change in T_e was indeed negligible [37]. Thus if the damage was indeed due to electron-shading effect, lowering of electron temperature was not the magic bullet that fixed it.

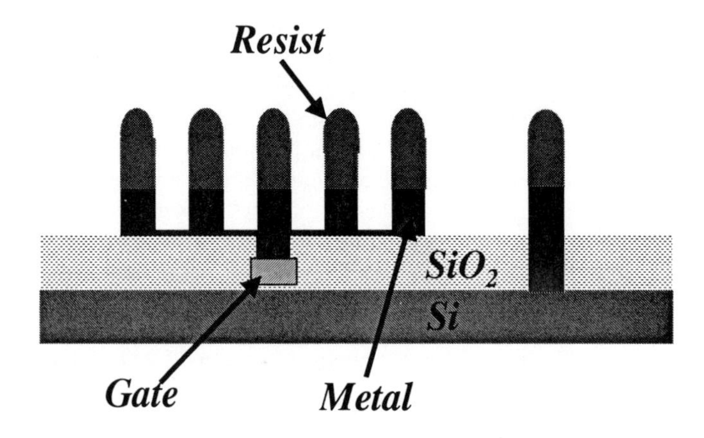

Figure 3.32. The "end point" of a metal etching process is shown. Isolated lines are cleared while dense-lines are still connected.

3.1.7 Negative Electron-shading Effect

In electron-shading effect, the potential of the shaded antenna is always positive. This is because ions are always more directional than electrons. In addition, reducing the aspect ratio either by increasing the spacing or lowering the mask height is supposed to always reduce shading damage. It was therefore a surprise to Hasegawa *et al.* [38] when they found that increasing the spacing between line actually increased damage to antenna capacitors. Figure 3.33 shows their experimental sequence. Antenna capacitors as shown were etched in a Cl_2/BCl_3 ECR plasma at 4mTorr pressure using 1000W of source power and 40W of bias power. After the metal/polysilicon stack that form the antenna was etched, resist was removed by ashing. The breakdown of antenna capacitor with antenna ratio of 10^6 was measured. The gate-oxide was 5.5nm thick and the line spacing varied from

0.4μm to 4μm. Starting thickness of the resist was 1.2μm so the maximum aspect ratio before etching was 3.

Figure 3.34 shows the measured breakdown behavior for 2 different overetch conditions. When the spacing is very narrow, the breakdown failure rate is very high, as expected from electron-shading effect. Increasing the spacing from 0.4μm to 0.5μm reduces the breakdown failure rate, again as expected. However, as spacing continue to increase, breakdown failure increases again and eventually levels off at large spacing.

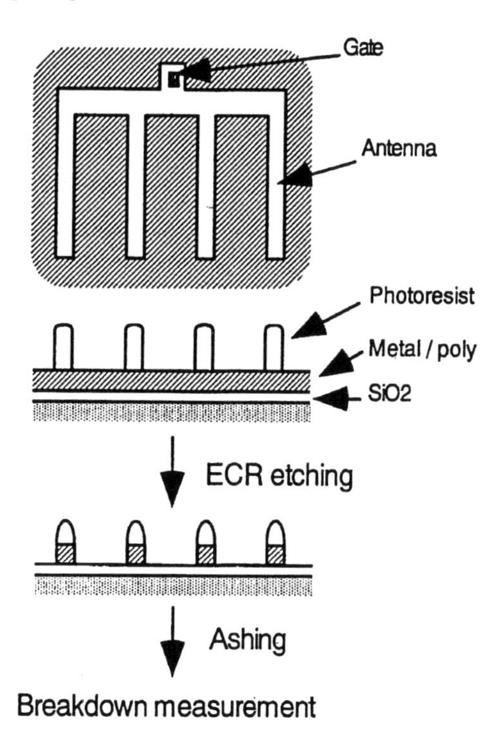

Figure 3.33. Antenna capacitor structure and experimental sequence in Hasegawa *et al.*'s [38] work that showed negative electron-shading effect.

To investigate the mechanism behind this unexpected phenomenon, Hasegawa *et al.* simulated the etching condition with several giant antenna probes (Figure 3.35). Probe A has dense, tall SiO_2 lines on top of a metal pad to serve as shading features. Probe B has thin SiO_2 mask on top of thick metal lines. Probe C is similar to probe B but with a larger spacing between lines. All three probes are referenced to a large open metal pad sitting next to the probe. Probe A simulates the condition at and slightly after "end point" where the resist is still thick and a thin layer of metal still exists in the space between lines. Probe B and C simulate the overetch situation where resist erosion has thinned down the resist and the metal in the space between lines are all etched away.

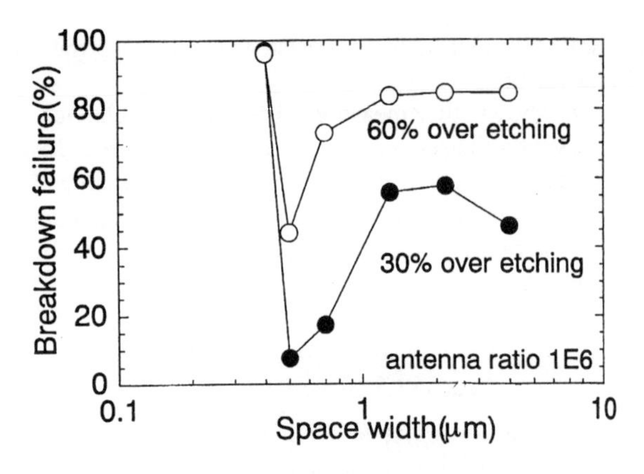

Figure 3.34. The antenna capacitor breakdown behaviors are shown as a function of spacing between lines and over etch time. Electron-shading damage go through a minimum as spacing increases. Taken from [38].

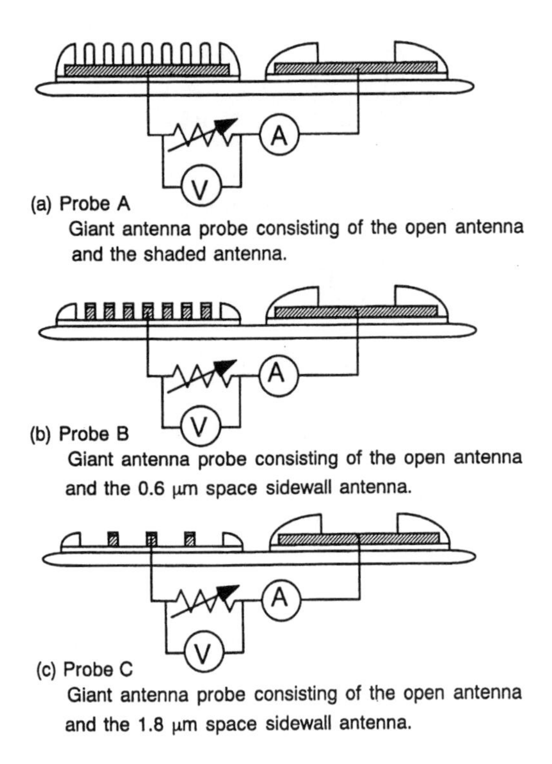

Figure 3.35. Giant antenna probes used by Hasegawa *et al.* [38] to find out the current and voltage behavior of electron-shading antenna with various spacing. Probe A has "tall" insulated dense-line mask on top of a metal pad. Probe B has thin insulating mask on top of the metal lines, no metal pad is underneath. Probe C is the same as probe B except the spacing is larger. All three probes are referenced to a large open metal pad.

The I-V characteristic of these probes in an ICP with 5mTorr Ar was measured using 1000W of source power and no bias. The result is shown in Figure 3.36. Probe A shows the normal electron-shading behavior of positive potential at the shaded antenna. Both probe B and C show negative potential at the antenna and, the large spacing probe C shows a larger negative potential.

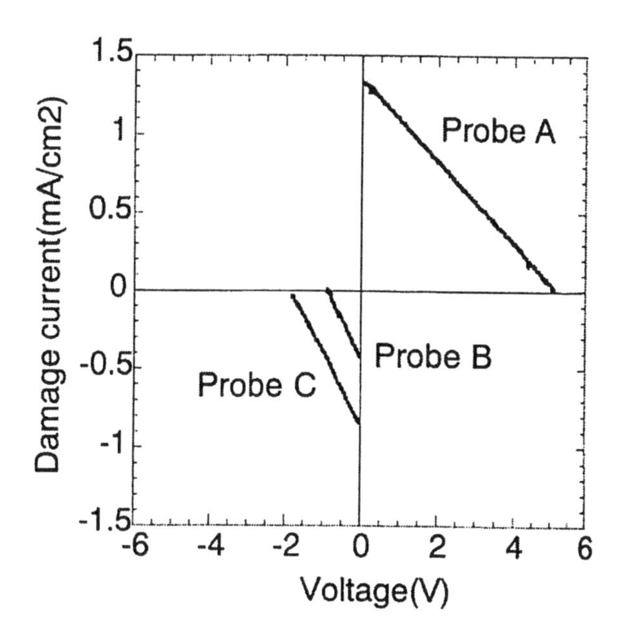

Figure 3.36. The measured I-V characteristics of the three probes shown in Figure 3.35 are shown. Probe B and C has an opposite sign compared to probe A. No bias was applied. Taken from [38].

To obtain further insight into the mechanism, Hasegawa *et al.* further studied the I-V characteristic of probe B and C as a function of electron temperature. Electron temperature was varied by changing pressure in the plasma while power was adjusted to maintain a constant ion current density. Figure 3.37 shows the response of probes B and C. Notice the T_e values shown are consistent with theoretical expectation of Figure 3.29. Contrary to electron-shading model, the negative current increases with electron temperature for both probes and the increase is larger for probe C.

Combining all the above results, Hasegawa *et al.* concluded that for probes B and C, the charging mechanism is the opposite of normal electron-shading, as represented by probe A. The charging mechanism can be understood with the help of Figure 3.38. Here the antenna type in probe A is named shading antenna while probes B and C are named sidewall antennae. The charging process in both antenna types is by the difference in directionality of ions and electrons. For shading antenna, the insulating mask intercepts more electrons while the metal antenna intercepts more ions. For sidewall antenna, the bottom insulating SiO_2 layer intercepts more ions while the tall metal sidewall intercepts more electrons.

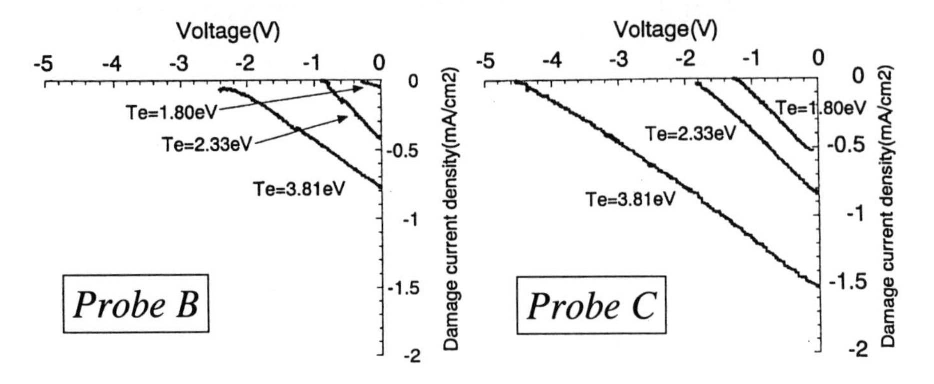

Figure 3.37. The I-V characteristics of Probe B and C as a function of electron temperature T_e are shown. No bias was used. Taken from [38].

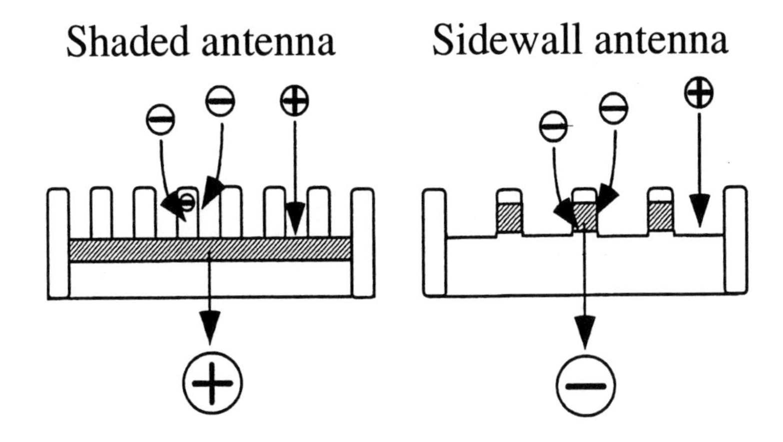

Figure 3.38. Proposed mechanism for the observed negative charging effect is illustrated. In the normal shaded antenna, electrons cannot reach the metal and excess ion flux charges the metal to a positive potential. In the sidewall antenna with thin insulated mask and large spacing, more electrons are collected by the metal line than ions, leading to a negative charging potential. Taken from [38].

The breakdown failure rates shown in Figure 3.34 are thus the combined result of both shading antenna damage and sidewall antenna damage. At "end point", resist is still thick and metal still exists in between lines, the antenna capacitors behave like a shading antenna. Shading damage happens during the short duration of "end point" until the bottom metal is cleared. Shading damage is expected to decrease as spacing between line increases. As etching progresses beyond "end point", resist thickness continues to reduce and metal between line etched away, the antenna capacitor gradually turns into sidewall antenna. Sidewall antenna damage increases with spacing between line. The combined damage therefore has a minimum at certain spacing. The higher damage observed for longer overetch can be understood by the combination of longer damage time and thinner resist, which is equivalent to a lower aspect ratio.

One must not forget that at the SiO_2 surface next to the metal lines of the sidewall antenna, positive potential develops due to electron-shading. This positive potential deflects low energy ions toward the sidewall as well as attracts electrons away from the sidewall. Both actions will reduce the negative charging of the metal line. If the metal potential has not been pinned to a less negative value by FN tunneling through a thin gate-oxide or by leakage through the variable resist of the giant antenna probe, it will float to a value that sustain charge balance. When the metal potential is pinned, fewer ions will reach the metal line while more electrons will. The dynamic balance is similar to the shading damage case with pinned bottom potential and is rather complex. It is actually rather remarkable that all the observed I-V characteristics are straight lines. This type of experiment offers an excellent opportunity to calibrate simulation models.

In the absence of deflected low energy ions, the metal line is expect to charge up until all electrons are turned away. This will mean many multiples of T_e. With the deflected low energy ions, the maximum voltage will be limited to 2 to 3 T_es depending on the bias condition. The maximum voltage found in Figure 3.37 for probe C is about 1.5 T_e. With bias, the value may go up somewhat. Given that T_e for most modern processing plasma are in the 2 to 4 eV range, sidewall antenna damage is not expected to occur for gate-oxide thicker than 10 nm. From the I-V characteristic of Figures 3.36 and 3.37, it is clear that both type of damage becomes more severe when the gate-oxide is thinner. Krishnan *et al.* [39] observed that the antenna damage during metal etching in ICP reactor is bi-directional in nature. The combined shading and sidewall antenna effect may explain their result [38].

3.1.8 Reduction of Electron-shading Damage Using Pulsed Plasma

As mentioned earlier, the most widely studied approach to reduce or eliminate electron-shading damage has been the use of pulse modulated plasma [22-29]. Most of these reports attributed the reduction of electron-shading damage to the reduction of electron temperature during the off cycle. Indeed, during the off cycle of pulsed plasma, T_e rapidly decreases due to preferential lost of high-energy electrons to inelastic collisions and to sheaths. Figure 3.39 shows the time dependence of T_e in pulsed Cl_2 plasma at zero bias [31]. Within 10μs of the onset of off cycle, T_e drops to a very low value. This rapid drop of T_e was observed by many other groups [40, 41] as well. It is commonly thought that during the off cycle, the lower T_e allows charges to be neutralized.

As discussed in previous section, lower T_e would help only in limited situations, such as low aspect ratio and low bias. Although the really low T_e during off cycle does push the transition to saturation regime at a higher aspect ratio or higher bias, the range is still limited. Recall that the reason low T_e help reduce shading damage is by increasing the amount of low energy ions and, when saturation has not set in, supply more low energy electrons that can be attracted by the bottom potential. When the bottom potential is pinned by tunneling, the supply of low energy ions is more critical than the supply of low energy electrons. Assume T_e =0, Equation 3.5 gives a V_{min} =0, so low T_e does help provide more low energy ions. However, since bias is typically not turned off in pulsed plasma, we still have a V_{max} given by Equation 3.6 and equal to $2V_{rf}$. A high bias still limits the supply of low energy ions. For pulsed plasma to be effective in suppressing shading damage at high aspect ratio or high bias, some other helping factors may be needed. So far, few pulsed plasma experiment were done at high aspect ratio or high bias, we do not know if the additional factors are really necessary in those situations. Nevertheless, there are indications that additional helping factors may exist.

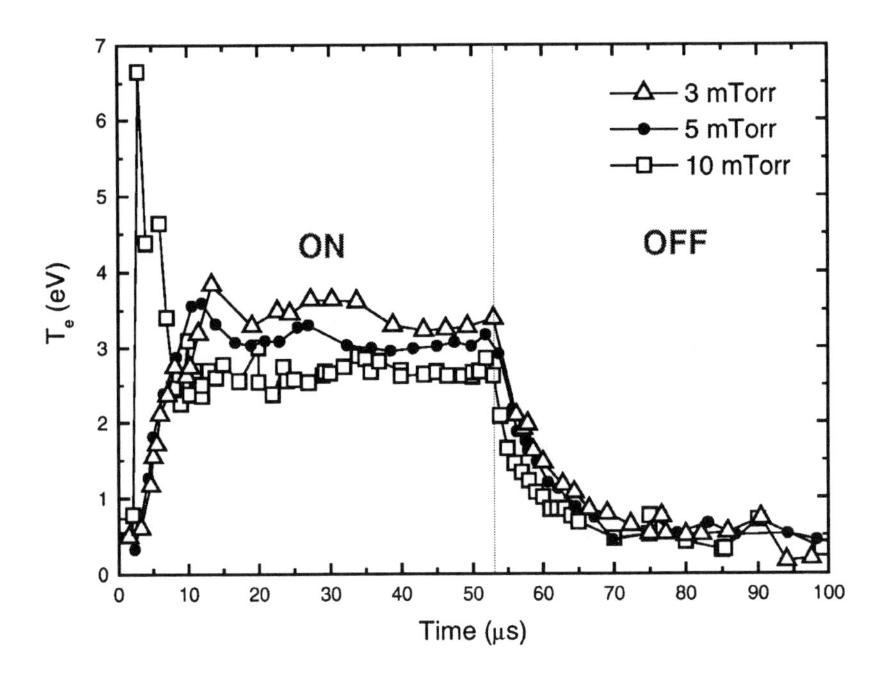

Figure 3.39. The electron temperature of a pulsed Cl_2 plasma is shown as a function of time. T_e drops off rapidly at the onset of off cycle. Taken from [31].

Fujiwara *et al.* [24] studied the effect of off cycle duration in a pulsed ECR plasma etching of polysilicon process. They found that the side-etch rate decreases with increasing off cycle duration, but the decrease is dependent on the discharge gas used (Figure 3.40). After analyzing the time dependent emission spectra, they concluded that the side-etch rate decrease is linked to the electron density decay rate. In other words, while the loss of high-energy electron to the sheath may cool the electron temperature fast, but the side-etch rate decreases with the decay of electron density through negative ion formation. Smaukawa *et al.* [42] also found that the threshold voltage shift in EEPROM devices exposed to pulsed ECR plasma decreases with increasing off cycle duration indicating a decrease of charging voltage (Figure 3.41). Note that for the same pulsed plasma condition that produced the very fast T_e decay in Figure 3.39, the electron density decay at a slower rate (Figure 3.42) [31].

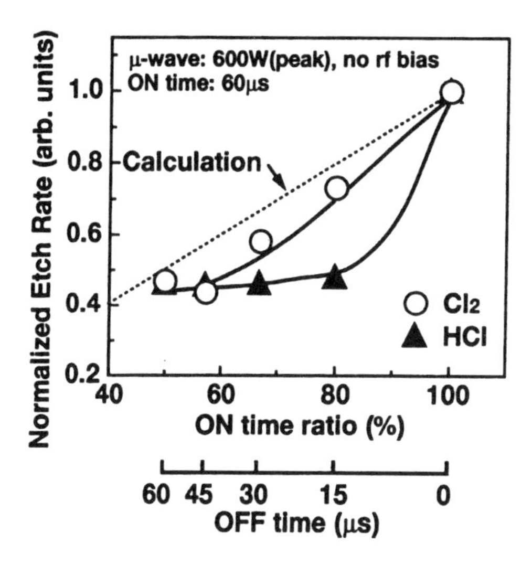

Figure 3.40. Side-etch rate for notch formation reduces with increasing off cycle duration. Taken from [24].

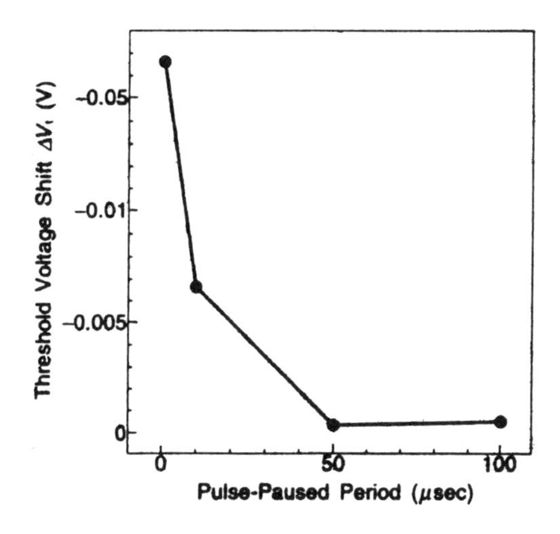

Figure 3.41. The EEPROM threshold voltage shift reduces with increasing plasma off duration, indicating lower charging voltage during plasma exposure. Taken from [42].

One obvious explanation for needing a longer than T_e decay time off cycle to further reduce shading damage is that the complete neutralization of accumulated charges needs a few bias cycles. However, that does not explain why the HCl discharge requires a shorter off cycle duration than Cl_2 discharge as shown is Figure 3.40. The link to electron density decay rate found by Fujiwara *et al.* suggests that the decrease in charging may have to do with negative ions. Normally, in continuous wave (CW) plasma, negative ions are trapped in the plasma and cannot reach the wafer. However, in pulsed plasma, negative ions may drift toward the wafer during off cycle. At the beginning of the off cycle, electron quickly cool and negative ions grow in density quickly. During the sheath contraction cycle, the collapsing sheath brings the plasma and therefore the negative ion density closer to the wafer. This movement of negative ion density behaves like a wave front (faster than the average thermal velocity of the negative ions). With the ultra-thin sheath associated with very low T_e, the negative ion front can reach the wafer as a pulse [43].

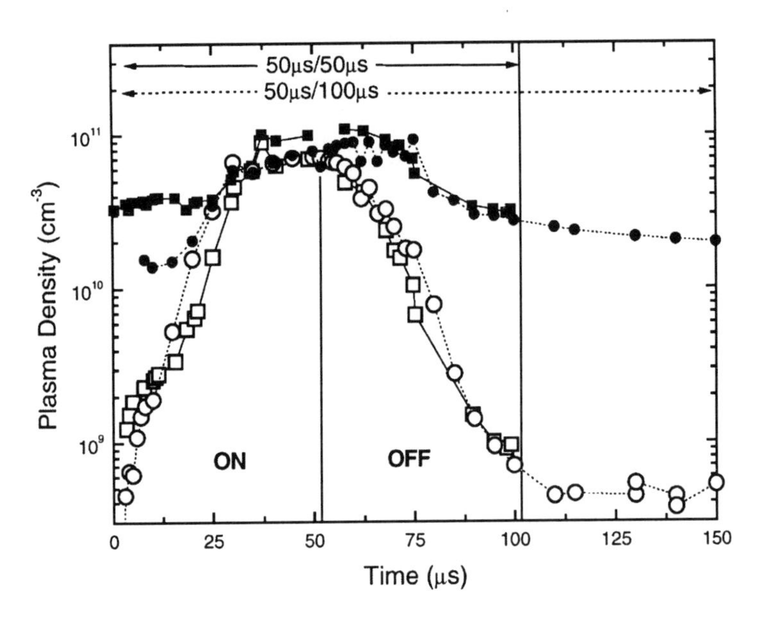

Figure 3.42. The electron density drops off at a longer time scale than the electron temperature shown in Figure 3.39 for the same pulsed plasma condition. Taken from [31].

This prediction was observed experimentally by Overzet *et al.* [44] using a set up shown in Figure 4.43. The discharge was an ICP. The ion energy analysis was through a pinhole. The discharge gas was SF_6 at 30mTorr. During the off cycle, they detected a negative ion flux that has three energy ranges, each associated with a distinct time scale (Figure 3.44). The highest energy one appears within a few tens of μs. The medium one appear ~200μs later. The low energy one starts to rise shortly after the first one and continue for a whole millisecond. The high-energy pulse may be due to the drift negative ion front moving along with the collapsing sheath. Note that Ocerzet *et al.* did not use a bias so the sheath simply collapses with T_e.

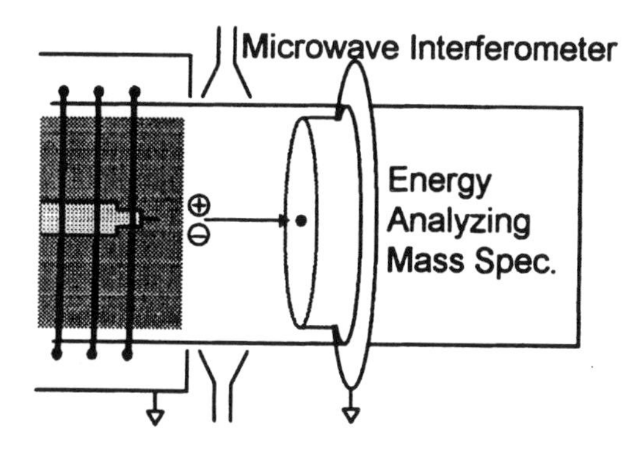

Figure 3.43. Pulsed ICP and ion energy analysis set up used by Overzet *et al.* [44]. Ion energy is measured through a pinhole.

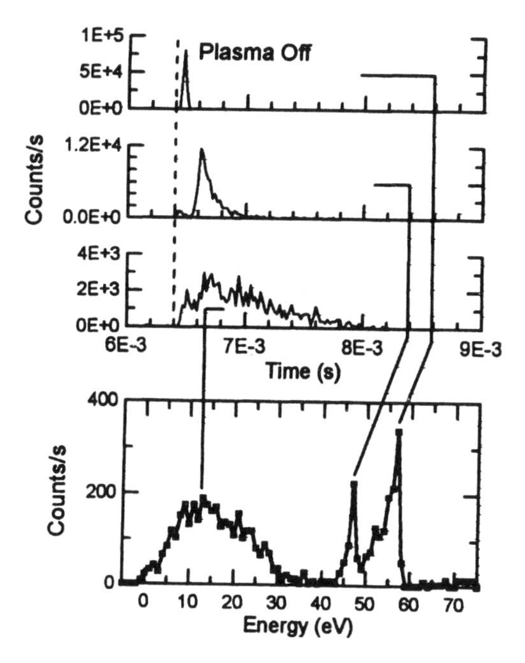

Figure 3.44. The measured negative ion energy and time through the pinhole during plasma off cycle are shown. Three groups of negative ions from the SF_6 discharge were detected. Each had its own distinct time mark. The highest energy ions appeared within 20µs of plasma off. The low energy group rise up soon after. The medium energy group showed up at about 200µs after plasma off. Taken from [44].

The low energy pulse of negative ions may be due to sheath reversal. As electron density continue to decrease during the off period, the electron flux can become lower than ion flux. The sheath voltage changes at that point from repelling electron to repelling ions. Such sheath reversal will also attract the now abundance negative ions toward the wafer. From the arrival time of the negative ion pulses, it is clear that these additional helping factors are only there when the bias frequency is very low or no bias at all. Under most etching condition, the bias frequency is too high for negative ions to have any impact.

In pulse modulated plasma, if the off cycle is too short, although one still gets a reduced T_e, it may not be enough for full damage elimination. Hashimoto *et al.* [11] did exactly that in their experiment and found that shading damage was not completely removed in some conditions. Figure 3.45 shows their shading damage study as a function of bias and bias frequency. This is the same experiment as those

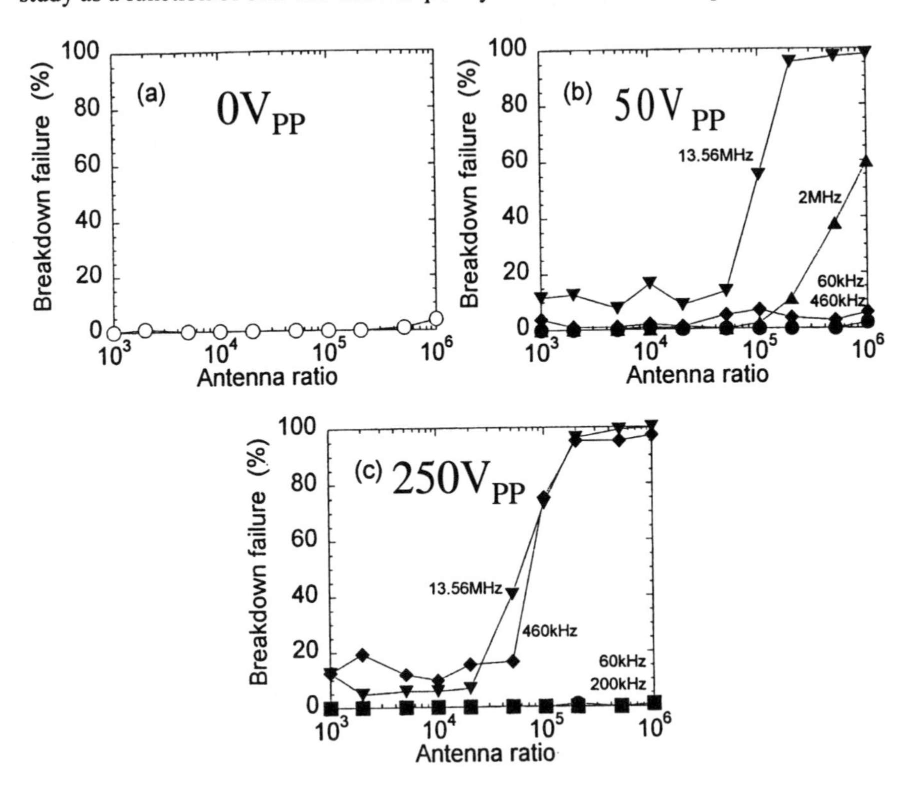

Figure 3.45. The antenna ratio at which breakdown failure increases significantly decreases with bias voltage in an electron-shading effect experiment using structures similar to Figure 3.4. The plasma was pulsed at 5μs/10μs on/off cycle. Damage is more severe with higher bias frequency. The bias frequency effect is stronger with higher bias voltage. Taken from [11].

shown in Figure 3.12, using the same antenna capacitors. The only difference is that this set of experiment was carried out with pulsed plasma. The plasma on/off period was 5μs/10μs. It is deem too short an off period to have significant negative ion growth. Unlike the CW case (Figure 3.12), $0V_{PP}$ bias produced no breakdown failures, showing the improvement due to pulsing of the plasma. When the bias was $50V_{PP}$, bias frequency up to 460kHz produced no breakdown failures. However, breakdown failures are observed for bias frequency of 2MHz. At 13.56MHz, the 50% breakdown antenna ratio is almost the same as CW case. With $250V_{pp}$ bias, even the 460kHz bias frequency is producing as much breakdown failure as 13.56Mhz. On the other hand, they are somewhat better than the CW case of the same bias.

Hashimoto *et al.*'s experiment clearly demonstrates that just lowering T_e is not enough to eliminate electron-shading damage in high bias (or high aspect ratio) etching. The frequency dependence is consistent with the ion energy effect. When the bias frequency increase, the transit time of ion becomes a larger fraction of a period of the bias RF. The result is that the energy distribution width reduces (see Section 3.1.5). The low-end of the ion energy increases while the high-end decreases. With less low energy ions available, the sidewalls are more negative and saturation sets in earlier. When bias voltage increases, the frequency effect is more pronounced because low energy ion density is reduced further.

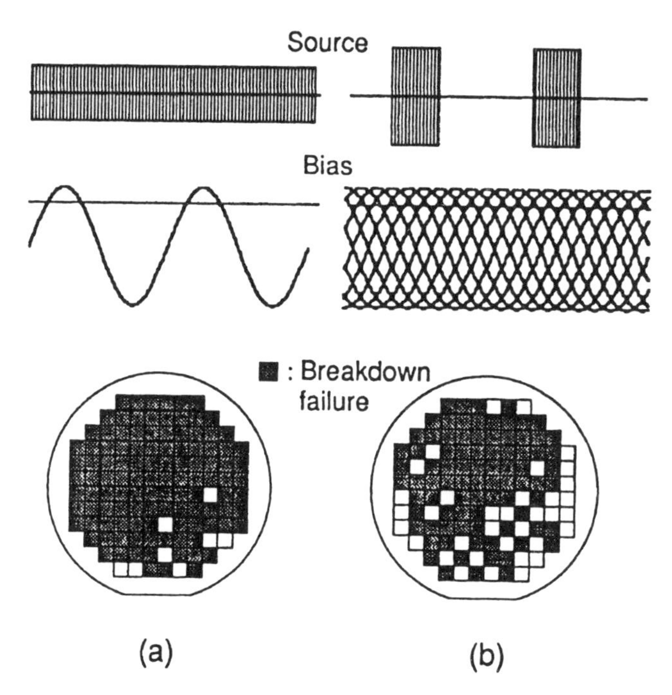

Figure 3.46. Pulsed plasma with asynchronous RF bias (60kHz, $1200V_{PP}$) is better than CW plasma with same average power and bias, but not by much. Dark sites signify broken down shaded antenna capacitor. Taken from [45].

In a separate experiment, Hashimoto *et al.* [45] showed that when an even higher bias voltage (1200V_{PP}) is used, very serious shading damage occurs even when the bias frequency is only 60kHz. In this case, the pulsed plasma is only slightly better than CW plasma (Figure 3.46). In Figure 3.46, the bias waveform as shown signifies that the bias RF is not synchronized with the pulsing of the plasma. When they try to synchronize the bias with the plasma pulsing (bias frequency changes to 66.7kHz to match the 5μs/10μs plasma on/off cycle), they found that the phase relationship between the bias and the plasma has significant impact on shading damage (Figure 3.47).

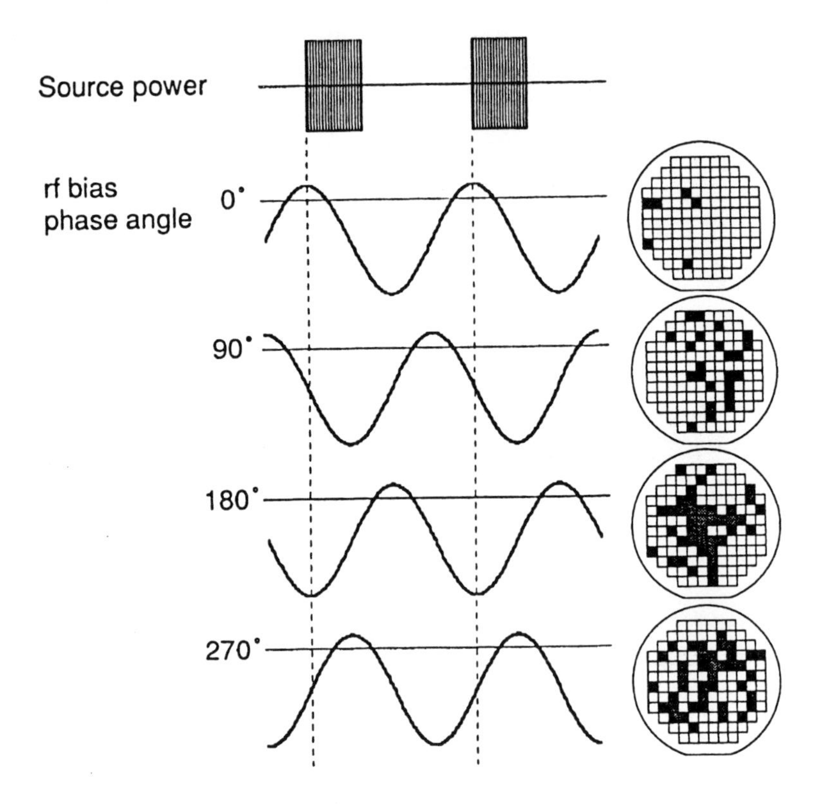

Figure 3.47. When making the bias (66.7kHz, 1200V_{PP}) to synchronize with the pulsing of the plasma, shading damage is found to be dependent on the phase relationship between the pulsing plasma and the bias. Taken from [45].

When the phase of the bias is such that its maximum is at the starting point of the on cycle of plasma (phase =0°), the damage is lowest. When the bias is at its minimum when the plasma begins its on cycle, the damage is highest. These interesting results can be understood as follows. At the peak of the bias cycle, the floating potential of the wafer comes closest to the plasma potential. At this point, the sheath thickness shrinks to a minimum, ions experience the least acceleration and electrons experience the least retardation. If this is coincident with the end of the plasma off cycle (phase =0°) when T_e is also at a minimum, the sheath is even thinner and ion energy is even lower. The net effect is to extend the low end of the ion energy distribution to lower energy. Thus the neutralization of the negative charges at the sidewalls is more effective and the shading damage will be reduced. If the valley of the bias voltage is coincident with the end of the plasma off cycle, the lowest T_e will only reduce the high end of the ion energy distribution. Low energy ions are not affected and therefore shading damage is not reduced as much. Some shading damage reduction is realized because of the asymmetric on/off cycle. When the minimum of the bias voltage is at the end of the plasma off cycle, the maximum is also within the off cycle where T_e cooling has already begun.

In Section 3.1.5, we saw that the low-energy peak of the ion energy distribution is larger than the high-energy peak. The reason was that the sheath shrinks faster than ion velocity. The effect of synchronizing the minimum sheath thickness with lowest T_e is to accentuate the low energy peak even more. Not only do we have lower ion energy at the low end, we also have more of these lower energy ions. Since the lowest energy ion are the most effective in neutralizing the negative charge at the top of the sidewalls. The synchronized biasing method derives the highest benefit of pulsed plasma. As can be seen in Figure 3.47, even with this benefit, shading damage cannot be completely eliminated with such a short off cycle.

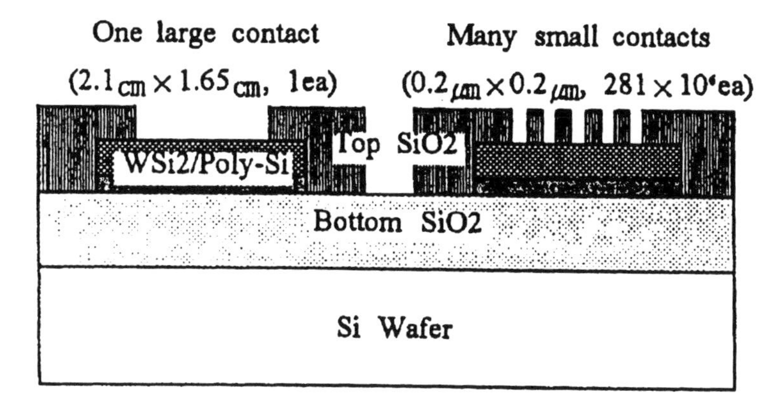

Figure 3.48. A giant antenna probe using the dense array of contact holes for shading. Taken from [29].

Recently, Shin *et al.* [29] proposed to go one step further than Hashimoto *et al.* by pulsing the RF bias synchronously with plasma. By keeping the plasma and the bias completely out of phase (180° phase shifted), the charging can be kept to a minimum. They used a charging probe similar to the one used by Hasegawa *et al.* (Figure 3.35a) except that instead of dense-line they used dense contact holes for shading (Figure 3.48). The probe was operating in open circuit mode so that the full bottom potential difference between the shaded antenna and open antenna is measured. Indeed, when the phase is 180°, the potential difference is lowest (Figure

3.49). Operating in this mode, they guarantee that the bias is on only during the low T_e period. The bias voltage used in the experiment appears to be $20V_{PP}$ (unspecified in their report). It is clear that pulsing the plasma and bias synchronously does not eliminate the charging at any phase angle even with the very low bias. This is due to shading is much stronger in a hole than in the spacing between lines. A surprising result is that phase angle does not seem to affect charging for the aspect ratio equals 1 case. On the other hand, the noise level in Shin *et al.*'s measurements is rather high. The variation may be buried in the noise.

3.1.9 Electron-shading Damage and Oxide Etching

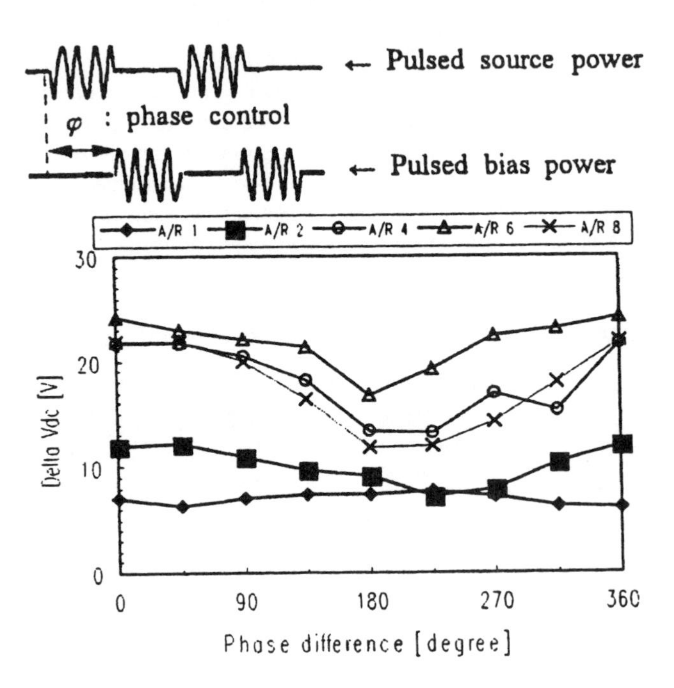

Figure 3.49. Pulsing both the plasma and the bias synchronously in a 180° out of phase manner can produce the lowest charging potential due to shading. Taken from [29].

Even with the method of pulsing both plasma and bias synchronously, significant charging is still observed for the low aspect ratio, low bias case (Figure 3.49). This clearly illustrates the shading problem encountered in oxide (contact or via) etching. Electrons are shaded in two dimensions in oxide (contact or via) etching while they are shaded in only one dimension for dense-line conductor etching. In addition, the aspect ratio encountered in oxide etching is higher than conductor etching in most cases. To make matter worse, oxide etching requires very high biases to achieve good etch-rate and profile control. It is of no surprise that ever since the introduction of high-density plasma for oxide etching, charging damage has been a serious problem. Okandan *et al.* [46] showed serious gate-oxide breakdown resulting from contacts etching using ICP source (Figure 3.50). Electron-shading was suspected to be the main culprit for the observed damage [47]. However, exactly how electron-shading effect causes charging damage in oxide etching is not clear. Hashimoto [48] tried to measure the charging characteristic in oxide etching using the giant antenna probe and found no signal. On the other hand, charging damage in oxide etching has been found in plasma that showed no non-uniformity related charging [46, 49-51]. One group tried to simulate shading damage in oxide etching numerically but found that the calculated voltage difference between top and bottom of the contact seemed too high to be realistic [52]. Here, we will examine the problem from the electron-shading effect point of view and try to explain some of the observed results.

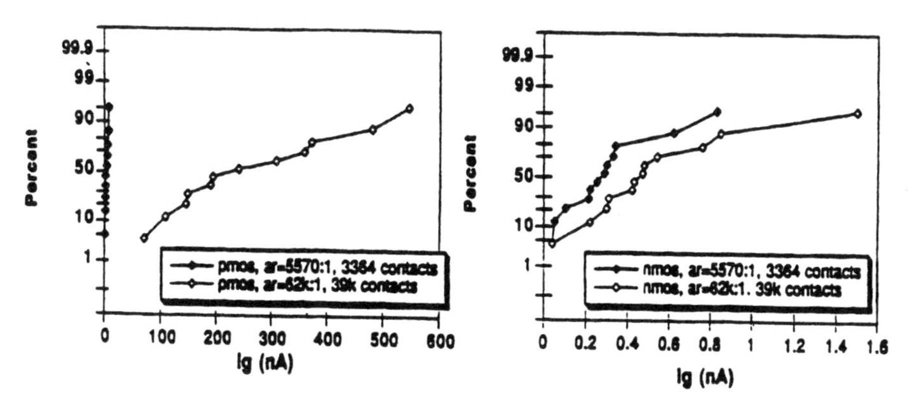

Figure 3.50. The cumulative gate leakage (I_g) at 3.3V (accumulation) for PMOS and NMOS transistors with medium and large contact antenna ratio are shown. Note that the abscissa scales are different for the two transistor types. Taken from [46].

In oxide etching, all windows are designed to have the same size (for lithography and etching control reason). There is no wide-open area to anchor the substrate's potential. Instead, substrate potential is determined by the difference in aspect ratio between windows connected to the gates and windows connected to the substrate. For example, in contact etching, both the contacts to gates and the contacts to source/drain/well are etched at the same time. Since the top of the gates is normally at higher position than source/drain/well, the aspect ratio of contacts to gates is lower than the aspect ratio of contacts to source/drain/well (Figure 3.51). In the absence of FN tunneling, the bottom potential of gate-contacts and substrate-contacts would both rise up to deflect enough ions to maintain charge balance. The bottom potential for the substrate contact is expected to be more positive than the bottom potential of gate contacts according to their aspect ratios. This is opposite to the giant antenna probes where the substrate potential from the normal electron-shading effect are always lower than the gate.

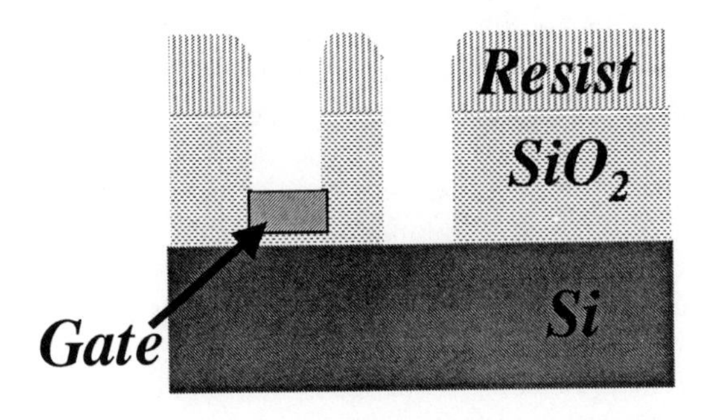

Figure 3.51. During the contact etching process, the contacts to the gate and to the substrate are etched together. Gate contacts tend to have lower aspect ratio than substrate contacts.

In the discussion of electron-shading damage of conductor etching, we concluded that lowering of the T_e would help reduce the damage for the low aspect ratio and low bias case. In general, methods that increase the supply of low energy ions would help to reduce shading damage. Does the same work for oxide etch? The answer is probably no. The best way to understand this is to use the extreme examples like we did in Section 3.1.5 when we discussed Figure 3.24c. In the absence of FN tunneling, the extreme case of very high aspect ratio (extreme shading) is that the bottom potential reaches V_{max}. The extreme case of wide-open space is, of course, zero potential rise. We already pointed out that shading is more severe in oxide etching. Because of 2 dimensional shading effect, the bottom potential must rise up quite high for both gate-contacts and substrate-contacts. Let us hypothetically choose a set of numbers to illustrate the situation. Lets assume that for the contact to gate, the shading situation is such that the bottom potential must rises high enough to deflect 65% of the ions, and for the contact to substrate, 75%. For a given ion energy distribution, the width of the distribution is determined by $2V_{rf}$ (see Equations 3.5 and 3.6) while T_e only shift the entire distribution up and down the energy scale (Figure 3.52). The potential difference (ΔV_{max}) between 65% and 75% of ion deflection is related to V_{rf}, not T_e.

During contact etching, contact to gate clears first. At that point, the depth of both contacts is the same, so is the bottom potential rise. As etching continues, the substrate contact hole becomes deeper. The bottom potential rises up higher. When the substrate contact begins to clear, substrate potential will be pulled up. As a result, a potential difference develops across the gate oxide. When this potential difference is high enough, FN tunneling increases rapidly. The FN current has two effects. It lowers the substrate's potential, which leads to fewer ions being deflected in the substrate contact hole. It increases the gate potential, which leads to more ions to be deflected as well as more electrons to reach the gate. Similar to the asymmetric response of electron flux from plasma to floating potential change discussed in Section 2.2.3, gate potential increase is going to be much less than substrate potential decrease if there are no other factors exist to hold the substrate potential high. If we assume that the gate potential changes very little, then the change in substrate potential must be ΔV_{max} - V_{ox} (neglecting band bending). The FN current involved must be the integrated ion flux within that energy range. If ΔV_{max} - V_{ox} ~ΔV_{max}, the FN current is simply the shading difference.

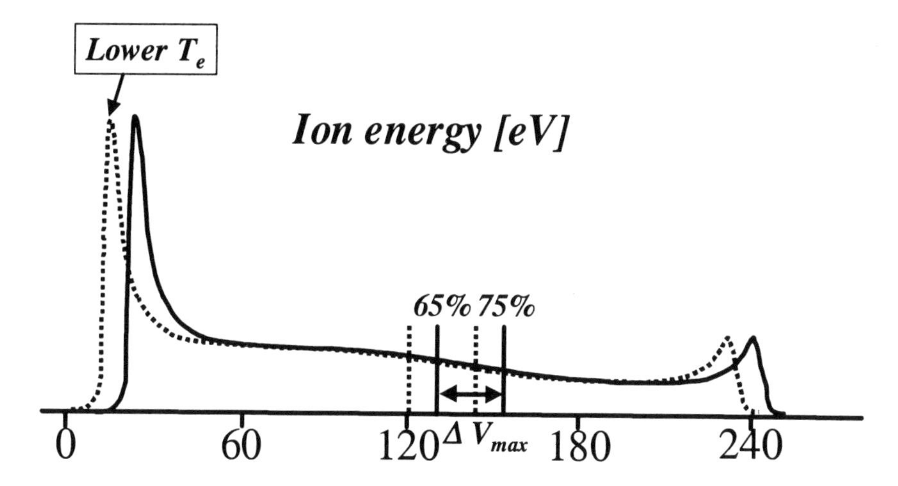

Figure 3.52. The bottom potential rise is determined by shading. The bottom potential difference is related to the width of the ion energy distribution. Lowering T_e can only shift the scale, the difference remains unchanged.

For the example at hand, the shading difference is 10% of the total ion flux. Notice that this is the ion flux collected by the substrate contact. In most antenna test device designs, large number of contacts are opened to gate antenna, the number of substrate contact is usually small. Thus, the damage current that flow across the gate-oxide should be very small. The structure used in Okandan *et al.*'s experiment [46] had 8 contacts to source and drain, the FN current should be very small regardless the number of contact to the gate. Even if the plasma density is very high, the FN current density should only be ~1mA/cm^2 (10% of ion current density times the total area of the 8 contacts divided by the gate area), hardly able to break the gate oxide for the short duration of oxide etching. Since the FN current is determined by substrate contacts only, the damage level should be independent of the number of contacts on the antenna. This conclusion is, of course, not in agreement with experimental observation.

In most cases, antenna test devices are implemented on a die where many other test devices or circuits exist. The number of contacts to the substrate from all devices and circuits can be very large. They can overwhelm the contacts to the gate. The end result is that substrate potential will not be lowered so easily. When substrate potential is pinned, gate potential must adjust to maintain charge balance. When the gate potential is forced to increase, more ions will be deflected and more electrons will reach the bottom of the gate-contact. Even with the shading geometry, electron flux can increase significantly. For example, Siu *et al.* [53] measured the I-V characteristic of line and space shaded and open pad. They found that although the electron flux collected with positive pad bias is almost two orders of magnitude lower than an open pad, it can still be a few time higher than ion flux (Figure 3.53).

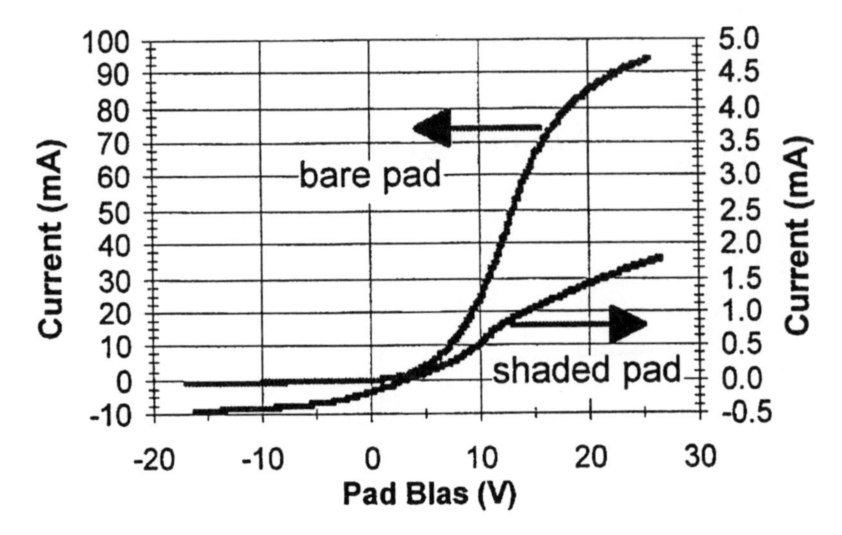

Figure 3.53. The I-V characteristic comparison between bare pad and shaded pad is shown. Although the electron flux collected by the shaded pad is much smaller, it still is a few times larger than ion flux. Taken from [53].

Of course, the shading is much more severe in oxide etching and we therefore should expect even lower electron flux density when the gate potential increases. Aritmoto *et al.* [21] made similar measurements with varying aspect ratio line and space and confirmed that more shading reduces the electron flux collected on the pad (Figure 3.54). Based on these observations, we can conclude that as the gate potential rises, more electrons are collected through the contact antenna to the gate. The electron flux increase is likely to be a weak exponential function of the potential rise. The magnitude of the electron flux should be from a small fraction of the total ion flux up to nearly equal to the total ion flux.

We can speculate that the substrate potential in Okandan *et al.*'s experiment was indeed pinned by all the devices and circuits on the die. Such a situation is actually rather common. If that were the case, we could expect that more number of contacts on the gate antenna would cause higher degree of damage, as they had observed. We could also expect that the degree of damage to be proportional to the plasma density. This was indeed the case for charging damage in oxide etching using another high-density plasma source [54].

If the interpretation proposed above were correct, lowering T_e would actually increase damage. This is because more electrons can be attracted to the bottom of the contact for a given potential rise.

We have assumed that the plasma is perfectly uniform in our treatment of the oxide etching damage problem so far. When the plasma is not uniform, one must combine the effect of potential driven by non-uniform plasma with the effect of potential driven by electron-shading. The situation gets complicated rather quickly.

In CMOS technology, one must take into account the effect of the well. The p-MOS well forms a pn-junction with the p-type substrate. This junction assures the p-MOS well to be at a higher or equal potential with the substrate. As a result, p-MOS devices are likely to be damaged more severely. This is indeed the case as shown in Figure 3.50.

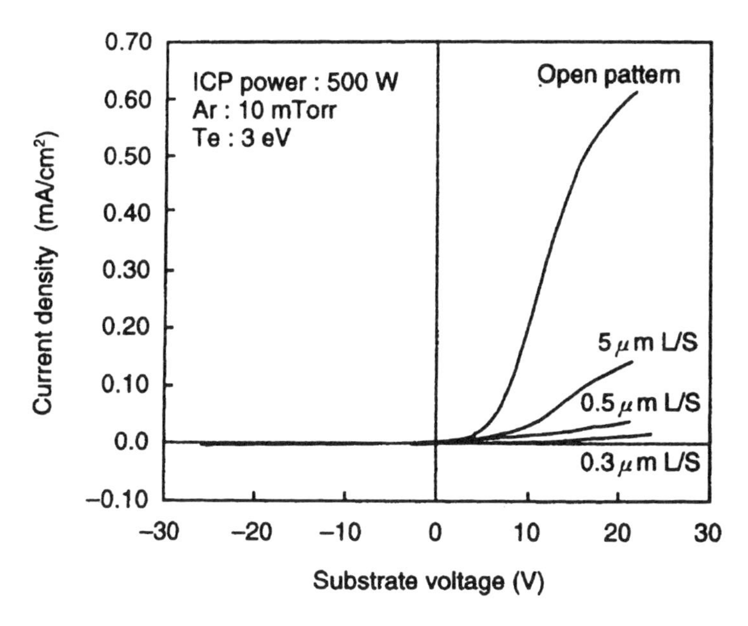

Figure 3.54. Increased shading would reduce the collected electron flux severely. Taken from [21].

For via etching, the source of aspect ratio difference between gate and substrate connection is not easily predicted. If the dielectric planarization process were perfect, there should be no aspect ratio difference between the vias connected to gate and the vias connected to substrate since both are making the connection through metal 1. No matter how high the aspect ratio is, the gate and the substrate are always driven to the same potential and no charging damage from shading effect should occur. On the other hand, if the substrate potential were pinned by some other means or aspect ratio variation does exist due to imperfect planarization, charging damage due to shading could occur. A good example was reported by Hook [55] recently. He found that electron-shading induced charging damage occurred during via etching only when the backside of the wafer was conductive. When the backside of the wafer had an insulating film, no damage was observed. He reasoned that when the backside was conductive, even though the wafer electrostatic chuck was more or less insulating, the substrate potential was not completely free from the influence of the chuck. Only when there was a backside insulating film, the substrate potential was driven by the shading effect of the vias.

Note that the bias in oxide etching is normally several hundred volts. The maximum width of the ion energy distribution has the same value. A 1% difference in ion flux due to shading will produce a few volts across the gate-oxide. As a result, a slight difference in aspect ratio can cause charging damage problem in thin gate-oxide. One way to reduce this sensitivity is to use very high frequency bias. Recall that the ion energy distribution is a function of the ion transit time across the sheath and the period of the RF bias. It will collapse into a single peak when the transit time is several RF bias periods. If the width of the ion distribution is only 20eV for example, a 10% difference in shading will produce only 2V across the gate-oxide. Such a low voltage during the short etching time will not damage even the thinnest gate-oxide. Unfortunately, high-density plasma makes narrowing of the ion energy distribution difficult. This is because the sheath thickness is inversely proportional to the square root of plasma density according to Equation 3.3. Coupling the thin sheath of high-density plasma with the very high bias, the average transit time is very short. Even at 13.56MHz bias, the ion energy distribution for oxide etching in high-density plasma will still be a relatively broad double peak structure.

Based on the above discussion, an obvious way to improve electron-shading damage in oxide etching is to reduce the plasma density. Lowering the plasma density would reduce the total charge flux arriving to the bottom of each contact and therefore would reduce damage even if the charging voltage were high. Lowering the plasma density would also reduce the voltage difference between different aspect ratio contacts by narrowing the ion energy distribution (for a given bias frequency). A smaller voltage difference further reduces the net charge flux reaching the bottom of gate contact. The combine reduction will drastically reduce the FN stress to the gate-oxide during etching. The best, of course, is to have lower density plasma with higher frequency bias. The trick is to have such plasma with perfect uniformity. In more than one case [51, 54], lowering the plasma power, and hence the density, was the only effective way to reduce damage in oxide etching.

3.1.10 Electron-shading Damage and RF-sputter Clean Process

A common process used to ensure low contact resistance in contacts or vias is RF-sputter clean before metal deposition. Such process has been found to cause charging damage [56-58]. In Ito *et al.*'s work, they clearly demonstrated that the charging damage observed was due to electron-shading effect. Figure 3.55 shows the antenna ratio (number of vias) dependent capacitor breakdown yield in Ito *et al.*'s experiment. Notice from the insert that their oxide was not planar. The contacts to gate and substrate may be differing in aspect ratio. In their report, they did not provide any information on how the substrate potential was determined. However, they observed antenna ratio dependent, as well as aspect ratio dependent breakdown failure (Figure 3.56). These are characteristic signals of electron-shading damage.

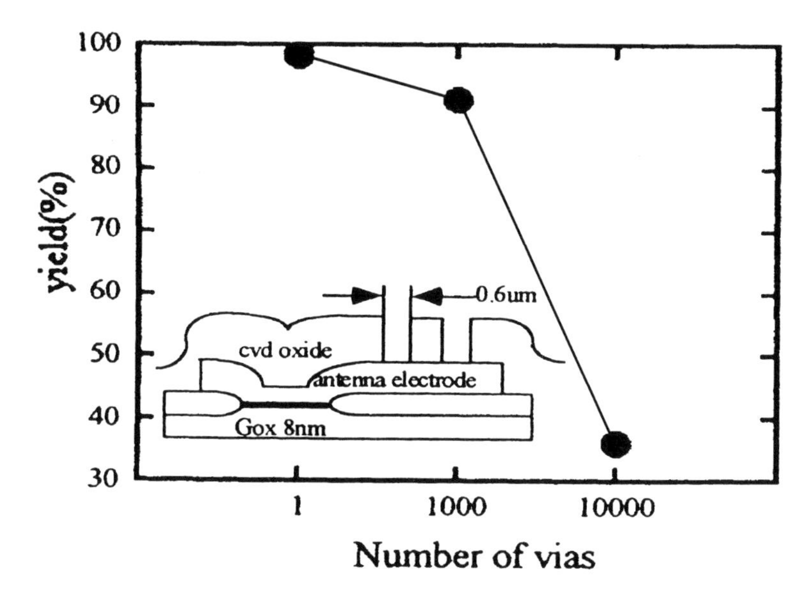

Figure 3.55. Via pre-clean using RF sputtering causes antenna size dependent breakdown yield, indicating charging damage. Taken from [57].

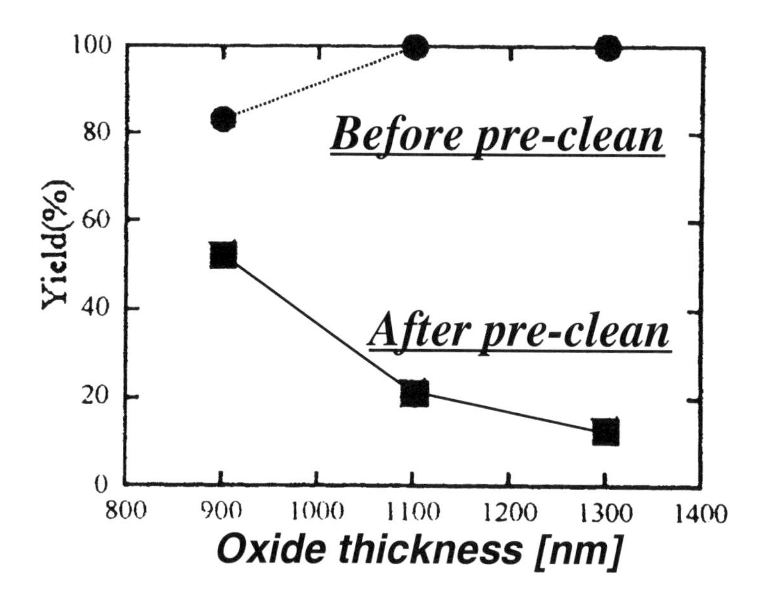

Figure 3.56. The breakdown yield of large antenna decreases with increasing oxide thickness, indicating electron-shading damage. Taken from [57].

In their experiment, high-density source (ICP) was used. They found that lowering the etch-rate (accomplished by lowering the power) was effective in lowering the damage (Figure 3.57), similar to oxide etching. Power reduction as a mean to reduce damage was reported by others as well [56, 58]. Ito *et al.* further showed that if there were a conducting liner on the sidewalls, damage were also reduced (Figure 3.58). Such result confirms that the observed damage was indeed due to electron-shading.

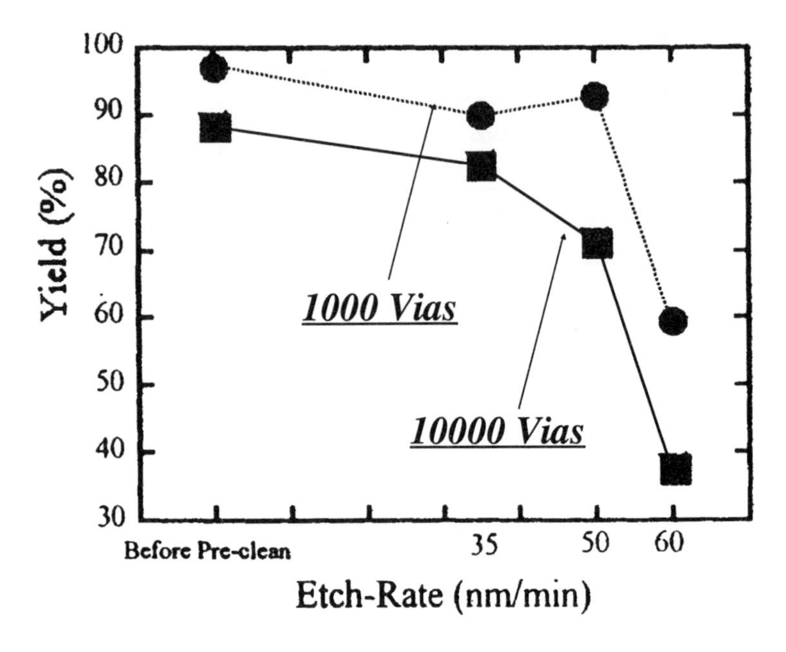

Figure 3.57. The breakdown yield is etch-rate dependent. Higher etch-rate using higher RF power causes more damage. Taken from [57].

Not all charging damage during RF-sputter clean are due to electron-shading. Simple diode-type low-density plasma is still commonly employed in many production machines. In these older style plasma systems, density tends to be too low to support significant charging damage due to electron-shading. Instead, charging damage from plasma non-uniformity is more common. For these systems, lowering the power may not be the solution. Increasing the power often improve plasma uniformity.

Before we leave the section on electron-shading effect, it should be pointed out that through out our discussion, an important factor has been neglected. This factor is secondary electron emission from energetic ions. Where secondary electron is emitted, the ion effectively carry two positive charges to the surface.

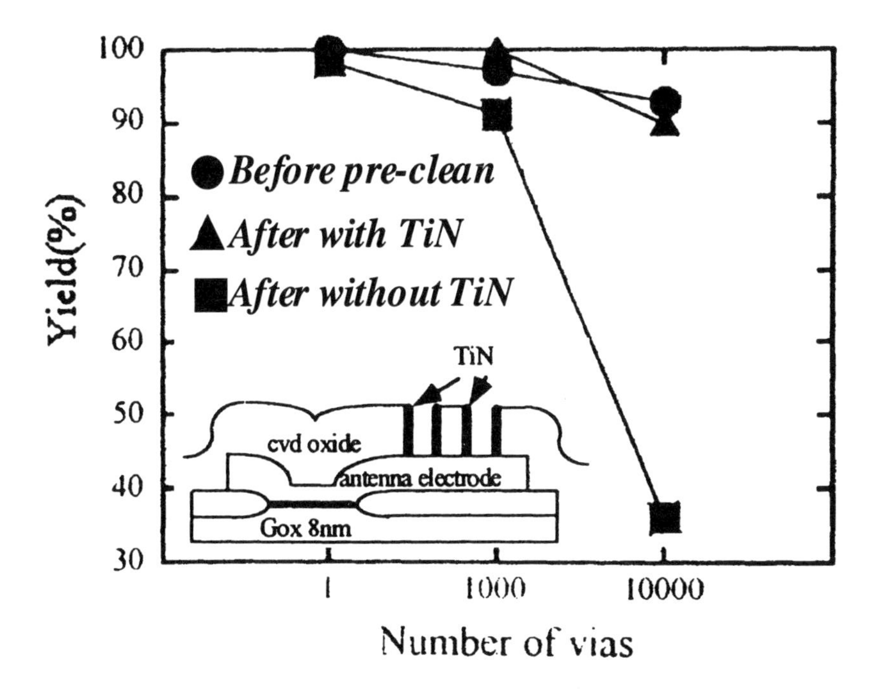

Figure 3.58. Using TiN to line the sidewall eliminated damage. Proofing that the damage is electron-shading induced charging. Taken from [57].

3.2 AC Charging Effect

3.2.1 Oscillating Oxide Field due to RF Bias

Charging damage that occurs in perfectly uniform plasma is not limited to electron-shading effect. AC effect can potentially do that as well. One AC effect may be from the capacitive coupling of the RF bias to the antenna capacitor. In Section 2.3.2, we saw that even when a very large AC voltage is applied to the wafer to create a very high bias potential for the ions, the voltage that appear across the gate oxide is very small. The explanation is that the applied voltage is divided among the equivalent capacitors in series and the voltage across each capacitor is inversely proportional to its capacitance. These capacitors are from the plasma sheath, the gate-oxide and the blocking capacitor. The gate-oxide thickness is so small and that its capacitance is very large comparing to the sheath capacitor. Consequently, its share of the applied voltage is also very small.

For high-density plasma, the sheath thickness can be 100μm or less. When a large antenna ratio is associated with some of the higher level metal layers, can one still assume that the voltage across the gate-oxide is very small? The problem can be treated quantitatively. For example, if the plasma sheath is 100μm thick and the total oxide thickness under the metal line antenna is 2.5μm, the voltage on the antenna would be 2.5/(3.9x100) ~ 0.64% of sheath voltage. Imagine etching this metal line with a $500V_{PP}$ bias in high-density plasma, the peak voltage between the metal line and substrate is somewhat higher than 3.2V. If the metal line is not an antenna (not connected to the gate), the voltage developed is too small to be of concern.

When the metal line antenna is connected to the gate, the antenna system can be regarded as two capacitors connected in parallel [59]. The total capacitance is the sum of the two. Let C_{sh} be the sheath capacitance per unit area, C_g the gate capacitance per unit area, and C_A the antenna capacitance per unit area. The voltage across the gate is:

$$V_g = V_{sh}\left(\frac{C_{sh}}{C_g + C_A}\right).$$

Using the parallel plate capacitor formula, we get:

$$V_g = V_{sh}\left(\frac{(A_g + A_A)/d_{sh}}{\varepsilon(A_g/d_g + A_A/d_A)}\right) = V_{sh}\left(\frac{d_A(1+R_A)}{\varepsilon d_{sh}(R_d + R_A)}\right) \tag{3.7}$$

Here A_g, A_A are area of gate and antenna respectively, d_{sh}, d_g, d_A are thickness of sheath, gate-oxide and oxide between antenna and substrate, respectively, R_A, R_d are antenna ratio (A_A/A_g) and oxide thickness ratio (d_A/d_g), respectively. In the limit of very large antenna ratio, Equation 3.7 reduces to the completely floating metal case:

$$V_g = V_{sh}\left(\frac{d_A}{\varepsilon d_{sh}}\right) \tag{3.8}$$

Thus, in the limit of large antenna ratio, the antenna voltage behaves as if it is not connected to the gate. When the antenna ratio is small, the gate capacitor will have a bigger influence and the total capacitance to sheath capacitance ratio will increase. The effect is to lower the voltage across the gate oxide. Thus, Equation 3.8 is the worst case estimate. Note that the voltage developed this way is not really an AC field across the gate-oxide. Figure 3.59 shows how both the substrate potential and antenna potential (large antenna limit) oscillate in relation to plasma potential when driven by a $500V_{PP}$ RF bias. In the figure, the antenna potential curve was artificially shifted up 50V so that one can see the two curves. Otherwise, they are too close to each other to be distinguished. Real potential curves are not sinusoidal as shown, but the two curves will still be closely following each other. The difference between the antenna and the substrate is shown in Figure 3.60. It is always positive. For the example considered ($V_g \sim 0.64\% \ V_{sh}$), a $500V_{PP}$ bias causes more than 3.5V peak voltage across the gate-oxide. By itself, this voltage is not of concern until gate-oxide is 30Å or less. One must remember, however, this voltage will be superimposed onto any other source of charging voltage.

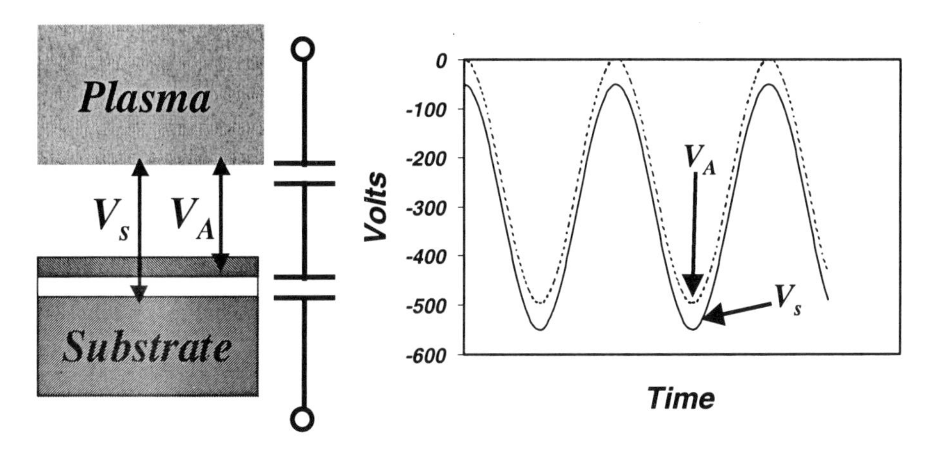

Figure 3.59. Sheath capacitor and oxide capacitor divide the voltage between plasma and substrate. The voltage between metal antenna and plasma is always somewhat smaller than voltage between plasma and substrate. The antenna voltage curve was shifted up 50V to separate it from the substrate curve. All voltage reference to the plasma which is taken to be zero.

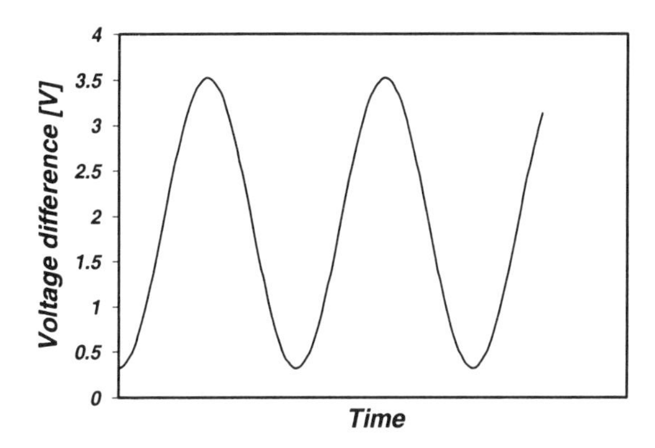

Figure 3.60. The potential difference between the antenna and substrate from capacitive division is always positive.

The etching condition of $500V_{PP}$ and 100μm sheath thickness is not commonly used. The only process that uses such condition is oxide etching. Even there the trend now is to reduce plasma density. Most other etching processes use lower density plasma and lower bias. Thus this AC effect is normally not a serious problem.

It is worth while to keep in mind some worst case situation that can occur potentially. It is not uncommon that up to 8 levels or more metal layers are used in deep submicron CMOS technology. The higher level metal layers can be as much as 5 to 6μm away from the substrate. While metal etching itself seldom uses very high-density plasma with very high bias, oxide etching does. The key is that this is a capacitively coupled phenomenon, the metal antenna does not have to be exposed to plasma to develop the voltage. Using the same numerical example above, the voltage across the gate-oxide can now be as high as 8V. This would not have been a problem if the metal were completely insulated from the plasma. The reason is that small amount of electron injected from the substrate to the gate that is connected to the metal antenna will bring down the potential, and FN tunneling will stop. The total amount of injected charge is whatever it takes to change the voltage across the capacitor. It is a very small quantity. The capacitance of a parallel plate capacitor with 5μm thick SiO_2 as dielectric is 0.7nF/cm^2. It takes 0.7nC/cm^2 to change the voltage by 1 volt. Assuming we need to lower the gate potential by 5V and that the antenna ratio is 100,000, the total injected charge across the gate-oxide is only 35mC/cm^2, not enough to cause any real problem.

Note that if the voltage were a true AC field across the gate-oxide, the problem would be extremely serious. The 35mC/cm^2 would become the charge tunneling across the gate-oxide per bias RF cycle. If the bias frequency were 13.56MHz, the AC charging current would be 474600A/cm^2. No gate-oxide can survive even a brief moment.

There are two factors that can get matters worse very quickly. One is that when oxide etching is opening vias down to the metal antenna. As the etching reaches "end point", the metal antenna can collect charges from the plasma. Even in the absence of electron-shading effect, the metal antenna will be able to sustain the FN current by neutralizing the electrons coming from the substrate with ion flux from the plasma. The FN tunneling will no longer be self-limiting and the damage can be much worse. The other factor is that oxide is not a good insulator in the presence of plasma. Most plasma emits large amount of vacuum ultraviolet (VUV) light. In the presence of VUV light, SiO_2 becomes conductive. The metal antenna thus can collect charges from the plasma to keep the FN tunneling going. We will discuss this conduction mechanism in more detail in the next chapter.

The worst case situation discussed above does not happen normally. There will almost always be low-level metals underneath the higher level metal antenna. The effective capacitance will be larger and the voltage will be lower. The only exception may be in the antenna test device design where no lower level metal underneath the antenna.

3.2.2 AC Charging from Pulsed Electron Flux

Another AC charging effect [60] due to the pulsed electron flux can be important under certain condition. Because electrons flow to the wafer in pulsed mode, the net charges on an antenna periodically go from positive to negative and back. This net charge can be translated into an alternating voltage across the capacitor formed by the antenna-gate electrode and the substrate. Figure 3.61 illustrates the pulsed electron flux and the steady ion flux from plasma, and the resulting potential on the antenna-gate electrode with respect to the substrate.

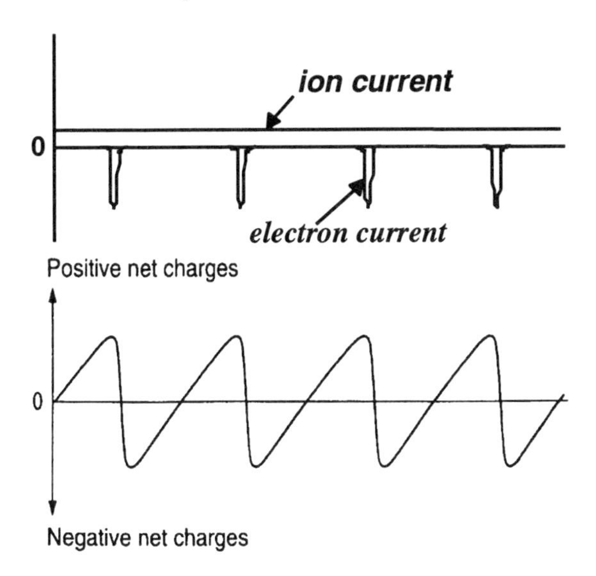

Figure 3.61. Continuous ion flux and pulsed electron flux from plasma causes the antenna-gate electrode's potential with respect to the substrate to have a substantial alternating value.

The magnitude of this AC voltage can be estimated from the ion current density multiplied by the RF period. For the antenna-gate capacitor:

$$V = \frac{Q}{C} = \left(\frac{Q}{C_g + C_A} \right) = \left(\frac{Q}{C_g \left(1 + {}^{R_A}\!/_{R_d} \right)} \right).$$

Charge per RF bias period Q is simply:

$$Q = 0.6 n \sqrt{\frac{kT_e}{M}} \left(A_g + A_A \right) \tau ,$$

where τ is the RF period. Combining the two expressions while keeping track of all the numerical factors and units, we get:

$$V = \left(\frac{0.6n\sqrt{\frac{kT_e}{M}}d_g}{(2.2\times 10^6)f} \right) * \left(\frac{R_d(1+R_A)}{R_d + R_A} \right) \tag{3.9}$$

where f is the RF bias frequency. Equation 3.9 can be interpreted as a charging factor times an antenna enhancement factor. When the antenna ratio R_A is much larger than the oxide thickness ratio R_d, the antenna enhancement factor reduces to R_d. In other words, the maximum antenna enhancement factor is the oxide thickness ratio. Coupling that with the dependent on d_g for the charging factor, the charging voltage is directly proportional to the oxide thickness under the antenna. Figure 3.62 shows the peak to peak voltage from this pulse electron flux effect for a variety of conditions (assuming d_g =100Å, M=40). As can be seen, the effect becomes important when the bias frequency is low and the oxide underneath the antenna is thick (R_d =200 implies a d_A of 2μm). Such behaviors make good sense because this is a simple case of charging a capacitor. The thicker the oxide under the antenna, the smaller the capacitance and therefore larger voltage change for a given quantity of charge change. The lower the RF frequency, the longer the period and therefore more total charge per period.

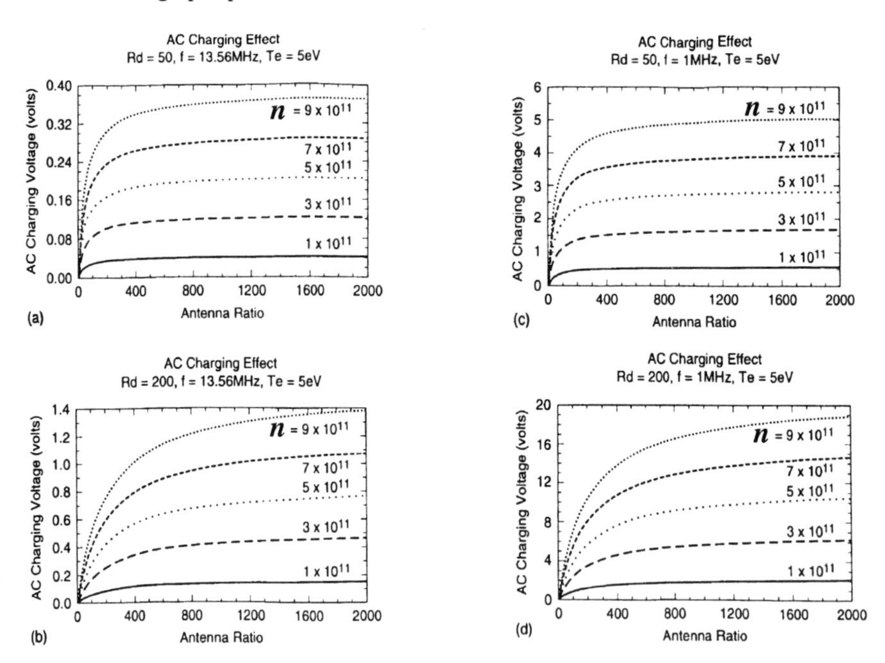

Figure 3.62. Peak to peak voltage across the gate-oxide due to pulsed electron flow from plasma for a variety of conditions. Each figure contains a family of curves for different plasma densities.

There has never been any experimental confirmation of Equation 3.9 reported in the literature. Bias frequency effect on damage was studied with ECR plasma source [61, 62]. The observed results were lower bias frequency had lower charging damage, opposite of Equation 3.9. However, the densities of the ECR plasmas in these experiments were only $1x10^{11}/cm^3$ or less. AC charging due to electron pulsing is not expected to be significant. In addition, Samukawa [61] reported that plasma uniformity was really poor in the high bias frequency regime. The observed more damage at higher bias frequency was most likely the result or poor uniformity.

In most etching conditions, AC charging due to electron pulsing should not be a problem by itself. It can, however, worsen any charging effect by superposition the AC voltage on other charging voltage. If they're charging due to plasma non-uniformity, the addition of AC charging can make the situation much worse. AC charging is directly proportional to plasma density while plasma potential is logarithmic to plasma density:

$$V_{P1} - V_{P2} = \frac{kT_e}{e} ln\left(\frac{n_1}{n_2}\right)$$

Thus, a moderate variation in V_P can cause a large variation in AC charging. Where the plasma potential peaks, AC charging also peaks. The combine effect is a highly localized severe damage region.

The discussion of AC charging due to electron pulsing has so far assumed that the substrate does not receive similar electron and ion flow. That is not likely to be the case in any real processing situation. The substrate is almost always able to collect charges directly from the plasma, either through direct exposure or by connecting to conductors that are exposed to the plasma. If it can collect exactly the same amount of charges as the antenna, then there should be no AC charging voltage at all. On the other hand, it is not likely that the substrate would collect exactly the same amount of charges per unit area as the antenna. It either collects more or less depending on the relative area of antenna to the gate and antenna to the substrate. There is always a residual AC charging effect, which is smaller than what is predicted by Equation 3.9. Unless one have the exact knowledge how much is the substrate's exposure to plasma, directly and indirectly, one has no way to predict how large would the AC charging effect be.

3.3 RF Bias Transient Charging Damage

Another potential source of charging damage with a perfectly uniform plasma is RF bias transient, particularly turn off transient. As discussed in Chapter 2, the self-bias on any object in contact with plasma is due to electrons accumulated on the object so that a repulsing potential is established to limit the flow of electrons to the object.

The self-bias become much larger when a RF voltage is applied to the object, meaning that the amount of accumulated electron is much larger. This larger self-bias serves to accelerate ions toward the object and completely eliminated the electron flow except for the brief moment when the RF voltage is near minimum. RF bias is always applied during etching. It is much larger for oxide etching, much smaller for polysilicon etching and metal etching is somewhere in between. At the end of an etching process, if the RF bias is turned off abruptly, the excess electrons accumulated on the wafer surface must be neutralized by collecting ions from plasma. The neutralization process for the antenna and the substrate will not be at the same pace. This is very similar to the situation in the AC charging due to pulsed electron flow. One cannot easily predict how large the transient charging will be.

One can, however, estimate what the worst case will be. Assuming that the substrate has negligibly small exposed area but not zero. If it were zero, the substrate is completely floating. A completely floating substrate can follow any changes in gate potential without current flow (assuming a uniform gate potential). When the RF bias is turned off abruptly, the gate antenna will rise in potential with respect to the plasma by collecting ions from the plasma and repel all electrons. The substrate lag behind in potential rise due to inability to collect as much ions. At some point, the gate potential will be so much higher than the substrate potential that FN tunneling occurs. The tunneling current will then drag the substrate potential along with the gate potential. The magnitude of the tunneling current will be the ion current times the antenna ratio, which can be very high if the antenna ratio is large. For example, if the ion current is $1mA/cm^2$ and the antenna ratio is 10,000 to 1, the peak tunneling current is $10A/cm^2$. Luckily, this current will last only a very short time, until the excess electrons in the substrate is neutralized. Assume that the sheath is 100μm thick during etching, the capacitance is $\sim 9pF/cm^2$. If the DC bias is 500V, the excess electrons will be $4.5nC/cm^2$. If 10% of the chip is covered by gate antenna and the antenna ratio is 10,000 to 1, then the area through which all the charges must tunnel is $10^{-5}cm^2$ per cm^2 of substrate area. The total charge through the gate-oxide will be $0.45mC/cm^2$. With a $10A/cm^2$ current density, it takes only 45μs to complete. The oxide field required to support the $10A/cm^2$ tunneling current density is around 14.5MV/cm. For very thick oxide, such high field will break the oxide even if the duration is very short. From 120Å on down, the oxide breakdown mechanism changes and the transient stress tends not to breakdown the oxide outright. Instead, it increases the breakdown probability (see oxide chapter).

The cleanest example of transient charging damage is the work reported by MacPhie *et al.* [63]. They were optimizing the machine parameter of an AMAT5000 MXP oxide-etching chamber to minimize charging damage. In their process, the RF plasma is shut down for a short duration (stabilization step) after endpoint before starting the overetch step. They found that the capacitor breakdown distribution improves with decreases in RF ramp down rate (Figure 3.63). At the lowest ramp down rate, the breakdown distribution, and hence the charging damage, is the same as without the stabilization step. At end point, the windows to the substrate are barely open. The substrate's ion current collection efficiency is poor and therefore unable to follow the gate voltage when the RF is ramped down quickly. Slowing down the ramp rate allow substrate to catch up and limit the voltage difference between gate and substrate. At the end of overetch, all windows are fully open, fast ramp down no longer has any effect.

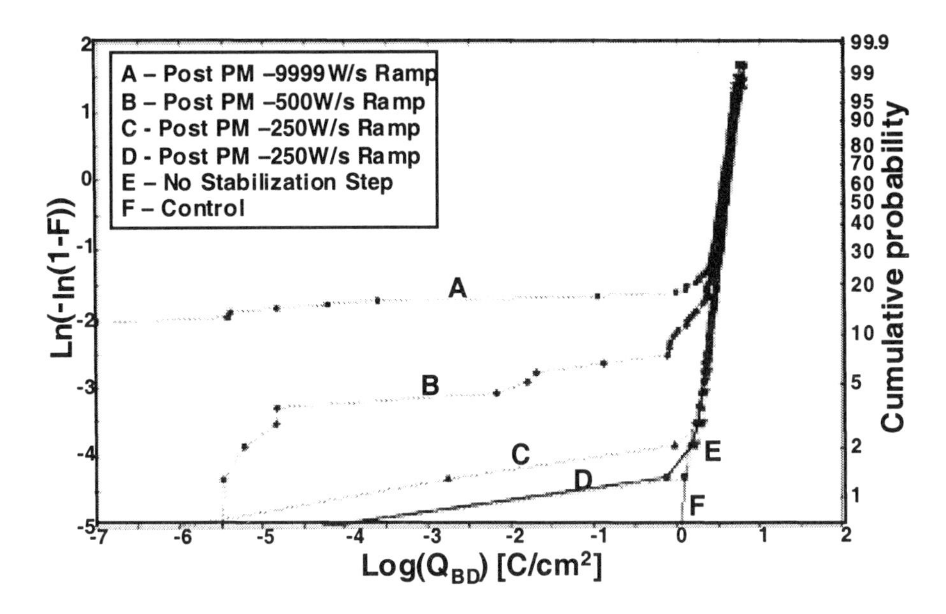

Figure 3.63. The capacitor charge to breakdown distribution improves as the RF ramp down rate reduces. At the lowest ramp down rate, the distribution is almost as good as the no stabilization case (no ramp down at end point). Taken from [63].

RF bias turn on does not post as serious a problem. This is because even for a very small exposed area of the substrate, it can collect a large electron current to help the substrate to keep pace with the antenna gate.

References

1. Hashimoto, K., *New phenomena of charging damage in plasma etching: Heavy damage only through dense-line antenna.* Jpn. J. Appl. Phys., 1993. **32, Part 1**(12B): p. 6109.
2. Ingram, S.G., *The influence of substrate topography on ion bombardment in plasma etching.* J. Appl. Phys., 1990. **68**(2): p. 500.
3. Arnold, J.C. and H.H. Sawin, *Charging of pattern features during plasma etching.* J. Appl. Phys., 1991. **70**(10): p. 5314.
4. Hashimoto, K., *Charging damage caused by electron-shading effect.* Jpn. J. Appl. Phys., 1994. **33**(part 1, no. 10): p. 6013.

5. Vahedi, V., N. Benjamin, and A. Perry, *Topographical dependent of plasma charging induced device damage.* in *International Symposium on Plasma Process Induced Damage (P2ID).* 1997. p. 41.
6. Chapman, B., *Glow Discharge Processes.* 1980, New York: John Wiley & Son.
7. Sudit, I.D. and F. Chen, *RF compensated probes for high-density discharges.* Plasma Source Sci. Technol., 1994. **3**: p. 162.
8. Bowden, M.D., *et al.*, *Study of the effect of a probe on the plasma in the source region of an electron cyclotron resonance discharge.* J. Vac. Sci. Technol., 1993. **A11**(6): p. 2893.
9. Fujiwara, N., T. Maruyama, and M. Yoneda, *Profile control of poly-Si etching in ECR plasma.* in *Dry Process Symp.* 1994. p. 31.
10. Hashimoto, K., *et al.*, *General dependence of electron-shading damage on the wafer bias power.* in *International Symposium on Plasma Process Inducted Damage (P2ID).* 1997. p. 49.
11. Hashimoto, K., *et al.*, *Response of electron-shading consistently indicating low-energy ion effect.* in *International Symposium on Plasma Process Induced Damage (P2ID).* 1998. p. 180.
12. Kinoshita, T., M. Hane, and J.P. McVittie, *Notching as an example of charging in uniform high density plasmas.* J. Vac. Sci. Technol., 1996. **B14**(1): p. 560.
13. Hwang, G.S. and K.P. Giapis, *On the origin of the notching effect during etching in uniform high density plasmas.* J. Vac. Sci. Technol., 1997. **B15**(1): p. 70.
14. Coburn, J.W. and E. Kay, *Positive-ion bombardment of substrates in rf diode glow discharge sputtering.* J. Appl. Phys., 1972. **43**(12): p. 4965.
15. Mizutani, N. and T. Hayashi, *Ion energy analysis throught rf-electrode.* Jpn. J. Appl. Phys., 1997. **36**(part 2, No. 11A): p. L1470.
16. Lieberman, M.A. and A.J. Lichtenberg, *Principles of Plasma Discharges and Materials Processing.* 1994, New York: John Wiley & Son.
17. Hwang, G.S. and K.P. Giapis, *Charging damage during residual metal overetching.* Appl. Phys. Lett., 1999. **74**(7): p. 932.
18. Kinoshita, T., *et al.*, *Simulation of topography dependent charging with pulse modulated plasma.* in *Dry Process Symposium.* 1996. p. 37.
19. Hwang, G.S. and K.P. Giapis, *The influence of electron temperature on pattern-dependent charging during etching in high-density plasmas.* J. Appl. Phys., 1997. **81**(8): p. 3433.
20. Kinoshita, T. and J. McVittie, *Reduction of topography dependent charging with low electron temperature plasma.* in *International Symposium on Plasma Process Induced Damage (P2ID).* 1998. p. 172.
21. Arimoto, H., T. Kamata, and K. Hashimoto, *Electron-shading effect during plasma process.* Fujitsu Sci. Tech. J., 1996. **32**(1): p. 136.
22. Samukawa, S., *Pulde-time-modulated electron cyclotron resonance plasma discharge for highly selective, highly anisotropic, and charge-free etching.* Appl. Phys. Lett., 1996. **68**(3): p. 316.

23. Ohtake, H. and S. Samukawa, *Charge-free etching process using positive and negative ions in pulse-time modulated electron cyclotron resonance plasma with low-frequency bias.* Appl. Phys. Lett., 1996. **68**(17): p. 2416.
24. Fujiwara, N., T. Maruyama, and M. Yoneda, *Pulsed plasma processing for reduction of profile distortion induced by charge buildup in electron cyclotron resonance plasma.* Jpn. J. Appl. Phys., 1996. **35**(Part 1, No. 4B): p. 2450.
25. Hashimoto, K., *et al.*, *Reduction of the charging damage from electron-shading.* in *International Symposium on Plasma Process Induced Damage (P2ID).* 1996. p. 43.
26. Sakamori, S., *et al.*, *Reduction of electron-shading damage with pulse-modulated ECR plasma.* in *Internationa Symposium on Plasma Process Induced Damage (P2ID).* 1997. p. 55.
27. Noguchi, K., *et al.*, *Suppression of topography-dependent charging damage to MOS devices using pulse-time-modulated plasma.* in *International Symposium on Plasma Process Induced Damage (P2ID).* 1998. p. 176.
28. Ohtake, H., *et al.*, *Pulse-time-modulated plasma etching for high performanace polysilicon patterning on thin gate oxide.* in *International Symposium on Plasma Process Induced Damage (P2ID).* 1999. p. 37.
29. Shin, K.-S., *et al.*, *Suppression of topography dependent charging using a phase-controlled pulsed inductively coupled plasma.* in *International Symp. Plasma Process Induced Damage (P2ID).* 1999. p. 155.
30. Fujiwara, N., *et al.*, *Effect plasma transport on etched profiles with surface topography in diverging field electron cyclotron resonance plasma.* Jpn. J. Appl. Phys., 1994. **33**(Part 1, No. 4B): p. 2164.
31. Malyshev, M.V., *et al.*, *Plasma diagnosis and its relation to damage.* in *International Symposium on Plasma Process Induced Damage (P2ID).* 1999. p. 149.
32. Chang, J.-S., *The inadequate reference electrode, a widespread source of error in plasma probe measurement.* J. Phys. D: Appl. Phys., 1973. **6**: p. 1674.
33. Malyshev, M.V. and V.M. Donnelly, *Trace rare gases optical emission spectroscopy: a nonintrusive method for measuring electron temperature in low pressure, low temperature plasmas.* J. Vac. Sci. Technol., 1997. **A15**: p. 550.
34. Hori, T., *et al.*, *Measurements of electron temperature, electron density, and neutral density in a radio-frequency inductive coupled plasma.* J. Vac. Sci. Technol., 1996. **A14**(1): p. 144.
35. Krishnan, S., *et al.*, *Inductively coupled plasma (ICP) metal etch dmage to 35-60A gate oxide.* in *International Electron Devices Meeting (IEDM).* 1996. p. 731.
36. Colonell, J.I., *et al.*, *Evaluation and reduction of plasma damage in a high-density, inductively coupled metal etcher.* in *International Symposium on Plasma Process Induced Damage (P2ID).* 1997. p. 229.

37. Malyshev, M.V., *Advanced plasma diagnostics for plasma processing*, Ph.D thesis, Department of Astrophysical Sciences, Princeton University, 1999.
38. Hasegawa, A., *et al.*, *Direction of topography dependent damage current during plasma etching*. in *Internation Symposium on Plasma Process Induced Damage (P2ID)*. 1998. p. 168.
39. Krishnan, S., *et al.*, *High density plasma etch induced damage to thin gate oxide*. in *International Electron Devices Meeting (IEDM)*. 1995. p. 315.
40. Ashida, S., M.R. Shim, and M.A. Lieberman, *Measurements of pulsed-power modulated argon plasmas in an inductively coupled plamsa source*. J. Vac. Sci. Technol., 1996. **A14**(2): p. 391.
41. Yokozawa, A., H. Ohtake, and S. Samukawa, *Simulation of a pulse time-modulated bulk plasma in Cl2*. Jpn. J. Appl. Phys., 1996. **35**(Part 1, No. 4B): p. 2433.
42. Samukawa, S., H. Ohtake, and T. Mieno, *Pulsed-time-modulated ECR plasma discharge for highly selective, highly anisotropic and charge-free etching*. NEC Res. & Develop., 1996. **37**(2): p. 179.
43. Makabe, T., *et al.*, *Toward charging free plasma processes: phase space modleing between pulsed plasma and microtrench*. in *Internationa Symposium on Plasma Process Induced Damage (P2ID)*. 1998. p. 156.
44. Overzet, L.J., B.A. Smith, and J. Kleber, *Time resolved measurements of pulsed discharges*. in *Dry Process Symp*. 1996. p. 9.
45. Hashimoto, K., *et al.*, *Reduction of electron-shading damage using synchronous bias in pulsed plasma*. Jpn. J. Appl. Phys., 1996. **35**(Part 1, No. 6A): p. 3363.
46. Okandan, M., *et al.*, *Soft-breakdown damage in MOSFET's due to high-density plasma etching exposure*. IEEE Electron Dev. Lett., 1996. **17**(8): p. 388.
47. Cheung, K.P., *VLSI technology workshop - plasma technology: etching and deposition*. in *VLSI Technology Symposium*. 1996.
48. Hashimoto, K., private communication, 1996
49. Marks, J., *et al.*, *Introduction of a new high density reactor concept for high aspect ratio oxide etching*. in *SPIE*. 1992. p. 235.
50. Tao, H.J., *et al.*, *Impact of Etcher Chamber Design on Plasma Induced Device Damage for Advanced Oxide Etching*. in *International Symp. Plasma Process Induced Damage (P2ID)*. 1998. p. 60.
51. Poiroux, T., *et al.*, *Study of the influence of process parameters on gate oxide degradation during contact etching in MERIE and HDP reactors*. in *International Symp. on Plasma Process Induced Damage (P2ID)*. 1999. p. 12.
52. McVittie, J.P., private communication, 1996
53. Siu, S. and R. Patrick, *Measurement of plasma parameters and electron-shading effects using patterned and unpatterned SPORT wafers*. in *International Sump. on Plasma Process Induced Damage (P2ID)*. 1998. p. 136.

54. Maynard, H.L., J. Colonell, and J. Werking, *Alternative interpretation of plasma processing damage data to facilitate comparisons between oxide etchers*. in *International SYmp. on Plasma Process Induced Damage (P2ID)*. 1999. p. 196.
55. Hook, T.B., *Backside Films and Charging During Via Etch in LOCOS and STI Technologies*. in *International Symp. Plasma Process Induced Damage (P2ID)*. 1998. p. 11.
56. Viswanathan, R., *et al.*, *Plasma damage-free multi-level metal backend for a sub-half micron CMOS process*. in *VLSI Multilevel Interconnect Conference (VMIC)*. 1997. p. 461.
57. Ito, T., *et al.*, *Plasma damage caused by electron-shading effect during sputter etch pre-cleaning of via contacts*. in *VLSI Multilevel Interconnect Conference (VMIC)*. 1997. p. 520.
58. Park, W.-J., *et al.*, *Charge-up damage of dual gate transistor during RF pre-cleaning of metal contact before barrier metal deposition*. in *International Symp. on Plasma Process Induced Damage (P2ID)*. 1999. p. 92.
59. Fang, S., A.M. McCarthy, and J.P. McVittie, *Charge sharing "antenna" effects for gate oxide damage during plasma processing*. in *International Symp. ULSI Science and Technology*. 1991. p. 473.
60. Cheung, K.P. and C.P. Chang, *Plasma-charging damage: A physical model*. J. Appl. Phys., 1994. **75**(9): p. 4415.
61. Samukawa, S., Jpn. J. Appl. Phys., 1991. **30**: p. 3154.
62. Noriji, K. and K. Tsunokuni, J. Vac. Sci. Technol., 1993. **B11**: p. 1819.
63. MacPhie, A., *et al.*, *Use of Recipe Parameters in an AMAT5000 Etcher to Reduce Qbd Failures in a 0.5um BiCMOC Process*. in *1st European Symposium on Plasma Process Induced Damage (ESPID'1)*. 1999. p. 35.

Chapter 4

Mechanism of Plasma Charging Damage III

The charging damage by non-uniform plasma of Chapter 2 and the charging damage by uniform plasma of Chapter 3 are basic mechanisms that cover a large class of damage events. There are other damage events that require somewhat different explanations. Some damage events reported in the literature are currently not understood. In this chapter, we will cover a number of these damage events. The purpose is to further develop the understanding of damage mechanisms and to illustrate how complicated plasma charging damage can be.

4.1 Plasma Charging Damage from Dielectric Deposition

Up to this point, all the plasma charging damage that we have discussed is concerned with potential on conductors driven by the charge balance characteristic of plasma. For damage to occur, a substantial amount of charge must tunnel across the gate-oxide. If an antenna collects electrons to sustain the tunneling current at one part of the wafer, the same amount of electrons must flow out of the wafer at another part and be neutralized by ions. Conservation of electric charge dictates that the current loop must be closed. If the current loop is incomplete, current flow will stop after a small initial transient and severe damage cannot occur. If a thin layer of insulator is deposited everywhere on the wafer, the plasma and wafer can no longer form continuous current loops. Accordingly, one would expect charging damage from dielectric deposition to only happen during the very beginning of deposition, before a completely insulating layer is formed. If charging can only happen for a very short time, it is not expected to cause severe damage.

Figure 4.1 shows an antenna structure halfway through a dielectric deposition process and the associated equivalent circuit (not including the plasma, just the floating potential at the surface of the wafer). The resistors across the capacitors represent the current leakage through the dielectrics. The leakage current across the deposited dielectric layer is expected to be extremely small. In fact, it is a technological requirement for the deposited dielectric to act as a good insulator. The leakage current across the gate-oxide is from tunneling processes and we therefore expect the resistance across the gate-oxide to be highly nonlinear.

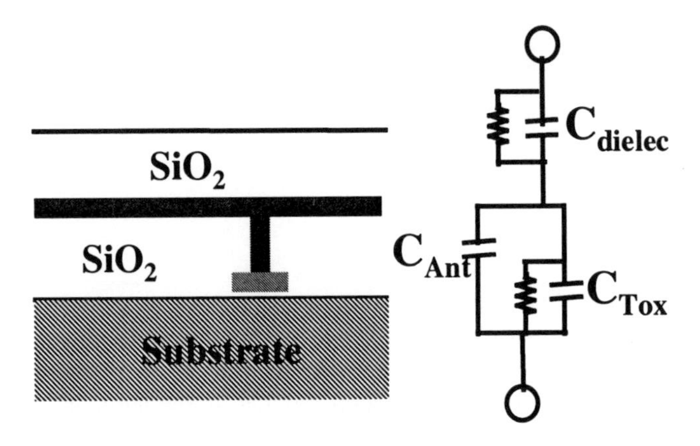

Figure 4.1. The schematic of an antenna structure halfway through dielectric deposition and the equivalent circuit.

In the absence of tunneling (or equivalently, the resistors have infinite resistance), the voltage across the gate-oxide is determined by the capacitor network. The combined capacitance of the antenna (with the substrate) and gate (with the substrate) in parallel ranges between the limits of gate capacitance at small antenna ratio to that of the antenna capacitance at larger antenna ratio. We therefore expect the voltage across the gate-oxide to increase with the antenna ratio.

With tunneling, the gate-oxide resistance decreases as voltage across it increases. From the capacitor network, it is clear that the increase in voltage across the gate-oxide with antenna ratio will stop when the tunneling current reaches the leakage current across the deposited oxide. Since the leakage of the deposited oxide is very low, the voltage across the gate-oxide will be very low as well.

Note that the current tunneling to or from the substrate must be balanced by opposite signed tunneling events somewhere else on the wafer to form a closed current loop with the plasma. Figure 4.2 illustrates the completed equivalent circuit. Here we show that the floating potential is higher at the center than at the edge of the wafer. Electrons flow from the edge of the wafer toward the center (current direction is opposite to electron). For charging damage to occur, current must be able to flow across the deposited oxide in both directions. Since the deposited oxide has very low leakage, charging damage to gate-oxide in principle cannot occur.

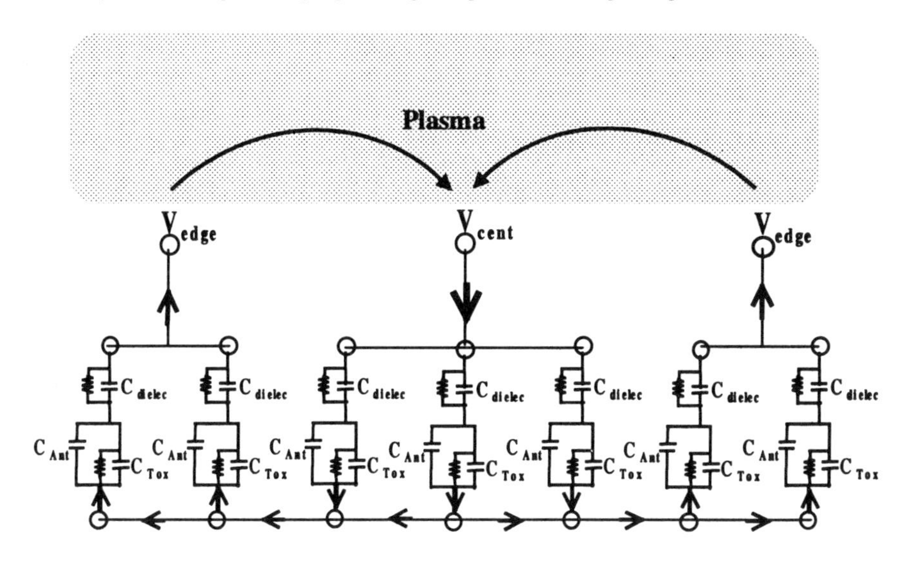

Figure 4.2. The completed circuit of many antennae on the wafer forming closed current loops with the plasma.

Given the above-discussed reasons, the detection of charging damage in dielectric deposition was a surprise to everyone. In fact, early reports are met with skepticism. It is suspected that for those observed cases, the plasma condition must be so bad that the initial few seconds of antenna exposure to plasma before the dielectric layer cuts off the current path is enough to cause all the damage. This belief is reinforced by the report that a thin layer of dielectric on top of the polysilicon gate can help reduce charging damage during polysilicon gate etching [1]. However, as time went by, more reports started to show up in the literature on charging damage in dielectric deposition [2-22]. These reports clearly show that plasma enhanced dielectric deposition is a major source of charging damage. In addition, some of these reports clearly show that charging damage did not happen only at the beginning of deposition.

The earliest report on plasma charging damage from dielectric deposition in the literature was by Shone *et al.* [2]. They did not provide much detail on the damaging oxynitride deposition condition. However, they did report that when a layer of non-damaging dielectric is put down first, the damage could be eliminated (Figure 4.3). This seems to support the notion that charging damage for dielectric deposition must be happening only at the very beginning of deposition. The severe damage they observed makes it hard to believe that it is all caused by the very first couple of seconds of plasma exposure, but it is not impossible.

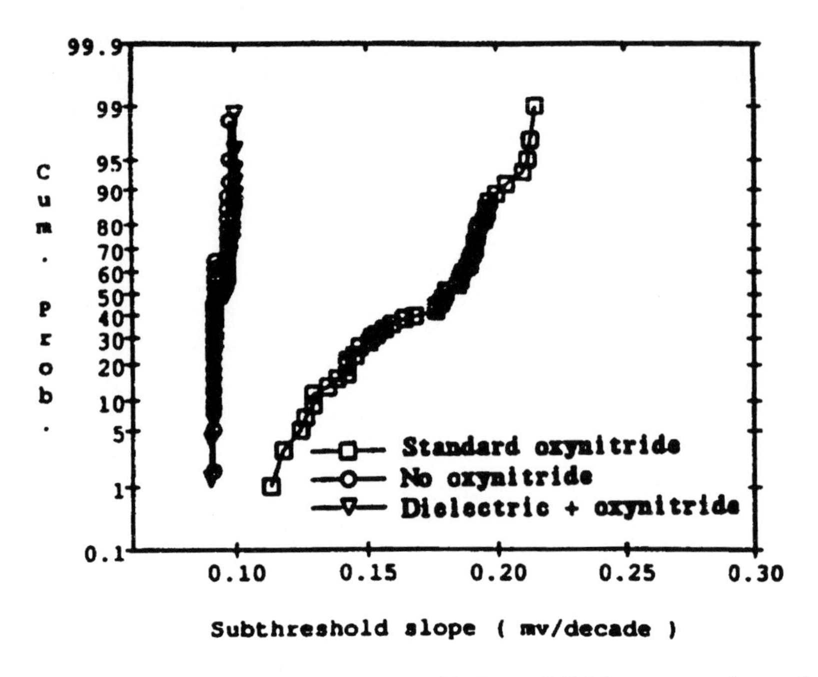

Figure 4.3. In-line measurement of sub-threshold slope of 400:1 antenna ratio transistors showing damage by plasma-enhanced oxynitride deposition. A dielectric liner layer helped to eliminate the damage. Taken from [2].

A more detailed study of charging damage from dielectric deposition by IBM groups a few years later [3, 4] provided the first evidence that the damage happens throughout the deposition. In their work, a 2μm thick PECVD (Plasma Enhanced Chemical Vapor Deposition) phosphorus-doped TEOS (tetraethoxysilane) SiO_2 (also called phosphorus-doped glass (PSG)) was deposited on a polysilicon gate antenna structure, which was pre-passivated with a 25nm thick PECVD Si_3N_4H layer. They found that the breakdown yield of their 128-kb DRAM (dynamic random access memory) array antenna structure was seriously degraded. They also found that the TEOS to oxygen ratio during PSG deposition strongly affected the breakdown yield (Figure 4.4). In addition, the transient response of the RF matching network employed in the PSG deposition also affected the breakdown yield (Figure 4.5). Both observations confirmed that PSG deposition was the source of charging damage. If the charging damage was all happening when the conducting surface of the antennae were exposed, the 25nm insulating nitride passivation layer should have prevented the damage. The fact that it did not suggested that the damage mechanism is not conventional.

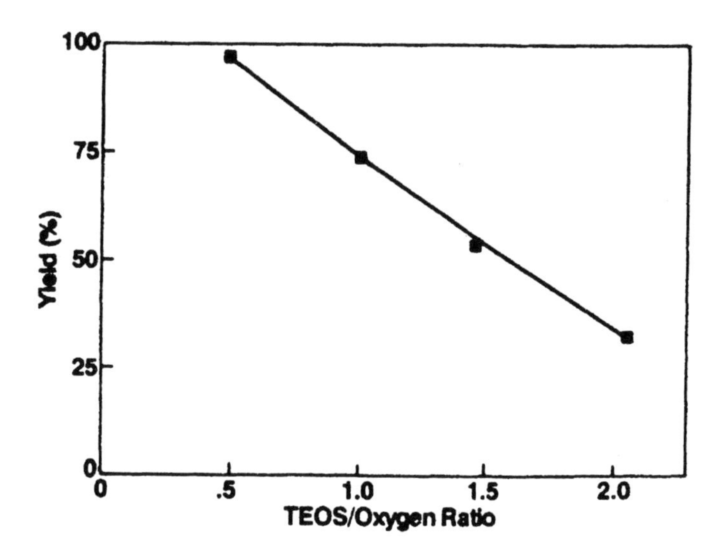

Figure 4.4. Yield of 128-kb DRAM antenna test structure is strongly dependent upon the TEOS/O_2 gas ratio in the plasma. Taken from [4].

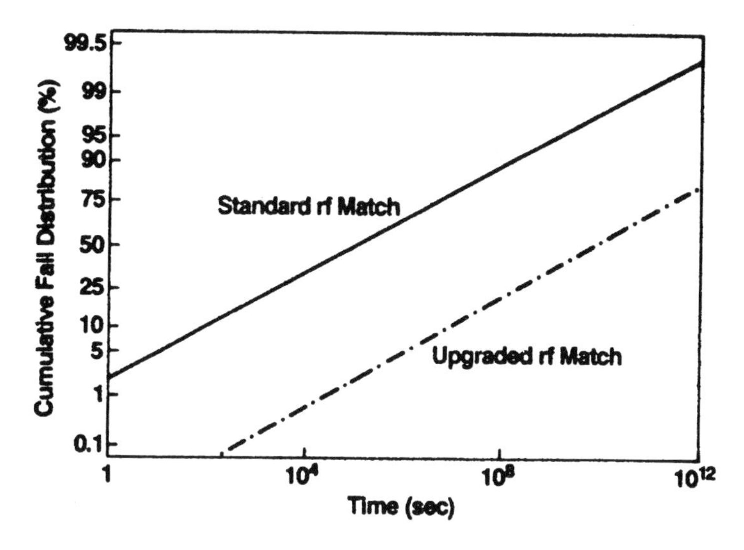

Figure 4.5. Modeled cumulative failure distribution of 16-Mb DRAM. The mean time to failure improved drastically when a better matching network was used in the plasma system. Taken from [4].

At about the same time, Hirao *et al.* [23] reported an interesting observation that provided some clues to the dielectric deposition charging damage puzzle. Hirao *et al.* was not looking for charging damage from dielectric deposition.

Instead, they reported the observation of charging damage to gate-oxide after metal antennae were exposed to Ar plasma. The interesting part of this experiment was that the metal antennae were covered by a 500nm thick interlevel dielectric layer. They found that the degree of damage depends on the type of interlevel dielectric (Figure 4.6). To explain their results (based on a similar argument given earlier in this chapter), they suggested that the dielectric layer must be leaky and that the two dielectric types must have different leakage levels. Fortuitously, they found that the I-V curve of the PSG (called P-TEOS in the figure) layer that allowed more damage to occur did have a higher leakage than the undoped glass (USG) layer (Figure 4.7). From the Q_{BD} degradation study, they estimated the amount of charge that must have flowed across the dielectric layer for the damage observed. Given the plasma exposure time, they found that to support the required leakage current the voltage difference between the dielectric surface and the metal antennae must be 100V or more. This means that the voltage between the dielectric surface and the substrate must have been over 200V (estimated from the antenna structure). However, such a high voltage, while not impossible, is not likely to exist.

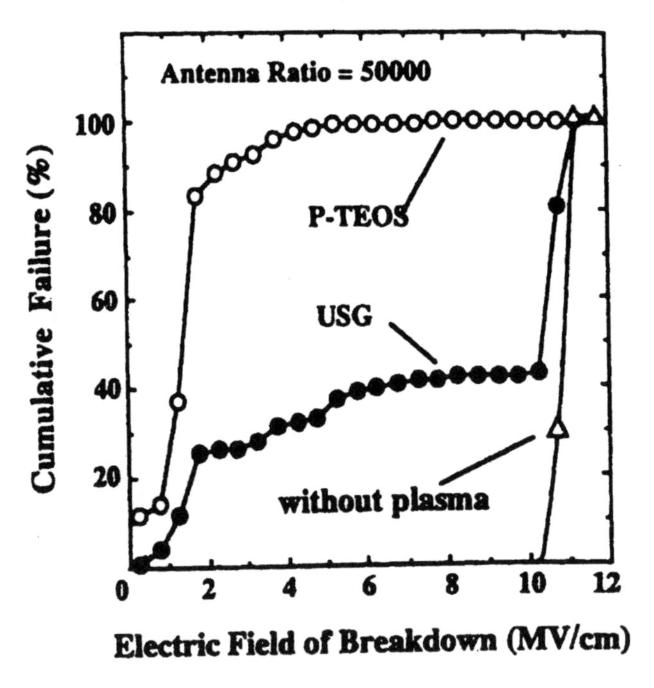

Figure 4.6. After Ar plasma exposure, gate-oxide of large antenna structure has increased low voltage breakdown failure. The P-TEOS covered testers have much higher failure rate than the USG covered testers. Taken from [23].

A more likely explanation of the observed difference in damage level between the two types of dielectric in Hirao *et al.*'s experiment is that they observed the combined damage of dielectric deposition and Ar plasma exposure. In their experiment, the P-TEOS film was deposited by a plasma enhanced process while the USG film was deposited by an atmospheric pressure CVD (APCVD) process. Since APCVD does not use plasma, no charging damage would be expected. The P-TEOS deposition, on the other hand, has been shown by the IBM group to cause serious charging damage. It is of no surprise, therefore, that the final damage level after Ar plasma exposure is higher for the antennae buried under P-TEOS.

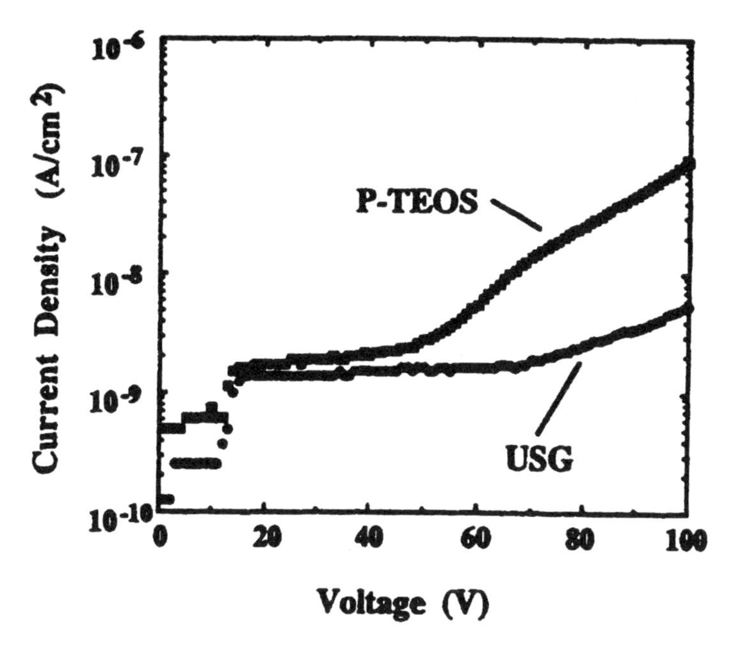

Figure 4.7. Under high voltage, the P-TEOS dielectric layer leaks more than the USG layer. Taken from [23].

Still, Hirao *et al.* were correct in thinking that current must flow across the dielectric layer in order for charging damage to occur. If the high voltage required to support the normal leakage mode were not realistic, one must then look for a conduction mechanism in the dielectric that does not require such a high field. In 1995, a photoconduction mechanism was proposed [6] to explain the observed increase in charging damage with deposited layer thickness (and hence deposition time) (Figure 4.8) in undoped TEOS deposition. This mechanism was based on three known factors. One, charging damage clearly did not happen only at the beginning of the deposition. Two, current must flow across the thick dielectric during plasma exposure for charging damage to occur, and three, plasma generates huge quantities of high-energy photons.

Photoconduction in SiO_2 has been studied by DiStefano *et al.* in the early 70's [24] as a means to pinpoint the band-gap of SiO_2. They showed that when photon energy is higher than 9eV, the conductivity of a 500nm thick SiO_2 layer rises up sharply by many orders of magnitude (Figure 4.9). Table 4.1 lists the resonance lines of some commonly used elements in plasma processing. Clearly, most processing plasmas including those used in dielectric deposition have very intense emission at energies above 9eV. It is therefore entirely reasonable to expect that photoconduction through the dielectric layer is a common occurrence.

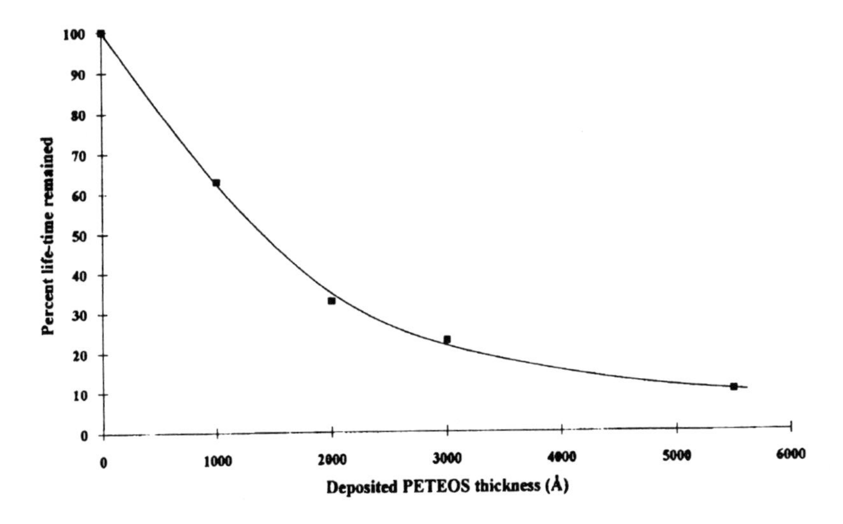

Figure 4.8. Hot carrier lifetime degradation of n-MOSFET (due to charging damage) as a function of deposited thickness of plasma enhanced TEOS (PETEOS) layer.

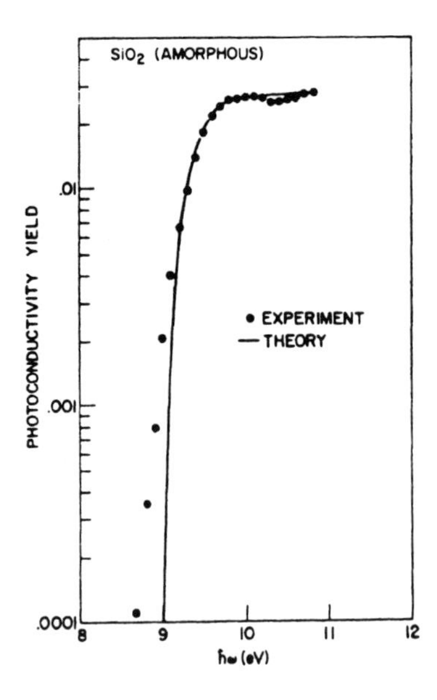

Figure 4.9. Relative phtoconductivity of amorphous SiO_2 as a function of photon energy. Solid line is predicted value from absorption coefficient. Taken from [24].

Table 4.1. Atomic emission lines for some commonly used elements in plasma processing.

Element	λ (Å) [eV]	λ (Å) [eV]
H	1216 [10.2]	1026 [12.08]
He	584 [21.23]	537 [23.09]
C	1657 [7.48]	1561 [7.94]
N	1200 [10.33]	1135 [10.93]
O	1356 [9.14]	1302 [9.52]
F	955 [12.98]	952 [13.03]
S	1900 [6.53]	1807 [6.82]
Cl	1380 [8.99]	1347 [9.21]
Ar	1067 [11.62]	1048 [11.83]
Br	1576 [7.87]	1489 [8.33]

The equation for photoconduction is

$$I = qF\frac{\tau}{\tau'} = qF\frac{\mu\tau V}{L^2} \tag{4.1}$$

where q is electron charge, F is photon flux, μ is electron mobility, τ is hole lifetime, τ' is electron transit time from cathode to anode, V is the voltage across the insulator and L is the distance between electrodes. The physics represented by Equation 4.1 is as follow: photons with energy larger than the band-gap of the insulator will create electron-hole pairs in the insulator. In the presence of an electric field, electrons will be swept toward the anode while holes will drift toward the cathode. Holes can also recombine with electrons. As long as holes are in the valence band, some level (depending on the hole concentration) of conductivity is maintained (an insulator, by definition, requires a filled band) and electrons can continue to flow from cathode to anode. Thus, for each absorbed photon, more than one electron can flow across the insulating layer, depending on the mobility of electrons and the distance they must travel. The number of electrons flowed across per incident photon is the gain of the photoconductor. Since L is only 500nm or less, the electron transit time will be so short that the gain can be as high as 10^6. In typical processing plasma, the power level is of the order of 1kW. The photon conversion efficiency can range from ~1% to 0.001% depending on the discharge type and photon energy. Even at the lower limit, there will be enough photons

(1mW of 10eV photon ~ 10^{15}photons/s, or ~10^{12}photons/cm^2-s at the wafer) to support a large photoconduction current. Although an extremely high-gain photoconductor can be made under special circumstance [25], it is usually difficult to achieve, particularly in the large signal (high-current) range. There are many factors that limit the photocurrent.

In order for electrons to flow from cathode to anode, they must be able to overcome the interfacial barriers. When making a photoconductive optical detector, great care is taken to ensure that no barrier exists between insulator and electrodes (Ohmic contact). Since the dielectric during deposition is not an ideal photoconductor, the interfacial barriers serve to limit the current level. The interfacial barriers are not symmetric at the two interfaces. The current will therefore be polarity dependent.

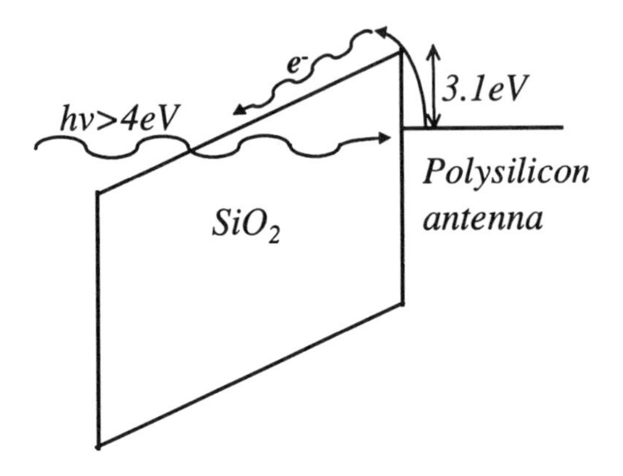

Figure 4.10. Photons with energy higher than the interfacial barrier can excite electrons from the electrode into the insulator by internal photoemission process.

When the potential at the surface of the oxide layer is higher than the substrate (and therefore the antenna), electrons must overcome the interfacial barrier between the antenna and the oxide to reach the oxide conduction band. If the antenna is made of polysilicon, the barrier is 3.1eV, a value few electrons can overcome without help. In this case, help comes in the form of internal photoemission (Figure 4.10). Photons arriving at the antenna electrode with energy higher than the interfacial barrier can excite electrons from the electrode into the conduction band of the insulator layer. This internal photoemission will flow regardless of whether or not the insulator has been made conductive by photons of energy higher than band gap. By itself, the efficiency of the internal photoemission process is quite low. Figure 4.11 shows the measured quantum yield for the SiO_2/Si system [26]. For a photon energy of 6eV, the quantum efficiency is 0.5%. The efficiency drops rapidly as photon energy is lowered and increases slowly when photon energy rises. The quantum efficiency is proportional to $(h\nu)^3$ [26]. Thus for the integrated photon flux, the average efficiency is about 1% (lower if the emission is mostly at the lower energy range). From this, we can estimate the injection current density as follows: If the RF power is 1kW, we can expect integrated photon flux to be around 10 watts (1% conversion efficiency assumed). For a $1000cm^3$ chamber, the intensity per square cm is about 10mW, which is about 10^{16} photons/s. If 1% of it turns into injected electrons, the current density injected into the SiO_2 by internal photoemission would be about $20\mu A/cm^2$. Depending on the discharge, this estimate can be higher or lower by an order of magnitude.

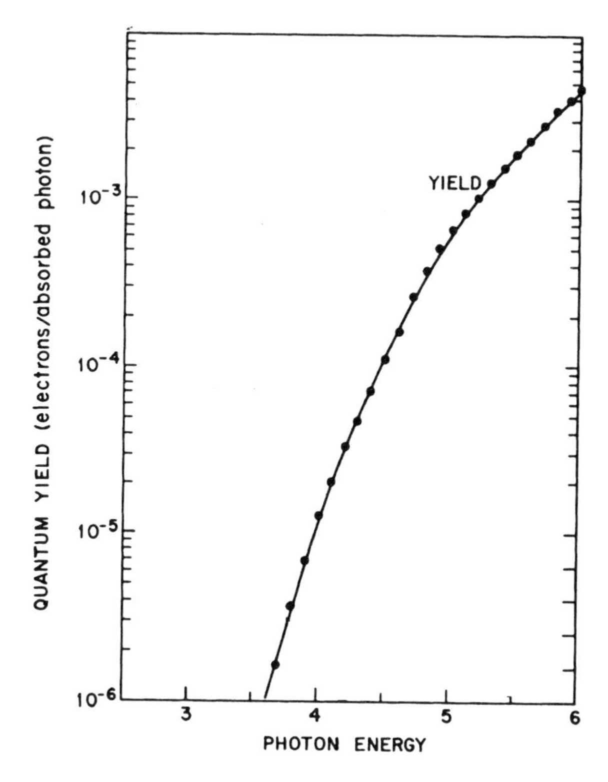

Figure 4.11. Quantum yield of internal photoemission in SiO_2/Si system as a function of photon energy. SiO_2 thickness was ~300nm. Taken from [26].

When photons with energy >9eV turn the oxide layer into a conductor, the band diagram of Figure 4.10 changes into Figure 4.12. However, the absorption of these high-energy photons by the oxide is so strong that the penetration depth is only 100Å. The intensity of the greater than 9eV VUV light drops by seven orders of magnitude at a depth of 1500Å. The band diagram of Figure 4.12 is valid only for thin oxides. It is the photons with energy in between 8 and 9eV that is most important for thicker oxide. Deposited oxides tend to have high level of defects that greatly increase the density of energy states just below the band gap. Photons at the sub-band gap level can generate electron-hole pairs that can be pull apart by the electric field. With sufficient intensity of light in the 8 to 9eV range, the band diagram of Figure 4.12 is valid for much greater oxide thickness. The flattening of the SiO_2 internal field causes the interfacial barrier to narrow. When the conductivity is high enough (large flux of photons), the barrier will become so thin that tunneling current becomes significant. How important this tunneling mechanism is during dielectric deposition is difficult to estimate without careful modeling.

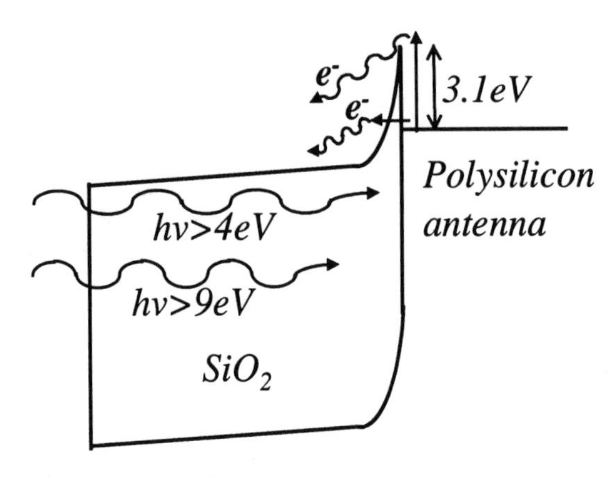

Figure 4.12. When photons with energies >9eV turn the SiO_2 layer into a conductor, the field inside the oxide flattens and the interfacial barrier narrows. The tunneling probability of electrons from Si into SiO_2 conduction band increases.

When the potential of the dielectric surface is lower than the substrate, the injecting electrode is the vacuum. If we think of the electrons at the floating surface as trapped charges at the vacuum/dielectric interface, we can draw the band diagram as shown in Figure 4.13. Internal photoemission with photon energies greater than 6eV can inject holes into the oxide valence band from the antenna electrode. The injected holes can then travel to the surface to neutralize the negative charges. This is not expected to be an efficient process and therefore cannot be counted on to produce significant current. On the other hand, the VUV light intensity is the highest at the surface. The barrier thickness at the surface can become very thin. The electrons accumulated at the oxide surface can be excited over the barrier as well as tunnel through the barrier into the oxide conduction band. The difficulty here is that we do not know the energy level of the trapped electrons at the surface and therefore cannot begin to estimate how efficient these two processes are. Photoconductive discharge of a freely floating surface has been studied extensively because of its technological importance in the photocopying industry [27]. In Batra *et al.*'s experiment [27], the discharge rate was extremely high and appeared to be limited only by the saturated photocurrent level. Thus we expect the conduction of electrons from the dielectric surface to the antenna be an efficient process as well.

In a recent experiment, Cismaru *et al.* [28] measured the resistance of a 900Å thick oxide layer when exposed to high-density Argon plasma. They found values ranging from $50k\Omega/cm^2$ to $200k\Omega/cm^2$. This value is much lower than the SiO_2 dark resistance of $>10^{12}\Omega/cm^2$. Translating to leakage current, the value would range from 5 to 20μA/cm^2-V. It is clear that their experiment measured only the case of electron flowing from the probe through the oxide to the plasma even though they scanned the voltage from negative to positive. The probe material may contribute to a lower internal photoemission efficiency with a larger barrier height than polysilicon. It appears that, at least in their case, electron tunneling across the barrier into the oxide conduction band is negligible.

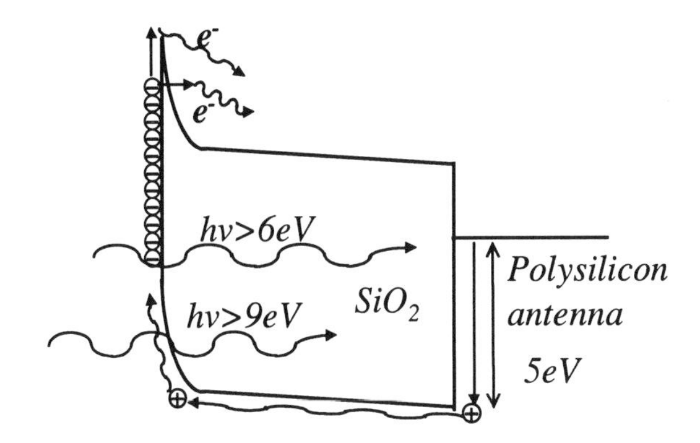

Figure 4.13. Internal photoemission of holes can travel to the surface to neutralize the negative charges. At the same time, electrons can be photo-excited as well as tunneling into the SiO_2 conduction band.

Other evidence of current must leak across the dielectric layer is the observation that the most damaging moment of dielectric deposition process is actually not during deposition. It is the "power-lift" process after deposition that causes the most significant charging damage [29, 30]. (Note that severe charging damage still exist after the "power-lift" step is eliminated in the dielectric deposition process). The "power-lift" process is just a short exposure to low-pressure oxygen plasma (no deposition is involved). It is for the purpose of neutralizing the residual electrostatic charges on the wafer before unloading. At this point, as much as 1000nm of dielectric has already been deposited on the wafer. For severe charging damage to occur in only a couple of seconds' worth of plasma exposure, the current that flows across the dielectric layer must be significant. The low-level current measured by Cismaru *et al.* seems to be at odds with this observation.

One possible explanation to this mystery is the deposition temperature. Plasma enhanced dielectric deposition, regardless of who makes the tool and what types of plasma source are employed, are all done at 350ºC to 400ºC range (occasionally even higher). At such a high temperature, the gate-oxide is particularly vulnerable, that is, has very low Q_{BD}. In Chapter 1 we discussed the temperature acceleration factor for trap generation during high-field stressing of gate-oxide. We have seen that the time to breakdown drops rapidly with increase in temperature. For a given stress level, shorter time to breakdown means smaller charge to breakdown. Apte *et al.* [31] measured directly the charge to breakdown (Q_{BD}) for 30Å thick gate-oxide for a range of temperature and stress current, their result is shown in Figure 4.14. Although they did not make measurements at 400°C, it is readily apparent from their data that the Q_{BD} will be even smaller. Assuming a leakage current of 20μA/cm^2 and an antenna ratio of 1000:1, the stress current density through the gate-oxide is 20mA/cm^2. At 400°C, the Q_{BD} would be somewhat less than 0.1C/cm^2, which means that a 5-second stress is enough to produce 50% failure. The actual stress current level during a high-density plasma enhanced dielectric deposition has been shown in Figure 2.42 as a function of antenna ratio. It is in full agreement with the low level of stress current discussed here.

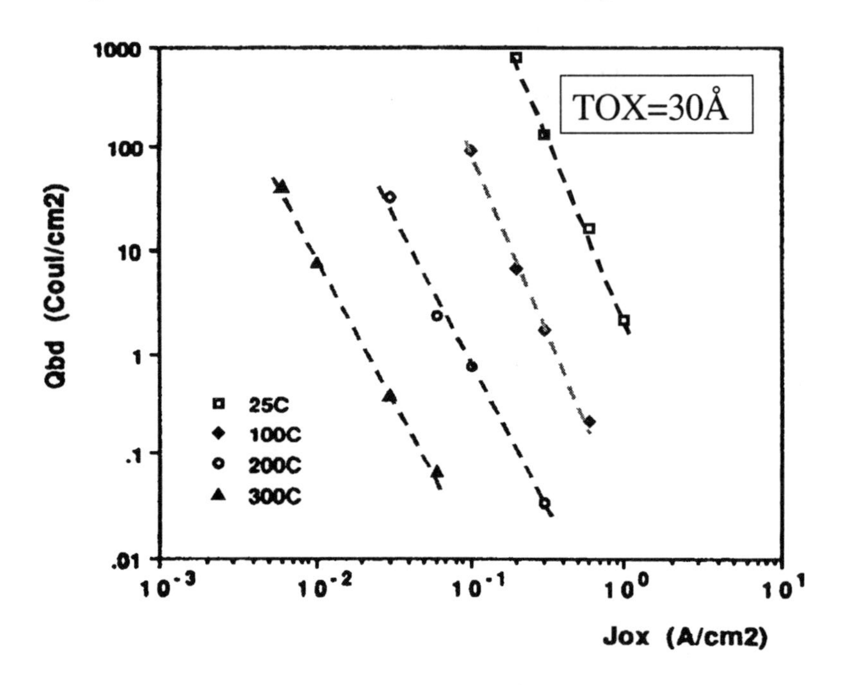

Figure 4.14. Charge to breakdown is a sensitive function of both temperature and stress current density. Taken from [31].

It is clear that charging damage during dielectric deposition requires the combination of photoconduction and very high temperature. This very high temperature effect is the main reason why dielectric deposition continues to be a major source of charging damage in advanced IC manufacturing. The rapid increase in temperature acceleration in thinner oxide as shown in Figure 1.30 suggests that the problem will only get worse.

While plasma enhanced dielectric deposition continues to be a major source of charging damage after all these years of studying the problem by so many groups, it is not to say that significant progress in damage reduction has not been achieved. As mentioned earlier, the post deposition "power-lift" was found to be the most damaging step. After the elimination of the "power-lift" step, significant charging damage still existed in the "standard" (vendor recommended) deposition process. Further improvements were achieved by fine tuning the processing recipe. Among the various changes, the deposition pressure turned out to have the biggest impact. Figure 4.15 [7] shows the impact of power and pressure on the antenna capacitor yield (left-hand-side) as well as the actual yield data for 100Kb DRAM cells as a function of deposition pressure. These advances allow the plasma enhanced dielectric deposition to continue to be used in IC manufacturing without seriously impacting the yield and reliability of products. However, as the scaling trend continues, the thinner gate-oxide used in deep sub-micron technologies is much more vulnerable to charging damage in dielectric deposition. This is due to two factors. One is that thinner oxide requires much smaller potential difference between gate and substrate to cause large tunneling current to flow. Thus even a lower variation in floating potential during plasma exposure can cause significant damage. The other is the significantly higher sensitivity to the temperature effect.

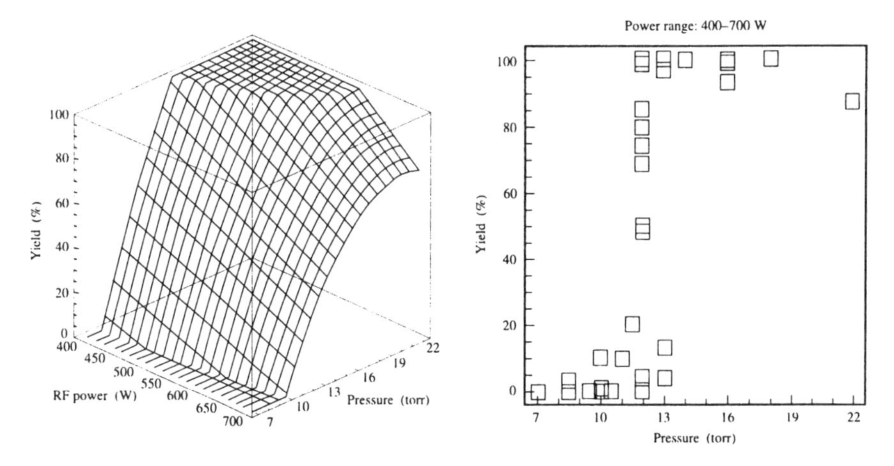

Figure 4.15. Left-hand-side shows the calculated response surface from a matrix of experiments. Yield is referring to antenna charging monitors. Right-hand-side shows 100Kb DRAM cell yield as a function of pressure. Each point represents a wafer lot. Taken from [7].Copyright 1995 by International Business Machines Corporation. Reprinted with permission of the IBM Journal of Research and Development (Vol.39, No.4).

4.2 Plasma Charging Damage from Magnetized Plasma

One of the best-established facts in plasma charging damage is that magnetized plasma tends to cause more severe damage. Perhaps the first to report magnetized plasma causing more severe charging damage was Greene *et al.* [32]. Soon after that, large number of reports from many groups [33-39] appeared in the literature.

The reason for this sudden rush of papers is two folds; the widespread adoption of magnetic field enhanced reactive ion etching (MERIE) system in the industry and the seriousness of the damage (therefore easy to detect).

The effect of a magnetic field on plasma is to modify the movement of charged particles and therefore increase the plasma density. Due to the huge mass difference, the impact on electron movement is much larger than on ion movement. In most cases, we can ignore the effect of magnetic field on ions and concentrate only on electrons. In the presence of a magnetic field, the electron mobility becomes anisotropic. The mobility along the magnetic field line is unaffected while the mobility crossing the field line is prohibited without the assistance of a collision event. This anisotropic mobility will have great impact on the floating potential distribution. Recall that sheath potential is set up by the difference in impingement frequency between ions and electrons. The change in electron mobility changes the impingement frequency and therefore the sheath potential. For uniform plasma potential, a varying sheath potential will result in a varying floating potential. Indeed, some of the early studies of plasma charging damage mechanisms relied on a non-uniform magnetic field to create highly non-uniform plasma to ensure charging damage can be observed [40].

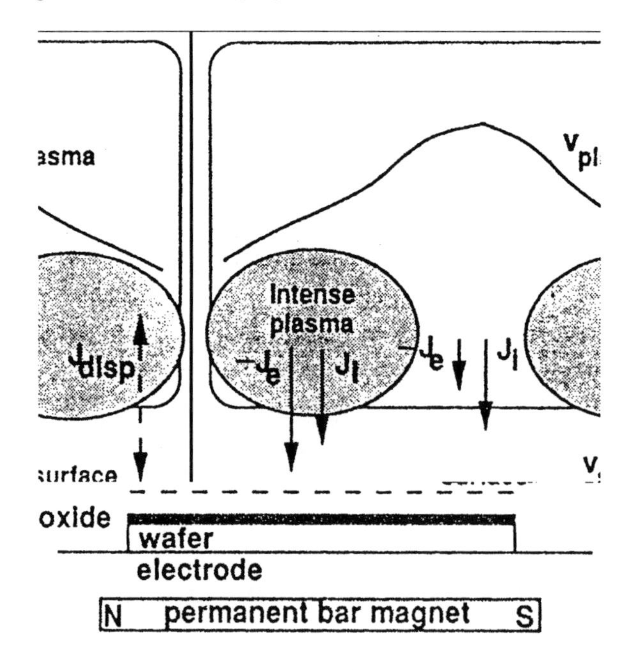

Figure 4.16. A single bar magnet creates a highly non-uniform magnetic field across the wafer. Where the field lines cross the wafer (both ends) the plasma are most intense. Taken from [36].

Figure 4.16 [36] shows the non-uniform plasma created by a single bar magnet. In this arrangement, the magnetic field lines cross the wafer near the two ends of the bar magnet and parallel with the wafer at the middle. Where the field lines cross the wafer the plasma is most intense. Electrons flow to the wafer along the field line without additional impedance due to the magnetic field. At the center where the field line is parallel to the wafer, electrons cannot easily reach the wafer. The floating potential at the center is therefore much less negative (w. r. t. plasma). Note that in Figure 4.16 the surface potential of the oxide (floating potential is drawn to be flat while the plasma potential is drawn to peak at the center. This is intuitively incorrect. One expects the plasma potential to be somewhat lower at the center because the plasma density is lower. The floating potential is expected to peak at the center because the sheath potential is the smallest here. It is the highly non-uniform floating potential at the oxide surface that causes charging damage to occur to the oxide. Highly non-uniform charging damage in this system was indeed observed [36], and the center region indeed showed more severe damage (Figure 4.17).

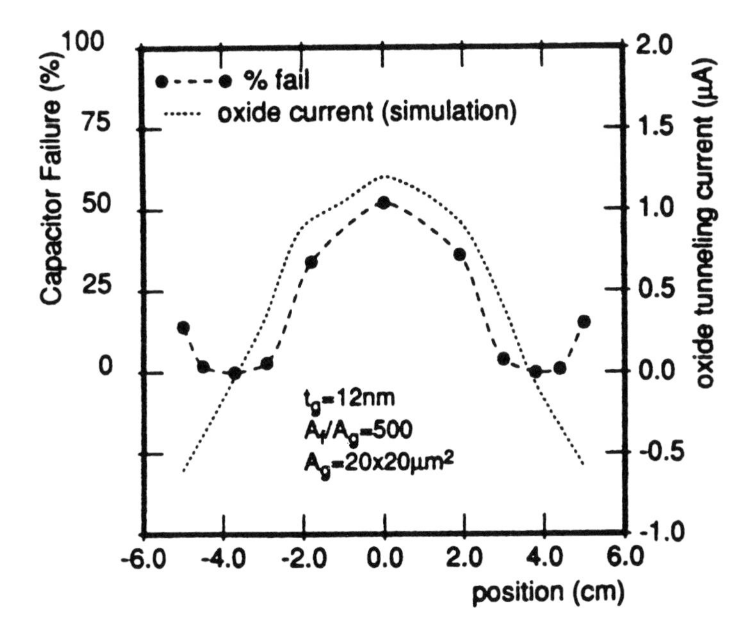

Figure 4.17. The antenna capacitor failure rate peaks at the same location as the surface floating potential. Taken from [36].

Normally, when designing a magnetic field enhanced plasma-processing system, great care is given to the uniformity of the magnetic field. However, even with a perfectly uniform magnetic field across the wafer, a non-uniform floating potential can still arise. This is because in addition to mobility modification, the presence of both a magnetic field and electric field produces a Lorenz force that is orthogonal to both fields. This is the so-called "**EXB**" drift of electrons flowing out of the plasma as well as secondary electrons produced by ions impinging on the wafer surface. The result is an exponential growth in plasma density in the direction of the drift.

Shin *et al.* [38] measured the effect of a uniform magnetic field in a commercial etcher where the plasma was designed to be uniform in the absence of the magnetic field. Figure 4.18 shows the result of their measurement when the uniform magnetic field is applied. Figure 4.18(a) shows the measured tunneling current level (damage level) experienced by the oxide during plasma exposure. Figure 4.18(b) shows the top view of the plasma showing that the hot (most intense) spot is off to the side in the direction orthogonal to the direction of the magnetic field as well as the electric field (perpendicular to the wafer surface). Figure 4.18(c) shows the cross section view and the tunneling current path.

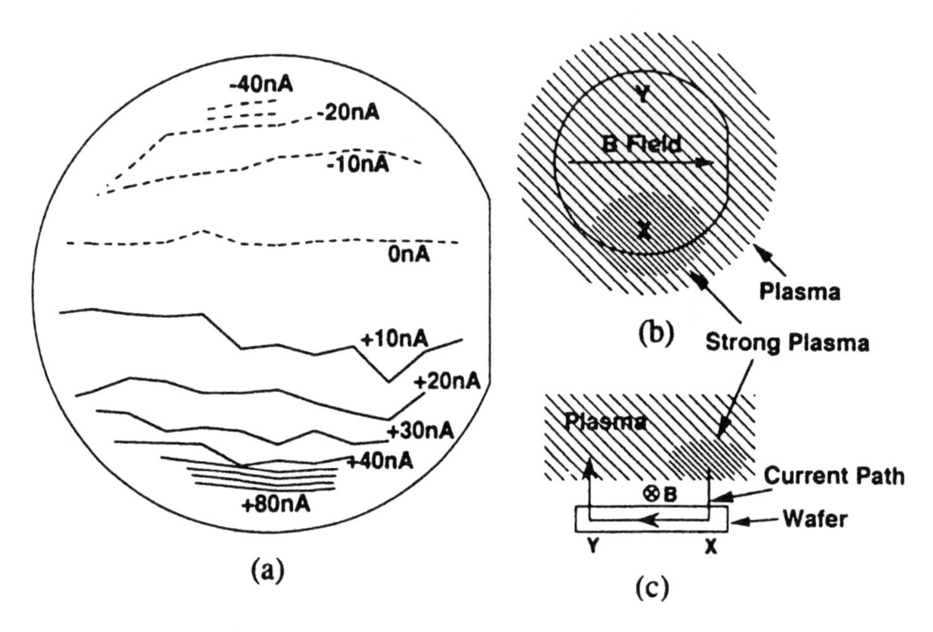

Figure 4.18. (a) Plasma charging current flowed through the oxide varies across the wafer during exposure to a plasma enhanced by uniform magnetic field. (b) Top view of plasma showing hot spot and magnetic field direction. (c) Cross sectional view showing the direction of charging current flow. Taken from [38].

Since magnetic field enhanced plasma offers the increased density that can shorten the processing time, there is a strong incentive to keep the magnetic field in the plasma. To combat the **EXB** drift induced non-uniform plasma problem, Nakagawa *et al.* [41] introduced a curvature to the magnetic field by offsetting the magnet. Figure 4.19 shows the principle of this concept. When the magnetic field lines are curved, the **EXB** drift will have a rotating component and therefore prevent the electrons from accumulating in one direction. The result is a more uniform plasma. They experimented with various degrees of curvature using the set up shown in Figure 4.20. Offsetting the alignment of the two magnets from the symmetric position generates a curved magnetic field. Using EEPROM transistor V_t shift as an indication of charging, they found that only a small offset angle is required to achieve damage free (as measured) (Figure 4.21). With such a small offset, the induced curvature in the magnetic field is very small. Figure 4.22 shows the measured magnetic field profile of the symmetric and 10° offset geometries. It is quite remarkable that such small changes can produce drastic results.

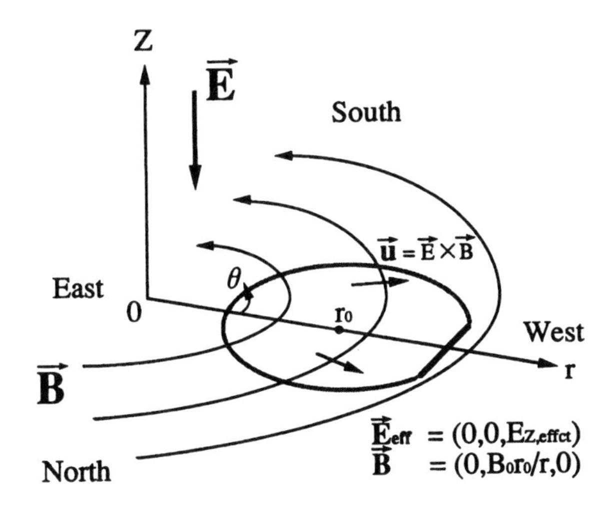

Figure 4.19. A curved magnetic field can introduce a rotating component to the **EXB** drift to avoid plasma density building up in one particular direction. Taken from [41].

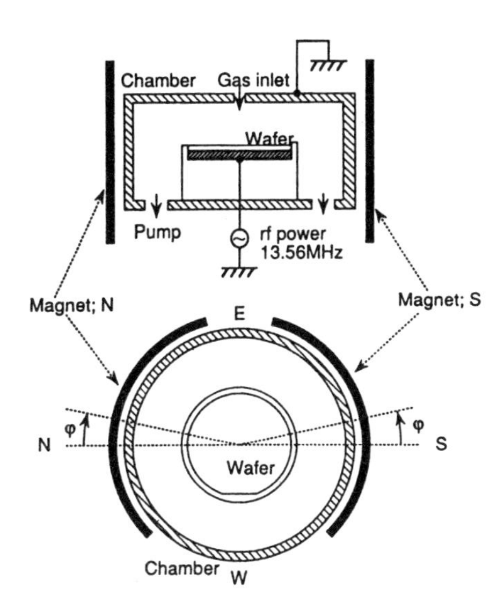

Figure 4.20. Experimental set up of Nakagawa et al. [41]. By offsetting the two permanent magnets from the symmetric position, they can introduce a curvature in the magnetic field.

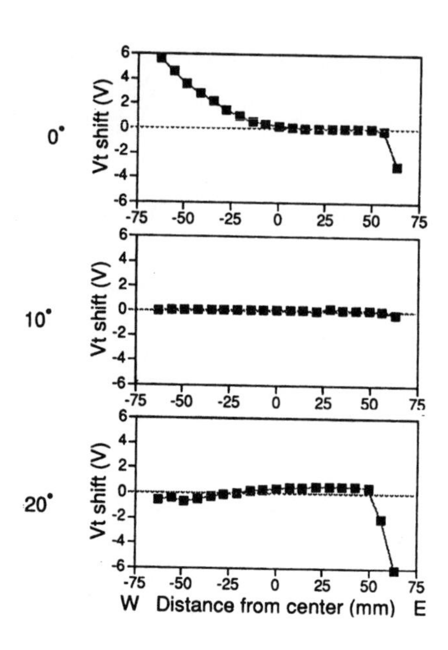

Figure 4.21. Transistor V_t shift distribution along the perpendicular direction of the magnetic field as a function of magnet offset angle shows that damage can be reduced by using a suitably curved magnetic field. Taken from [41].

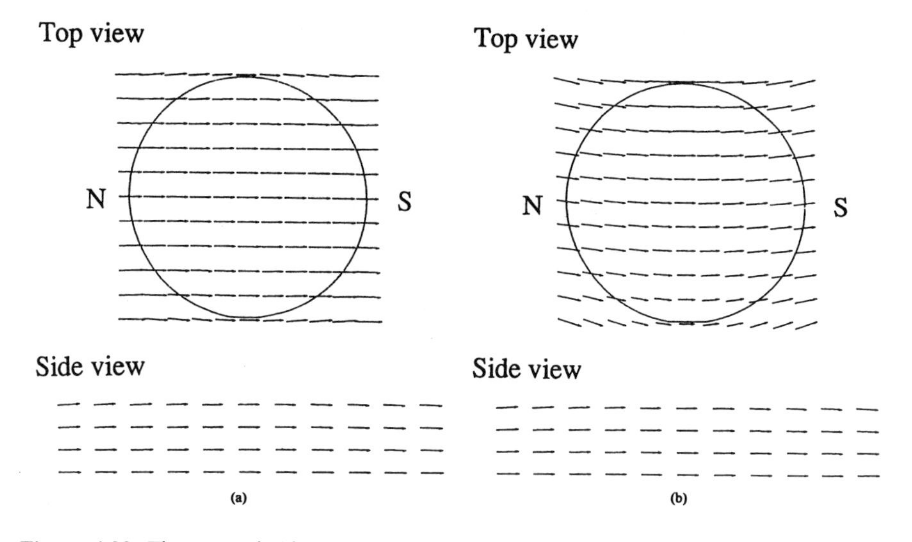

Figure 4.22. The top and side view of the magnetic field with and without the 10° magnet offset angle. Taken from [41].

A commercial etching system has taken this concept one step further [42][43] by introducing a magnetic field intensity gradient in addition to the curvature in field direction (Figure 4.23). In this system, the magnetic field curvature is not uniform as is shown in Figure 4.23.

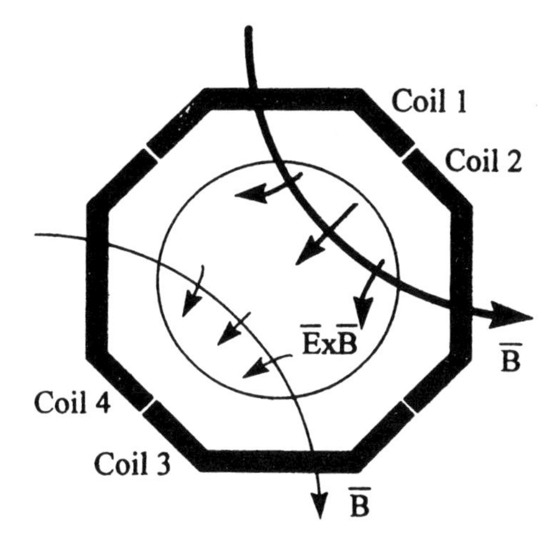

Figure 4.23. Curved magnetic field as well as controlled gradient of field strength are used to reduce charging damage in a commercial etching system. Taken from [43].

The method of curved magnetic field or curved plus gradient magnetic field requires a careful balance between rotational force and **EXB** force, thus requiring careful tuning for each processing condition. While it is a powerful technique for controlling damage, such sensitivity to processing condition details makes it rather complex to optimize processes.

4.3 Plasma Charging Damage at the Transistor Channel's Edge

The transistor channel's edge has been shown by many groups [44-51] to suffer more damage than the rest of the channel. Many methods have been used to study this phenomenon, including Gate-Induced-Drain-Leakage (GIDL) current [45], charge pumping [45, 46, 51], channel-length dependent threshold voltage shift [44, 45], drain-bias dependent threshold voltage shift [48], over-etch-time dependent effective channel length [47], edge length dependent gate leakage current [49], and localized charge injection through the gate-drain overlap region [50]. While all of these works demonstrated conclusively that gate polysilicon etching processes tend to create more damage at the edge of the channel, none actually shows that the damage is charging related.

As mentioned at the beginning of this book, plasma is a rather hash environment for the wafer to be exposed to. There are all sorts of radiation including ions, electron and photons bombarding the wafer. Most of the above mentioned channel edge damage can be attributed to ion damage or photon damage. For example, the gate-oxide thickening at the edge of the channel reported by Brozek *et al.* [50] could easily be the result of silicon lattice damage by ions. The damaged silicon at the channel edge leads to faster oxide growth during the re-oxidation step. The edge length dependent leakage current reported by Kang *et al.* [49] was clearly shown to be due to photons.

While there is no conclusive evidence that at least some of the reported edge damage is related to or worsened by the presence of charging, one cannot conclude that charging did not play a role. It is well established that, for example, damage to oxide by energetic photons is much stronger in the presence of an oxide field. Figure 4.24 is an example of how the gate bias on a capacitor during radiation affects the level of damage (trapped charge generation) [52]. Clearly, the presence of charging during plasma exposure causes the radiation damage to be significantly worsened.

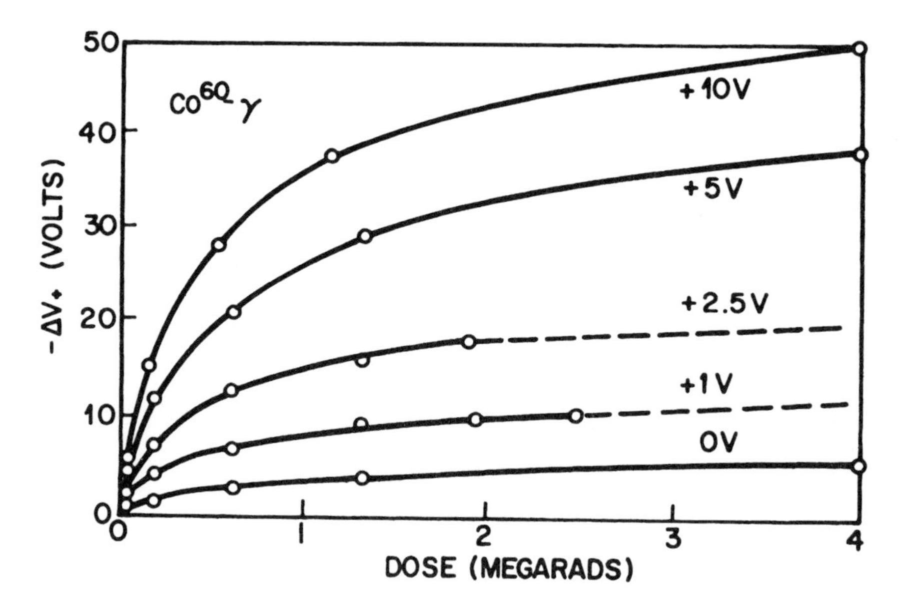

Figure 4.24. Capacitor flat-band voltage shift as a function of radiation dosage and gate bias. The oxide was 1600Å thick. Taken from [52].

The photon energy required to produce damage in oxide is not very high. For example, Ling [53] reported interface state generation by 4.9 eV photons. Although the photon energy is below the band gap of SiO_2 and should not generate electron-hole pairs in the oxide, it is above the threshold of internal photoemission at the Si/SiO_2 interface (see Figure 4.11). When photon energy is above the 9 eV SiO_2 band gap, damage to SiO_2 is due to electron-hole pair generation. If the transistor gate is made of 0.5μ thick polysilicon, only about 1% of the photon flux can reach the gate-oxide through the gate. While 1% of a large flux can still generate a lot of damage in the channel, significantly more damage can be expected at the edge of the channel where the photons can reach directly. If the gate is silicided or made of metal, almost none of the plasma-generated photons can reach the gate-oxide through the gate. In that case we will have only edge damage.

In our discussion of charging damage during dielectric deposition, we already indicated that plasma could generate large quantities of energetic photons. Most of the commonly used chemicals in plasma processing have strong emission lines in the range of 5 to 25 eV. Thus charging enhanced photon damage is a real problem, particularly during the gate-etching step. Fortunately, there are high temperature anneal steps after the gate etching step in the normal manufacturing process. Much of the damage can be annealed except the most severe cases.

4.4 Plasma Charging Damage in Very Short Range

In Chapter 2, we discussed the charging damage mechanism due to nonuniform plasma. In general, we tend to think of the variation of the plasma potential as a smooth function across the dimension of a wafer. The plasma potential at two locations not too far from each other on the wafer is expected to be quite small. This simple picture might not be justified in some cases. Figure 4.25 shows the charging damage induced transistor threshold voltage shift on a 6" wafer. Clearly, the plasma that produced this damage distribution has relatively large density variation in relatively short range. The damage distribution shown in Figure 4.25 is from a commercial etcher that produces a highly uniform etching rate across the wafer. It is therefore expected that on the average the plasma density is quite uniform. Note that electron-shading type of charging damage should produce a uniform (or random) distribution of damage. The distribution of damage in Figure 4.25 is definitely caused by non-uniform plasma.

Why would uniform plasma (measured by etching rate) produce non-uniform plasma type of charging damage? One possibility is that the damage distribution is simply a reflection of non-uniform gate-oxide quality distribution. If the charging is uniform and gate-oxide quality is not, the resulting damage distribution will follow the distribution of gate-oxide quality. However, gate-oxide quality can be checked by other means and was found to be uniform for the wafers that produced the damage map of Figure 4.25. Another possibility is that the plasma has large local fluctuations. While these fluctuations would not show up in measurements that can only detect the time-averaged property of the plasma, they would cause transient charging to occur that leads to damage. Figure 4.26 [54] illustrates the plasma fluctuation and its time-averaged uniformity. Aum *et al.* [54] discussed how to use different antenna lay outs in testing devices to detect such a transient charging problem. Unfortunately, they did not provide measured results as an example of such a fluctuation.

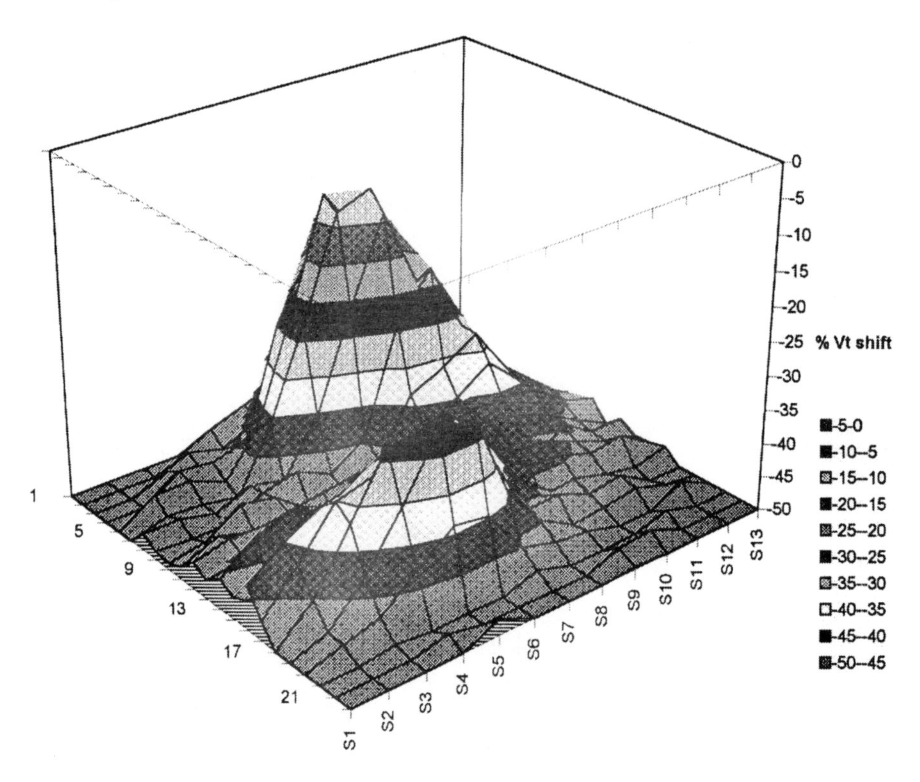

Figure 4.25. An example of highly localized charging damage indicating significant variation of plasma potential in a relatively short range.

A possible example of the kind of random fluctuation induced charging problem might have been the hot-carrier damage reported by El Hassan *et al.* [55]. In El Hassan *et al.*'s experiment, they used the same type of transistor with three different interconnect layouts (Figure 4.27). Type A layout shorted the gate to the source. Type B layout has independent pads for all four terminals of the transistor. The pad for the gate was enlarged to serve as a charging antenna. Type C is the control, all four terminals are shorted together using fuse links during processing. All devices have 90Å thick gate oxide and were annealed in forming gas after metal 1 definition and before testing.

Figure 4.28 shows the measured charge pumping current for all three types of devices. Compared to the control, the type B device has a somewhat higher interface-state density (high charge pumping current). This suggests that some level of charging damage has occurred during metal etching. Since the antenna ratio in type B devices is the largest, it is surprising to see that type A devices show a much higher level of damage. In addition, the damage level is clearly higher for the shorter channel device, suggesting that the damage may be concentrated at the edge of the channel.

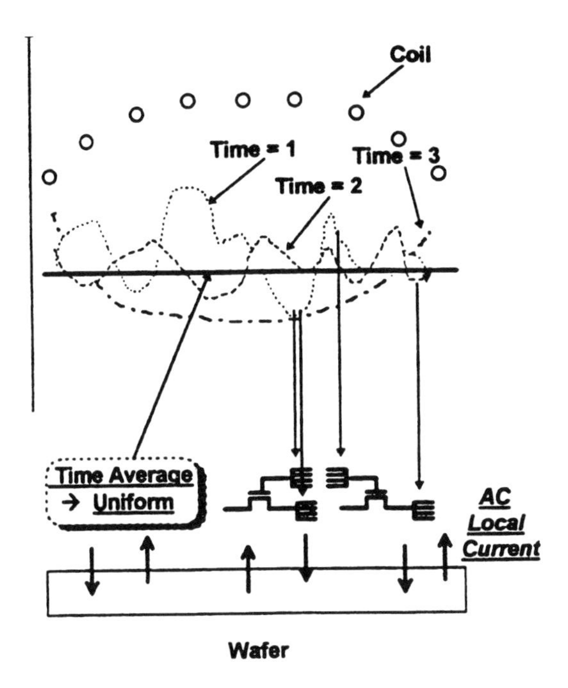

Figure 4.26. Plasma fluctuation can cause transient charging locally. Different device lay outs may have different responses. Taken from [54].

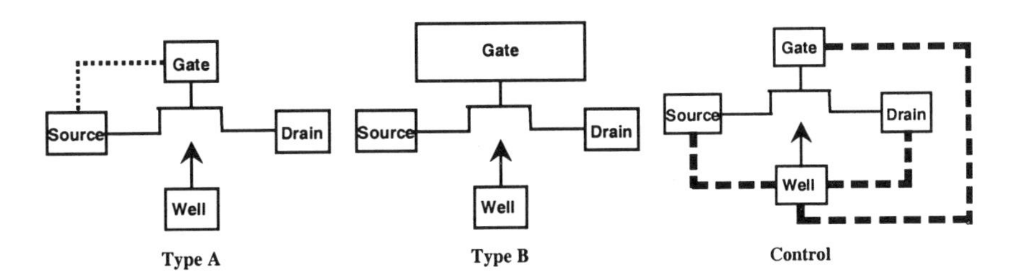

Figure 4.27. Type A devices have gate and source shorted together. Type B devices have a large area gate antenna. Reference device has all 4 terminals shorted together using a fuse-link. After [55].

Figure 4.29 shows the threshold voltages for all the devices after a fixed amount of hot-carrier stress. All the devices show similar behavior except type A devices under forward bias stress. This result suggests that not only is the damage concentrated at the channel's edge, it is asymmetric. Only the drain end of type A devices received additional damage. The only logical explanation for the observed result seems to be that forward hot-carrier stress has happened to type A devices during plasma etching of metal 1. Since these are 3.3V devices, in order to cause that much hot-carrier damage during the relatively short duration of metal 1 etching,

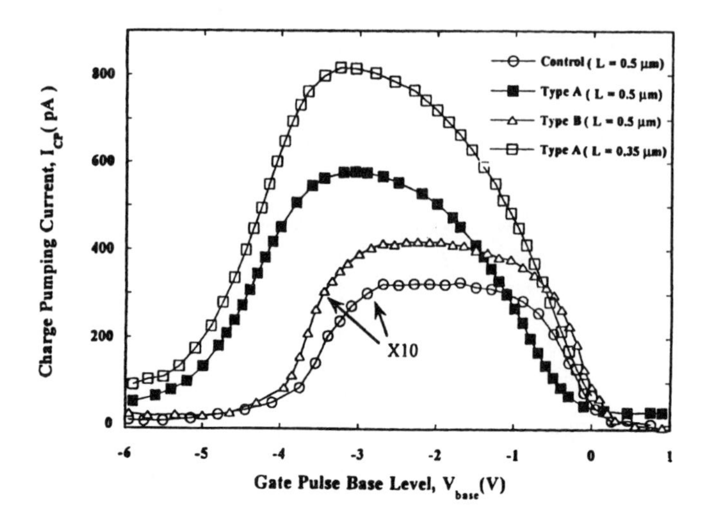

Figure 4.28. Charge pumping current indicates that while type B devices have a slightly higher interface state density than control, the density is far less than the type A devices. In addition, type A devices show a strong channel-length-dependent interface state density. Taken from [55].

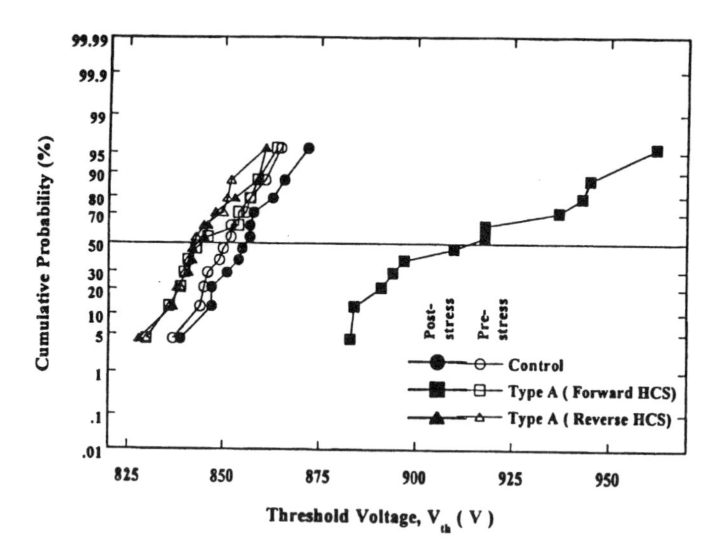

Figure 4.29. After hot-carrier stress, only type A devices under forward stress show an abnormal level of degradation. All other devices, including type A devices under reverse stress behave similarly. Taken from [55].

the voltage difference between source and drain is at least 5V. The layout of these devices is such that the source and drain pads are separated by less than 1mm. In other words, the observed result seems to suggest that over the length scale of 1mm, the floating potential on the wafer surface varied by as much as 5V. Such a drastic floating potential change over a short range is only possible during plasma instability.

The kind of random plasma instability suggested in Figure 4.26 probably happen more often in plasma processing systems than most people realize. A tell-tale sign is that most measured plasma properties such as light emission are far noisier than the RF power that sustains the plasma. An extreme form of instability is micro arcing which contributes to the formation of particles. In a recent report, Lassig *et al.* [21] found an unexpected large and rapid (30kHz) floating potential fluctuation which is pressure sensitive (Figure 4.30). While the root cause for the observed floating potential fluctuation is not yet known and therefore should not be generalized to other plasma conditions, it does demonstrate that there are various types of plasma instabilities that can cause layout dependent charging damage.

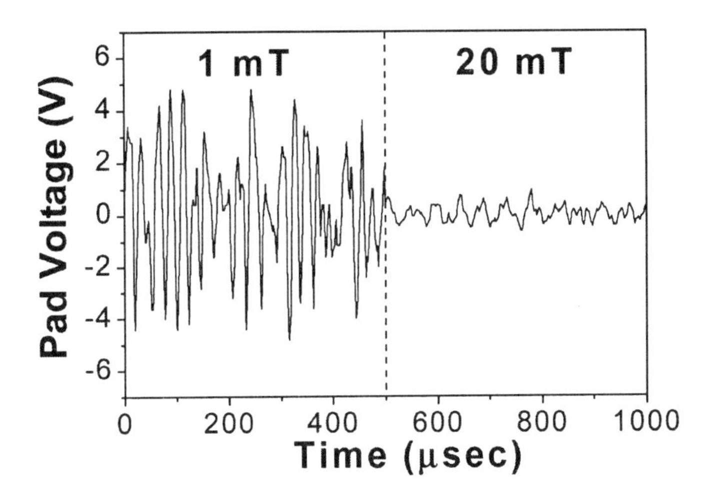

Figure 4.30. Pad potential as a function of time measured using a SPORT probe shows unexpected high frequency (~30kHz) fluctuations in plasma potential which is diminished at high pressure.Taken from [21].

The plasma fluctuation that produced the damage distribution of Figure 4.25 is different from the plasma fluctuation discussed by Aum *et al.* [54]. The observed distribution is highly reproducible from wafer to wafer, which suggests that the fluctuation is not random like it is illustrated in Figure 4.26. Rather, it is highly localized. This type of localized transient plasma charging damage is probably even more common than the random fluctuation type.

To understand why localized transient charging damage happens, we need to remember that plasma is in steady state very far from equilibrium, and that the plasma distribution is the result of a dynamic balance of generation and loss at every point in time and location. Loss happens mostly at the boundary, which is the chamber wall. The wafer is always part of the boundary. The main different between the wafer and the wall is that the wall is a passive boundary while the wafer is an active boundary. When we apply bias to the wafer, that bias will drive the plasma potential. As the plasma potential on top of the wafer is being driven up and down at the frequency of the RF bias, the plasma potential at the wall may or may not follow. If the wall is a conductor anchored at ground, its potential will not change with the plasma. If the wall is a perfect insulator, its potential will follow the plasma potential as long as the time constant associated with charging up the capacitance of the wall (capacitor formed by the insulating layer) is short compared to the rate of change in plasma potential.

In a recent paper, Malyshev *et al.* [56] showed how an insulating chamber wall allows the plasma potential to be driven by the bias applied to a Langmuir probe. Figure 4.31 shows the current collected by the Langmuir probe as a function of bias from plasma in a chamber with insulating wall. Also shown is what happens when a solidly grounded electrode is inserted into the plasma. Even though the area of the Langmuir probe is tiny compared to he area of the chamber walls, the bias applied to the probe is clearly able to push the plasma potential up. As a result, the potential difference between plasma and the probe is not as large as the applied bias. Thus, current collected by the probe as a function of bias is much less than expected. This reduced current is clearly seen when compared with the case of an anchored plasma potential (with grounded electrode).

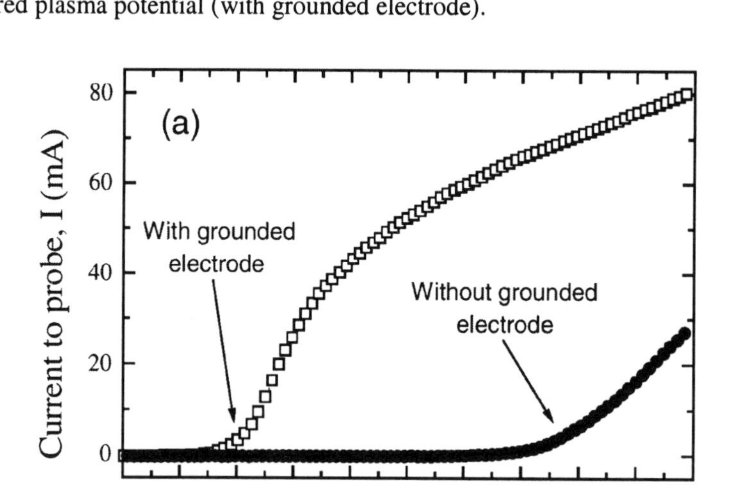

Figure 4.31. Current collected by a Langmuir probe as a function of bias is much lower in plasma with insulating chamber walls. When a grounded electrode is introduced, the plasma potential is anchored and the collected current is much higher. Taken from [56].

Since there will be a plasma deposited film at the wall in most processing chambers, the wall is never perfectly insulating or conducting. The wall potential will partly follow the plasma potential in almost all cases. Figure 4.32 illustrates how the plasma potential distribution changes with bias applied to the wafer. The solid curve represents the plasma potential distribution when the bias is off. The plasma potential is drawn to be fairly flat over the wafer to represent a uniform average plasma density. The potential drops rapidly beyond the wafer chuck near the wall, as is normally the case. The dotted curves represent the plasma potential distributions at the two extremes of the applied AC bias. Note that the potentials at the wall shift toward the same direction but are much smaller in magnitude. The potential drop from the plasma to the wall is much larger, and the rapidly changing part of the potential distribution now extends deeper into the plasma and over the wafer. The potential at the center of the wafer is now very different from the potential at the wafer edge. Thus we can expect charging damage to occur.

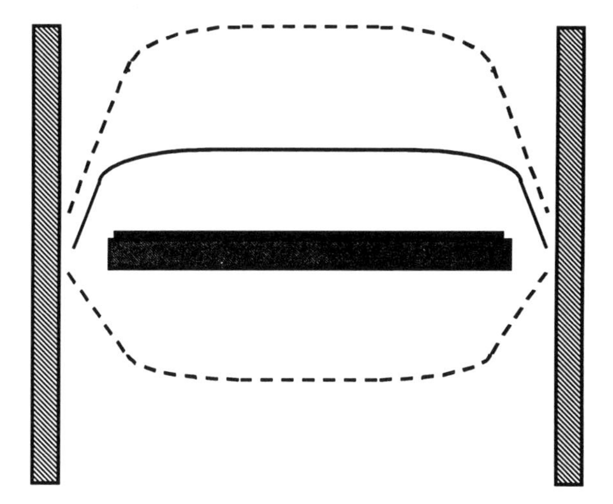

Figure 4.32. Plasma potential with (dotted lines) and without (solid) applied RF bias. The with bias distributions are the extremes of the applied AC bias.

Lassig *et al.* [21] recently reported that the floating potential across the wafer is far more non-uniform when wafer bias is applied. Since their measurement method is sensitive to the maximum excursion of the potential, the distribution they observed is the distribution during the extremes of bias. Their measured potential distribution is shown in Figure 4.33, which is quite similar to that of Figure 4.32. Their measured distribution is much more irregular than the one shown in Figure 4.32. This is not surprising because in Figure 4.32 we have only considered the side chamber wall. In a real chamber, we must take into account the top chamber wall as well. We also need to take into account the fact that no chamber can be perfectly symmetric, no plasma excitation source is perfectly uniform and the deposited film on the chamber wall will not be perfectly uniform. When all of these factors are considered, we can expect at times the resulting transient plasma potential distribution is even more irregular than the one observed by Lassig *et al.* Thus, the type of damage distribution shown in Figure 4.25 can easily have been the result of transient charging damage.

In most commercial plasma processing systems, the average plasma uniformity is very good or they will never be accepted into production. Since plasma density increases exponentially with plasma potential, the uniform etch rate or deposition rate of plasma processing systems implies that the plasma potential cannot vary too much across the wafer. Accordingly, there have been arguments in the literature that plasma charging damage due to non-uniform plasma is unimportant in modern processing tools. Thus there is a widespread belief that electron-shading is the only important charging damage mechanism. However, charging damage due to electron-shading should result in a uniform damage distribution across the wafer. In practice, most of the measured charging damages are not uniform across the wafer. While a combination of mildly non-uniform plasma and electron-shading damage would still produce a non-uniform damage, the electron-shading effect cannot explain many of the non-uniform damage observed. This suggests that charging damage due to transient plasma non-uniformity is an important source of damage.

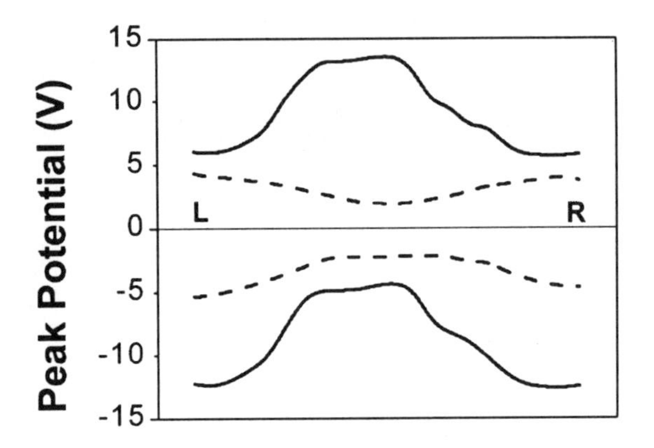

Figure 4.33. Peak potential distribution of a high-density plasma system with (solid line) and without (dotted line) RF bias. Upper plots are positive peak potentials while the negative peak potentials are in the lower plots. The potential distribution in the absence of bias varies only slightly, in agreement with a uniform deposition rate. Taken from [21].

4.5 Hidden Antenna Effects

191-2It is a common practice in the semiconductor IC industry to use antenna design rules to limit the impact of plasma charging damage to product yield and reliability. These design rules are established based on antenna test devices. However, there are damage mechanisms that these design rules cannot account for. One of these mechanisms is the transient existence of hidden antenna ratio due to processing conditions. It is well known that the etch rate of fine patterns depends on feature width [57]. This phenomenon is called Reactive-Ion-Etch (RIE) lag. It is due to difficulties of transporting reactance down to the narrow spacing (Knudsen transport) as well as shadowing effect. Figure 4.34 shows an example of RIE lag in the etching of a narrow trench into silicon.

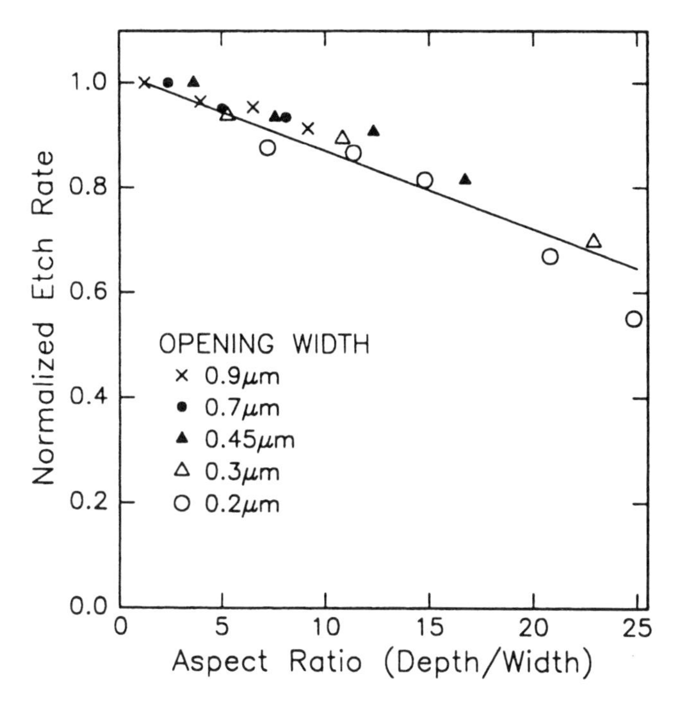

Figure 4.34. Due to reactive ion etch (RIE) lag effect, etch rate is feature width dependent. Narrower openings are etched slower. Taken from [57].

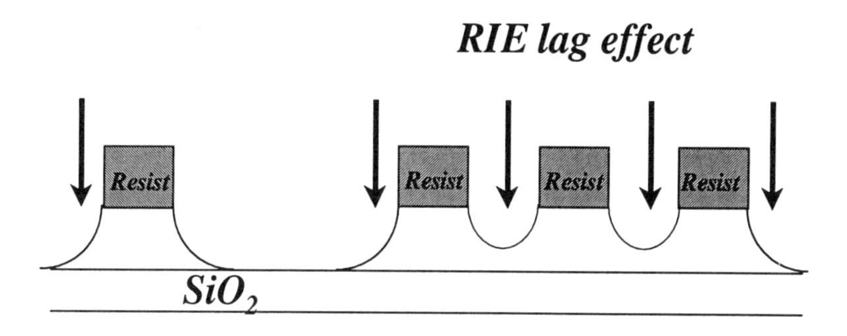

Figure 4.35. Due to RIE lag,larger spacing starts to clear (etching end-point) first. From this point on, until the small spaces are also cleared, dense lines remain connected while isolated lines are separated.

Usually, etching-engineers work hard to minimize RIE lag. However, as technology advances, the required smaller spacing in fine features makes it difficult to eliminate RIE lag completely. When RIE lag exists, we will encounter situations where densely packed lines take longer time to etch through than isolated lines. During the time span between the clearing of larger spacing (end-point of etching) and the narrowest spacing, we have to treat all the dense lines as connected together even though they were not intended to be by design. Figure 4.35 illustrates the problem graphically. During this time period, we effectively have a very large antenna. A very nice example of this hidden antenna (also called giant antenna) effect was reported by Krishnan *et al.* [58] and is shown in Figure 4.36. In their experiment, two types of perimeter intense antenna devices are compared with a control that was protected by a diode. The two types of antenna are made of many parallel lines. One type has all the lines connected together while the other type does not. The charging damage on both types of devices as measured by threshold voltage shift from control is the same, clearly demonstrating the RIE lag induced hidden antenna effect. This result also implies that the majority of damage occurred during the clearing step and not the overetch step. (See Section 2.5.3).

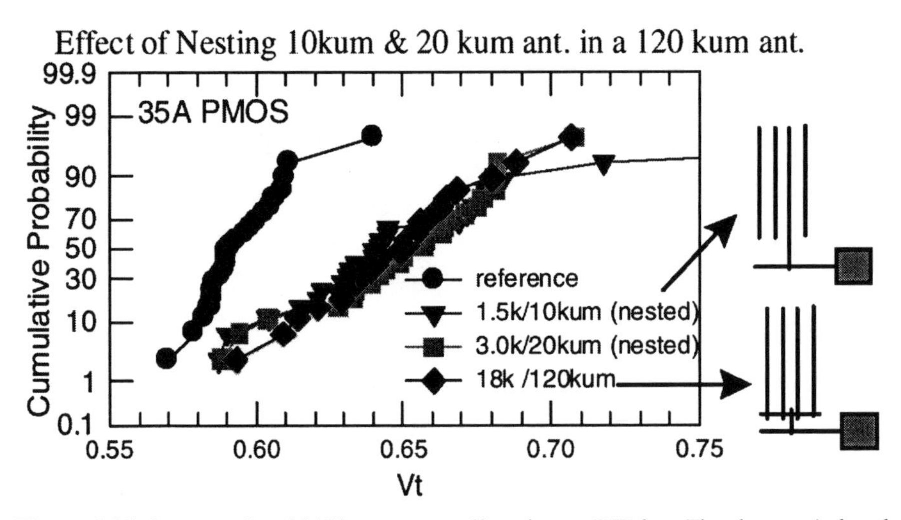

Figure 4.36. An example of hidden antenna effect due to RIE lag. The damage induced threshold voltage distribution indicates that the nested antenna behaves as if all the lines are connected. Taken from [58].

References

1. Gabriel, C.T. and M.G. Weling, *Gate Oxide Damage Reduction Using a Protective Dielectric Layer.* IEEE Electron Dev. Lett., 1994. **15**(8): p. 269.

2. Shone, F., *et al.*, *Gate oxide charging and its elimination for metal antenna capacitor and transistor in VLSI CMOS double layer metal technology*. in *VLSI Technology Symp.* 1989. p. 73.
3. Stamper, A.K. and S.L. Pennington, *Passivation-layer charge-induced failure mechanism in 0.5um CMOS technology*. in *IEEE VLSI Multilevel Interconnection Conference (VMIC)*. 1992. p. 420.
4. Strong, A.W., *et al.*, *Gate dielectric integrity and reliability in 0.5-um CMOS technology*. in *IEEE International Reliability Physics Symp. (IRPS)*. 1993. p. 18.
5. Stamper, A.K., J.B. Lasky, and J.W. Adkisson, *Plasma-induced gate-oxide charging issues for sub-0.5um complementary metal-oxide-semiconductor technologies*. J. Vac. Sci. Technol. A, 1995. **13**(3): p. 905.
6. Cheung, K.P. and C.S. Pai, *Charging Damage from Plasma Enhanced TEOS Deposition*. IEEE Electron Dev. Lett., 1995. **16**(6): p. 220.
7. Cote, D.R., *et al.*, *Low temperature chemical vapor deposition processes and dielectrics for microelectronic circuit manufacturing at IBM*. IBM J. Res. Develop., 1995. **39**(4): p. 437.
8. Machida, K., *et al.*, *Charge build-up reduction during biased electron cyclotron resonance plasma deposition*. J. Vac. Sci. Technol. B, 1995. **13**(5): p. 2004.
9. Bothra, S., *et al.*, *Control of Plasma Damage to Gate Oxide during High Density Plasma Chemical Vapor Deposition*. J. Electrochem. Soc., 1995. **142**(11): p. L208.
10. Cote, D., *et al.*, *Process-induced gate oxide damage issues in advanced plasma chemical vapor deposition processes*. in *International Symp. Plasma Process Induced Damage (P2ID)*. 1996. p. 61.
11. Krishnan, S. and S. Nag, *Assessment of charge-induced damage from high density plasma (HDP) oxide deposition*. in *International Symposium on Plasma Process Induced Damage*. 1996. p. 67.
12. Roche, G.A. and J.P. McVittie, *Application of Plasma Charging Probe to production HDP CVD tool*. in *International Symp. Plasma Process Induced Damage (P2ID)*. 1996. p. 71.
13. Hook, T.B., A. Stamper, and D. Armbrust, *Sporadic charging in interlevel oxide deposition in conventional plasma and HDP deposition systems*. in *International Symp. Plasma Process Induced Damage (P2ID)*. 1997. p. 149.
14. Xie, J. and S. Jun, *A simple and effective electrical monitor do detect plasma induced damage in deep submicron VLAI process technology*. in *VLSI Multilevel Interconnection Conf. (VIMC)*. 1997. p. 214.
15. Lin, Y.M., *et al.*, *Plasma charging damage and water-related hot-carrier reliability in the deposition of plasma-enganced tetraethylorthosilicate oxide*. J. Electrochem. Soc., 1997. **144**(7): p. 2525.
16. Lin, Y.M., *et al.*, *Improvement of water-related hot-carrier reliability by optimizing the plasma-enhanced tetra-ethoxysilane deposition process*. J. Electrochem. Soc., 1997. **144**(8): p. 2898.

17. Lie, D.Y.C., *et al.*, *Hot-carrier degradation for deep-submicron n-MOSFETs introduced by backend processing*. in *SPIE*. 1997. p. 258.
18. Moens, P., *et al.*, *Plasma process-induced charging during PECVD overlay nitride deposition*. in *International Symp. Plasma Process Induced Damage (P2ID)*. 1998. p. 68.
19. Trabzon, L., O.O. Awadelkarim, and J. Werking, *The effects of interlayer dielectric deposition and processing on the reliability of n-channel transistors*. Solid-State Electronics, 1998. **42**(11): p. 2031.
20. Shih, H.H., *et al.*, *The prevention of charge damage on thin gate oxide from high density plasma deposition*. in *International Symp. Plasma Process Induced Damage (P2ID)*. 1999. p. 88.
21. Lassig, S.E., *et al.*, *Effects of processing pressure on device damage in RF biased ECR CVD*. in *International Symp. Plasma Process Induced Damage (P2ID)*. 1999. p. 96.
22. Shiba, K. and Y. Hayashi, *Investigation of plasma-induced charging damage for nMOSFETs with conventional or damascene Al interconnects*. in *International Electron Device Meeting (IEDM)*. 1999. p. 101.
23. Hirao, S., *et al.*, *Plasma Damage of Gate Oxide through the Interlayer Dielectric*. in *1993 International Conference on Solid State Devices and Materials (ISSDM)*. 1993. p. 826.
24. DiStefano, T.H. and D.E. Eastman, *The band edge of amorphous SiO2 by photoinjection and photoconductivity measurements*. Solid State Communications, 1971. **9**: p. 2259.
25. Capasso, F., *et al.*, *Effective mass filtering: Giant quantum amplification of the photocurrent in a semiconductor superlattice*. Appl. Phys. Lett., 1985. **47**(4): p. 420.
26. Powell, R.J., *Photoinjection into SiO2: Use of Optical Interference to Determine Electron and Hole Contributions*. J. Appl. Phys., 1969. **40**(13): p. 5093.
27. Batra, I.P., K.K. Kanazawa, and H. Seki, *Discharge Characteristics of Photoconducting Insulators*. J. Appl. Phys., 1970. **41**(8): p. 3416.
28. Cismaru, C., J.L. Shohet, and J.P. McVittie, *Plasma Vacuum Ultraviolet Emission in a High Density Etcher*. in *Internation Symposium on Plasma Process Induced Damage*. 1999. p. 192.
29. Cheung, K.P., *et al.*, *Is n-MOSFET hot-carrier lifetime degraded by charging damage?* in *International Symp. Plasma Process Induced Damage (P2ID)*. 1997. p. 186.
30. Cheung, K.P., *et al.*, *Impact of plasma-charging damage polarity on MOSFET noise*. in *International Electron Device Meeting (IEDM)*. 1997. p. 437.
31. Apte, P.P., T. Kubota, and K. Saraswat, *Constant Current Stress Breakdown in Ultrathin SiO2 Films*. J. Electrochem. Soc., 1993. **140**(3): p. 770.
32. Greene, W.M., J.B. Kruger, and G. Kooi, *Magnetron etching of polysilicon: Electrical damage*. J. Vac. Sci. Technol., 1991. **B9**(2): p. 366.

33. Namura, T., *et al.*, *Charge Buildup in Magnetized Process Plasma.* Jap. J. Appl. Phys., 1991. **30**(7): p. 1576.
34. Sekine, M., *et al.*, *Gate Oxide breakdown Phenomena in Magnetized Plasma.* in *Dry Process Symp.* 1991. p. 99.
35. Hoga, H., *et al.*, *Charge build-up in magnetron-enhanced reactive ion etching.* Jap. J. Appl. Phys., 1991. **30**(11B): p. 3169.
36. Fang, S. and J.P. McVittie, *A model and experiments for thin oxide damage from wafer charging in magnetron plasmas.* IEEE Electron Dev. Lett., 1992. **13**(6): p. 347.
37. Fang, S. and J.P. McVittie, *Charging damage to gate oxides in an O2 magnetron plasma.* J. Appl. Phys., 1992. **72**(10): p. 4865.
38. Shin, H., *et al.*, *Spatial Distributions of Thin Oxide Charging in Reactive Ion Etcher and MERIE Etcher.* IEEE Electron Dev. Lett., 1993. **14**(2): p. 88.
39. Fang, S. and J.P. McVittie, *Model for oxide damage from gate charging during magnetron etching.* Appl. Phys. Lett., 1993. **62**(13): p. 1507.
40. Fang, S., S. Murakawa, and J.P. McVittie, *A new model for thin oxide degradation from wafer charging in plasma etching.* in *IEEE International Electron Device Meeting (IEDM).* 1992. p. 61.
41. Nakagawa, S., *et al.*, *Charge Build-Up and Uniformity Control in Magnetically Enhanced Reactive Ion Etching Using a Curved Lateral Magnetic Field.* Jap. J. Appl. Phys., 1994. **33**(Part 1, No 4B): p. 2194.
42. Lindley, R., *et al.*, *Advanced MERIE technology for high-volume 0.25-um generation critical dielectric etch.* Solid State Technonlgy, 1997. **1997**(August): p. 93.
43. Lindley, R., *et al.*, *Magnetic field optimization in a dielectric magnetically enhanced reactive ion etch reactor to produce an instantaneously uniform plasma.* J. Vac. Sci. Technol. A, 1998. **16**(3): p. 1600.
44. Gadgil, P.K., T.D. Mantei, and X.C. Mu, *Evaluation and control of device damage in high density plasma etching.* J. Vac. Sci. Technol. B, 1994. **12**(1): p. 102.
45. Balasinski, A. and T.P. Ma, *Impact of radiation-induced nonuniform damage near MOSFET junctions.* IEEE Trans. Nucl. Sci., 1993. **40**(6, pt . 1): p. 1286.
46. Gu, T., *et al.*, *Impact of polysilicon dry etching on 0.5um NMOS transistor performance: The presence of both plasma bombardment damage and plasma charging damage.* IEEE Elect. Dev. Lett., 1994. **15**(2): p. 48.
47. Li, X., *et al.*, *Effect of plasma poly etch on effective channel length and hot carrier reliability in submicron transistors.* IEEE Elect. Dev. Lett., 1994. **15**(4): p. 140.
48. Gu, T., *et al.*, *Degradation of submicron N-channel MOSFET hot electron reliability due to edge damage from polysilicon gate plasma etching.* IEEE Elect. Dev. Lett., 1994. **15**(10): p. 396.
49. Kang, T.K., *et al.*, *Effects of polysilicon electron cyclotron resonance etching on electrical characteristics of gate oxides.* Jap. J. Appl. Phys., 1995. **Part 1, 34**(5A): p. 2272.

50. Brozek, T., *et al.*, *Localized charge injection through the gate oxide over gate-drain overlap region: mechanism, device dependence, and application for device diagnostics.* in *International Electron Device Meeting (IEDM).* 1996. p. 869.
51. Chung, S.S., *et al.*, *Charge pumping technique for the eveluation of plasma induced edge damage.* in *International Reliability Physics Symposium (IRPS).* 2000. p. 389.
52. Nicollian, E.H. and J.R. Brew, *Radiation effects in SiO2*, in *MOS Physics and Technology.* 1982, John Wiley & Son: New York. p. 549-576.
53. Ling, C.H., *Trap generation at Si/Sio2 interface in submicrometer metal-oxide-semiconductor transistor by 4.9 eV ultraviolet irradiation.* J. Appl. Phys., 1994. **76**(1): p. 581.
54. Aum, P.K., *et al.*, *Controlling plasma charge damage in advanced semiconductor manufacturing - Challenge of small feature size device, large chip size, and large wafer size.* IEEE Trans. Elect. Dev., 1998. **45**(3): p. 722.
55. Hassan, M.G.E., O.O. Awadelkarim, and J.D. Werking, *The impact of metal-1 plasma processing - induced hot carrier injection on the characteristics and reliability of n-MOSFET's.* IEEE Trans. Elect. Dev., 1998. **45**(4): p. 861.
56. Malyshev, M.V., V.M. Donnelly, and J.I. Colonell, *Plasma diagnosis and charging damage.* in *International Symp. Plasma Process Induced Damage (P2ID).* 1999. p. 149.
57. Gottscho, R.A., C.W. Jurgensen, and D.J. vitkavage, *Microscopic uniformity in plasma etching.* J. Vac. Sci. Technol. B, 1992. **10**(5): p. 2133.
58. Krishnan, S., *et al.*, *Inductively coupled plasma (ICP) metal etch dmage to 35-60A gate oxide.* in *International Electron Devices Meeting (IEDM).* 1996. p. 731.

Chapter 5

Charging Damage Measurement I - Determination of Plasma's Ability to Cause Damage

The measurements of plasma charging damage can roughly be separated into two categories. One category determines plasma's propensity for causing charging damage. The other determines directly the extent of damage that actually happened in devices or circuits. In this chapter, we concentrate only on the first category. The second category will be discussed in next chapter.

5.1 Direct Plasma Property Measurement with Langmuir Probe

One of the most common methods of characterizing plasma is to use a Langmuir probe. It is commercially available and is widely used by plasma processing equipment vendors as well as process-development engineers. The Langmuir probe, when used correctly, can in principle measure electron temperature T_e, electron density n_e, positive ion density n_i^+, plasma potential V_p as well as floating potential V_f. It can measure all of these quantities as a function of position and therefore map the distribution directly. A clear discussion of how to use Langmuir probes to measure all of these quantities and the potential pitfalls was given by Donnelly [1] recently. The discussion given below is taken mostly from that work but skips over the more intricate details.

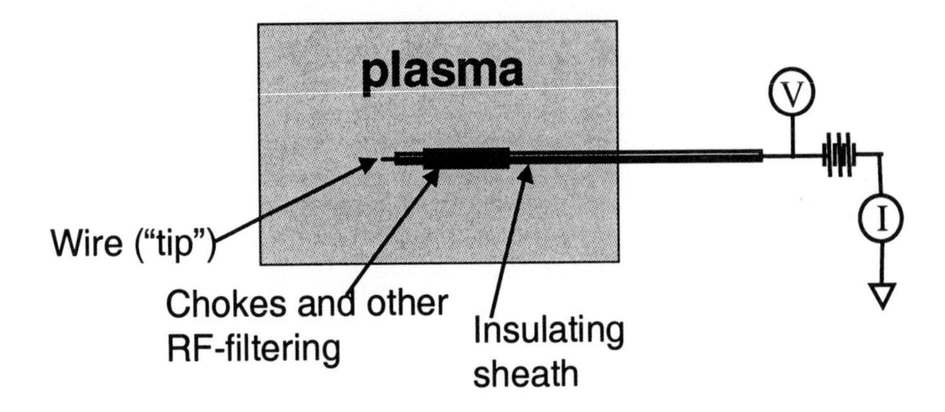

Figure 5.1. A single Langmuir probe. Taken from [1].

Figure 5.1 shows a simple Langmuir probe in plasma. The actual sensor is at the tip of the probe, which is really a small piece of bare wire sticking out of the insulating sheath into the plasma. This wire is connected to an external voltage source as well as a current sensor. To avoid picking up the large RF signal either from the power that drives the plasma or the bias, a RF choke or low pass filter is employed between the sensor wire and the DC source. The operation of the probe is to measure current collected by the sensor wire while the DC voltage is being scanned. Figure 5.2 shows an example of the measured current as a function of DC voltage on the probe (with respect to ground). When the voltage is negative, the probe collects ions from plasma. When the probe is positive, the probe collects electron from plasma. The shape of the measured current-voltage curve is similar to the one shown in Figure 2.12 of Chapter 2. Since ion flux saturates easily, the ion current density measured remains almost flat for all the negative voltage range. Electron flux, on the other hand, increases exponentially with the applied positive voltage until the plasma potential V_p is approached.

The applied voltage at which the collected current density is zero is the floating potential V_f. By definition, the electron flux balances the ion flux exactly at floating potential. Figure 5.3 shows the same I-V curve of Figure 5.2 in expended scale to highlight the zero crossing point. For the particular plasma under study, the floating potential at the position of the probe is shown to be 1.5V. In this expanded scale, the ion current as a function of negative voltage is no longer flat. This is an unavoidable artifact that the sheath thickness expands around the probe with the applied voltage (see discussions in Chapter 2). As a result, the effective probe area also increases with the applied voltage. A larger effective probe area would of course collect more ions. The saturated ion flux on a flat wafer would not suffer from this artifact.

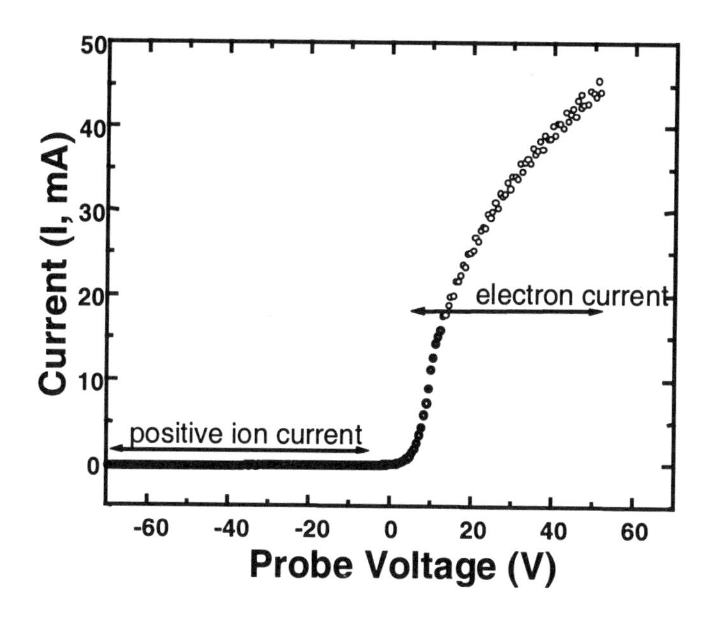

Figure 5.2. Typical current-voltage (I-V) characteristics of a Langmuir probe. Taken from [1].

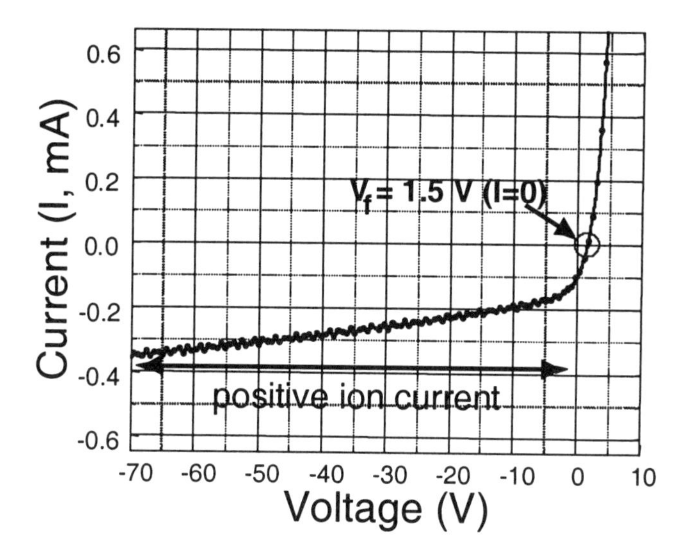

Figure 5.3. Expended view of the Langmuir probe I-V characteristics to show floating potential at which ion current equal electron current.. Taken from [1].

As mentioned above, the electron flux increases exponentially with positive applied voltage until the plasma potential is approached. Plotting the collected current in semilog scale produces a straight line that saturates at plasma potential V_p (Figure 5.4). This is how V_p is measured with the Langmuir probe. There are two ways to extract V_p from the I-V curve as shown in Figure 5.4. One the intercept method and the other is the maximum derivative method. As shown in Figure 5.4, the two methods do not produce exactly the same result. The difference is probably due to the noise level in the I-V curve that gets amplified by the derivative method.

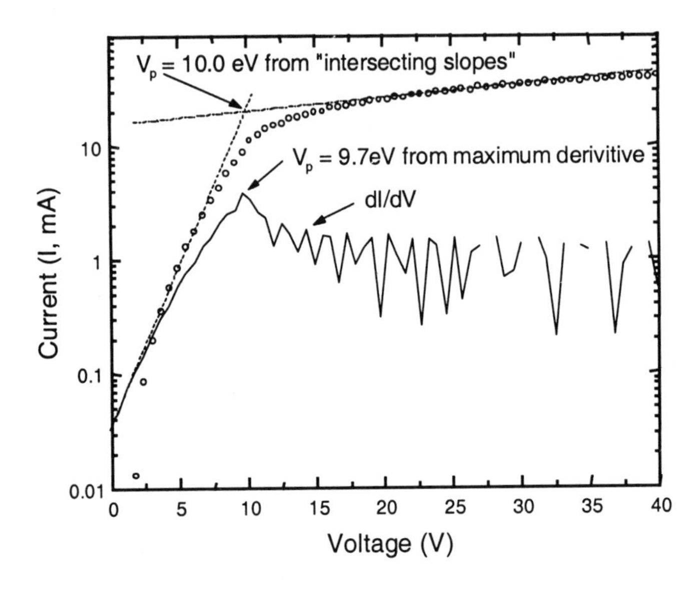

Figure 5.4. Methods to extract the plasma potential from the probe I-V curve. Taken from [1].

Recall from Chapter 2 that electron flux from plasma is:

$$I_e = \frac{1}{4} n_e \sqrt{\frac{8kT_e}{\pi m_e}} \exp\left(\frac{eV_{sh}}{kT_e}\right) \tag{5.1}$$

We can rewrite it as:

$$I_e = I_{e,p} \exp\left(\frac{V - V_p}{kT_e}\right) \tag{5.2}$$

Thus, plotting the electron current density as a function of applied voltage far from V_p will produce a straight line in semilog scale. The slope of this straight line will give $1/kT_e$, the reciprocal of electron temperature in eV. Figure 5.5 shows the electron current density as a function of probe voltage in semilog plot. The

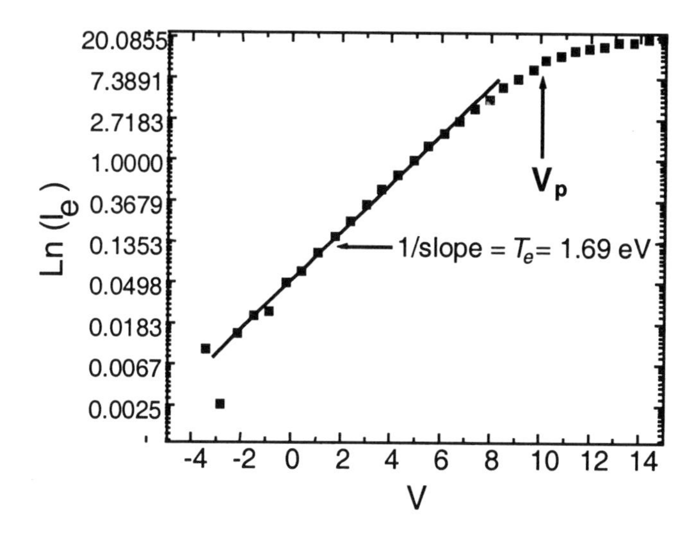

Figure 5.5. Electron temperature can be extracted from the slope of electron current as a function of applied voltage in semilog plot. Taken from [1].

extracted electron temperature for this particular case is 1.69eV. Note that the electron current shown in Figure 5.5 is not the probe current at positive voltage. Rather, it is the probe current minus the ion current which is a function of applied voltage. When the (collisionless) probe sheath is >> probe radius,

$$I_i^+ \propto \sqrt{V - V_p} \tag{5.3}$$

The ion current part of the probe current can be obtained by plotting the probe current at negative voltage as a function of $\sqrt{V}$ and extrapolate to V_p as shown in Figure 5.6.

Clearly, when used properly, a Langmuir probe can provide much information about plasma that is important for accessing its propensity for charging damage. In addition to V_f, V_p, n_e, T_e and I_i^+, one can in principle obtain n_i^+ as well as electron energy distribution function using more sophisticated analysis. In principle, when all of these parameters are known, one can predict whether charging damage will occur or not. Unfortunately, life is seldom so simple. All of these measurements rely on one very important assumption: that the application of a voltage to the probe does not disturb the plasma potential. We already discussed in last chapter that most real plasma processing systems do not provide enough anchoring for the plasma potential. When a voltage is applied to the probe, it pushes the plasma potential to follow. How well the plasma potential follows the applied voltage depends on how well the plasma potential is anchored. Figure 5.7 illustrates the effect of poorly anchored plasma on the I-V characteristic of the probe. With a distorted I-V curve, all the parameters discussed above, with V_f the only exception, would be deviated from the real value.

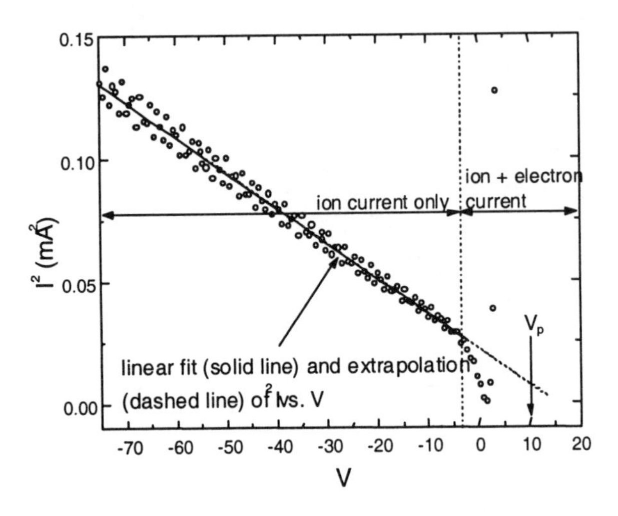

Figure 5.6. Ion current at voltage where electron current dominate can be extracted by extrapolating the plot of ion current square versus applied voltage. Taken from [1].

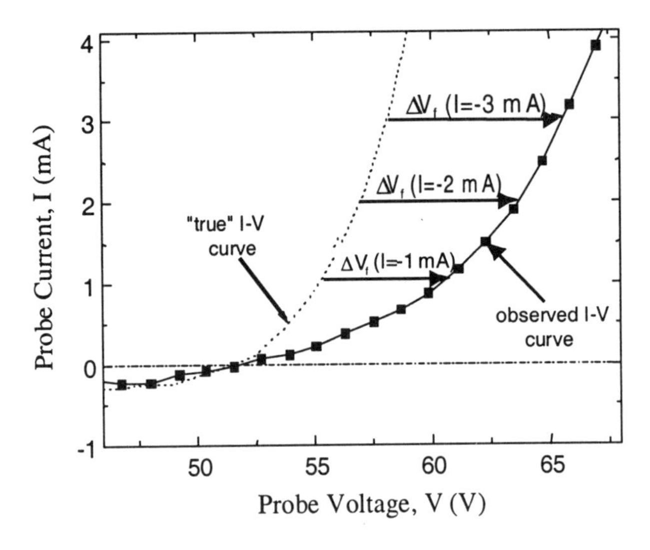

Figure 5.7. When the plasma potential is not anchored, it can be influenced by the applied bias. As a result, the measured I-V would be distorted by stretching out. Taken from [1].

In the last chapter, we saw that by introducing a grounded electrode into the plasma the I-V curve can be restored. Another way to overcome this problem is to use double probe. Figure 5.8 shows an example of a double probe. The reference probe is for measuring the V_f at each voltage applied to the sensor probe. Since measuring V_f involves no net current, the reference probe does not disturb the plasma. It only senses how much the plasma potential has been changed by the sensor probe. This information can then be used as correction factor in the I-V curve as shown in Figure 5.9. Once the correct I-V curve is obtained, the real plasma parameters can again be extracted.

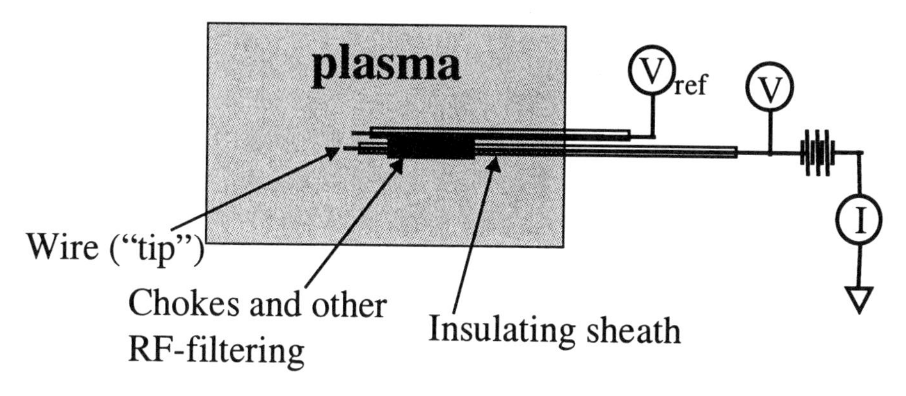

Figure 5.8. An example of a double Langmuir probe. Taken from [1].

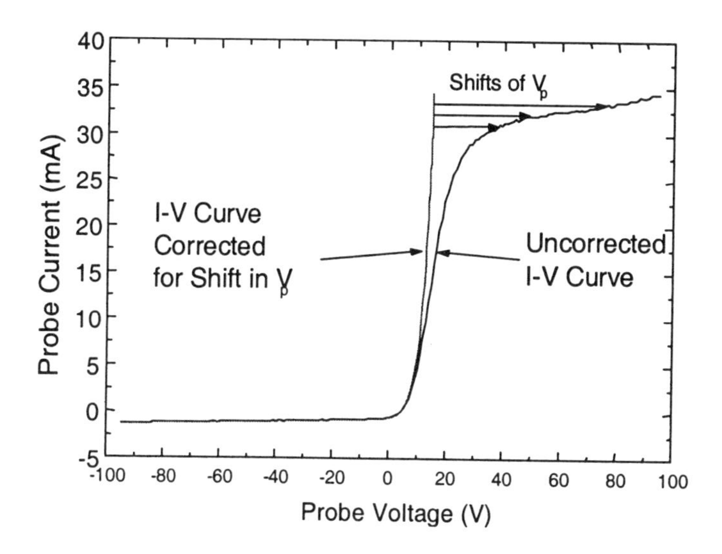

Figure 5.9. The floating potential shift measured by the reference probe can be used to correct for the plasma potential shift due the voltage applied to the sensing probe. Taken from [1].

The use of a double probe does not guarantee that accurate plasma parameters be obtained. There are other factors at work that cause errors in the measurement. For example, while the reference probe can provide a correction factor for the plasma potential, the sensor probe can have difficulty collecting current without applying an uncomfortably large voltage that the hardware of the probe may not support. In addition, accurate probe measurements require a clean probe tip. Cleaning of the probe tip can be accomplished by applying a large voltage to collect enough electrons to heat the probe tip to red-hot. When the probe cannot collect enough current, it may not be possible to clean the probe tip, in-situ.

Furthermore, there are subtle details such as the probe tip size and geometry can influence the parameter extraction. For example, the subtraction of ion current to obtain the true electron current may not be so simple if the condition for Equation 5.3 to be valid is not satisfied. Such a condition is often encountered when the pressure is not low enough or the plasma density is very high.

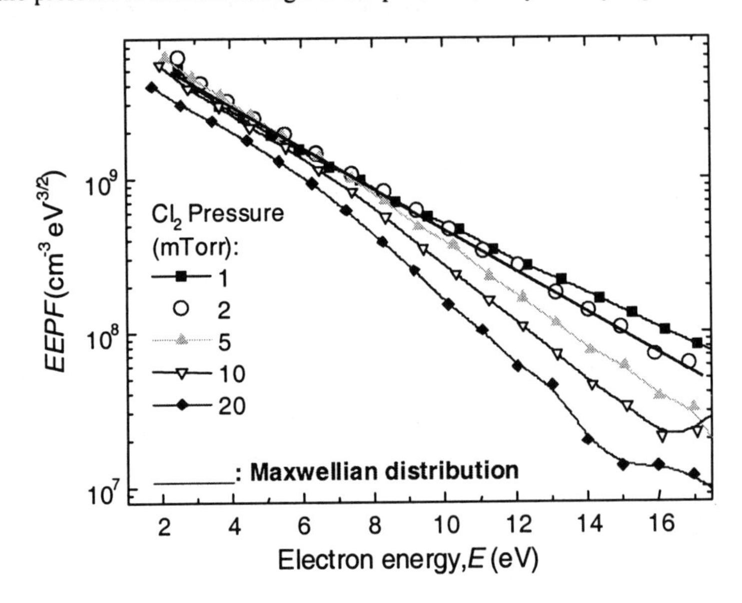

Figure 5.10. Measured electron energy distribution (EEDF) function for a wide pressure range of chlorine plasma is not Maxwellian. Only the data at 2 mTorr fit the Maxwellian behavior. Taken from [2].

Even in the best of circumstances, the measurement may still be in error. Inherent in the method of T_e extraction is the assumption that the velocity distribution of electron in plasma is Maxwellian. This assumption is only an approximation. Figure 5.10 shows the measured electron energy distribution function (EEDF) in chlorine plasma as a function of pressure [1] [2]. The EEDF should be a straight line in the semilog plot if the velocity distribution is Maxwellian. As can be seen, only the 2 mTorr data fit the straight line. EEDF at all other pressures deviates from Maxwellian. The non-Maxwellian distribution would obviously affect the T_e extraction from the probe I-V curve. Using a non-intrusive trace rare gases optical emission spectroscopy (TRG-OES), Malyshev *et al.* demonstrated that indeed the actual T_e differs from the extracted T_e using Langmuir probe when the EEDF is not Maxwellian [3] [4]. Figure 5.11 shows the comparison between T_e measured by the double Langmuir probe and the T_e measured by the TRG-OES method for the same plasmas in Figure 5.10. When the EEDF deviated from Maxwellian by having more high-energy electrons (1 mTorr), the Langmuir probe found a lower T_e than TRG-OES. When the EEDF deviated from Maxwellian by having less high-energy electrons, the Langmuir probe found a T_e higher than TRG-OES. The larger the deviation, the bigger the difference.

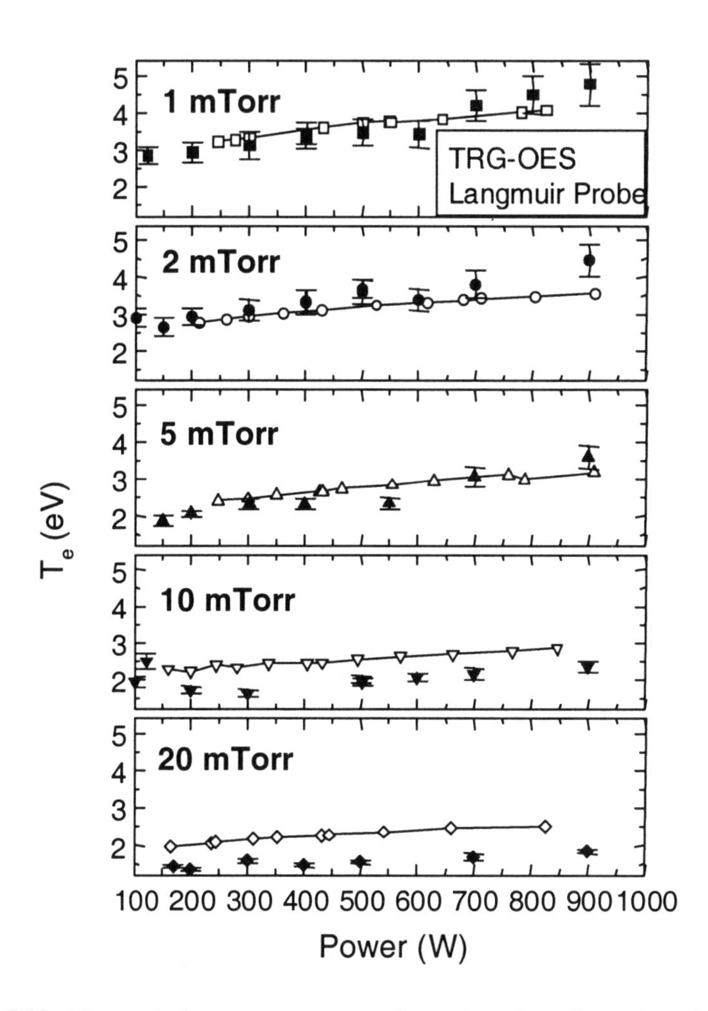

Figure 5.11. Measured electron temperatures from Trace-Rare-Gases-Optical-Emission-Spectroscopy (TRG-OES) differ from those measured from double Langmuir probe. Taken from [4].

Strictly speaking, we can only define a temperature when the velocity distribution is Maxwellian. Therefore, we cannot say that the T_e from the TRG-OES method is more accurate. The differences shown in Figure 5.11 only serve to highlight that to simply assume a Maxwellian distribution and extract T_e from the Langmuir probe I-V curve can be misleading. Since T_e play an important part in the electron-shading damage model, the erroneously measured T_e using Langmuir probe has created a lot of confusion. The use of Langmuir probes to measure T_e is even more problematic as the velocity distribution becomes anisotropic. We have seen how far off the measured value can be from the real value in Chapter 3.

Finally, before we leave the discussion of using Langmuir probe to measure plasma parameters, we should recall the discussion of transient charging at the end of last chapter. To avoid the RF voltage overwhelming the probe, all of them employ RF choke or low pass filters to remove high frequency signals. In doing so, it cannot provide information of plasma parameters that will affect transient charging.

5.2 Stanford Plasma-On-Wafer Real-Time (SPORT) Probe

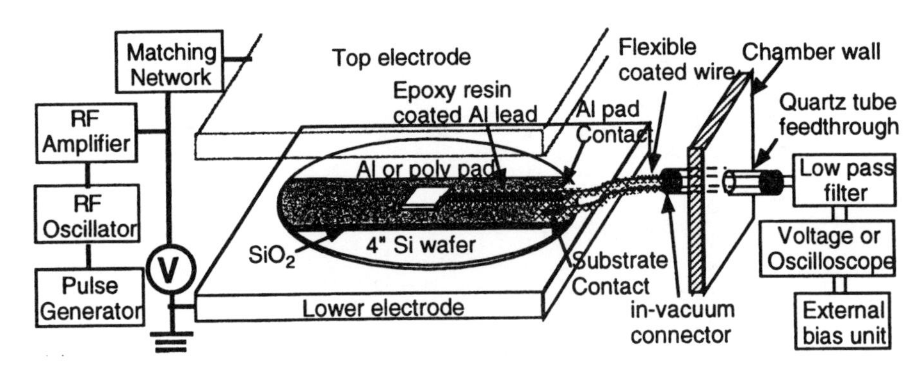

Figure 5.12. Details of the construction of a SPORT probe. Taken from [5].

The SPORT probe is like a specialized Langmuir probe. Instead of a movable probe, it is directly built on a wafer. Figure 5.12 shows how the SPORT probe is constructed [5]. In its simplest form, a conductor (aluminum, for example) pad sits on top of an oxide layer on a silicon wafer, and serves as the probe. It is connected to the outside world through an in-vacuum coaxial cable and feed-through on the chamber wall. Part of the silicon wafer is not covered by oxide. The exposed silicon wafer serves as a reference electrode. The connection of the reference electrode to the outside world is the same as the probe pad. All the leads are coated with insulating layer to keep them from contacting the plasma directly.

The most important advantage of the SPORT probe is that it can measure the voltage difference between the floating probe and the silicon substrate directly. This voltage difference, in principle, is the voltage that set up by the plasma to drive charging damage. The voltage measurement can be accomplished with the SPORT probe completely in passive mode. No driving voltage or current need to be supplied from outside.

Although the probe pad is fixed at one location on the wafer, it is possible to put large number of probe pads on the same wafer to provide location dependent information. Figure 5.13 shows an example of such a layout [5]. Clearly, there is a limit on how many probe pads one can put on the wafer. Since each probe pad requires a coaxial cable and a feed through on the chamber wall, more than a few would be quite difficult in practice. In fact, most commercial plasma processing systems do not even have access ports. It is near to impossible to use the SPORT probe in real processing systems. This limitation is perhaps the most serious one for using SPORT probe to characterize plasma.

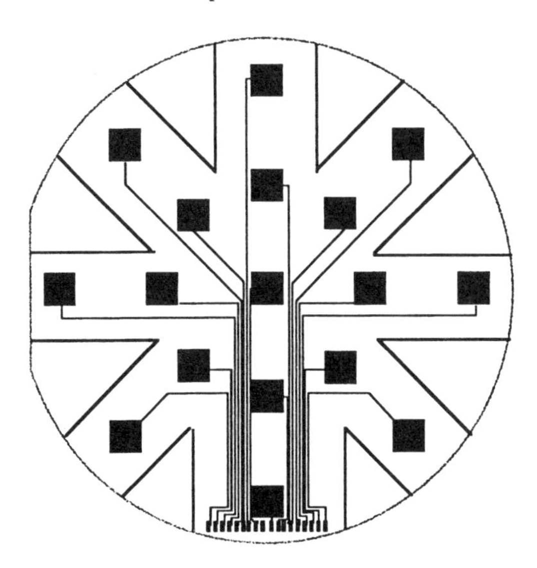

Figure 5.13. A multi-probe arrangement example. Taken from [5].

Owing to its real-time capability, much useful information can be obtained once the probe is installed. Figure 5.14 shows an example of measuring the charging voltage at five locations across a wafer [5]. Also shown are the charging voltage of the same locations when a bar magnet is placed under the wafer (see Figure 4.16 for the physical arrangement of the magnet and the discussion of its impact.). The increase in charging damage is immediately apparent.

Being a specialized Langmuir probe, the SPORT probe can also be operated like Langmuir probe and the I-V characteristics of the plasma can be measured. Figure 5.15 shows the change in I-V characteristics of the center location with and without the bar magnet [5]. Notice the voltages at zero current for both curves. These are the floating potential for with and without magnet cases. The difference between the two is the increase in charging voltage. One can certainly obtain the same information with a Langmuir probe.

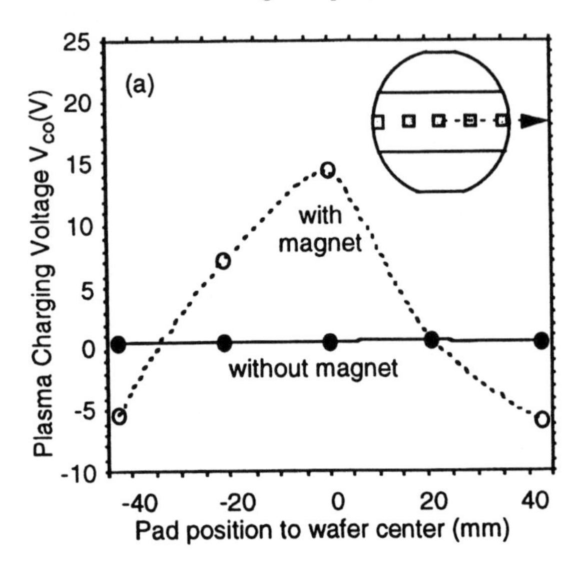

Figure 5.14. Measured charging voltage at five location across a wafer. The impact of a magnet under the wafer is demonstrated. Taken from [5].

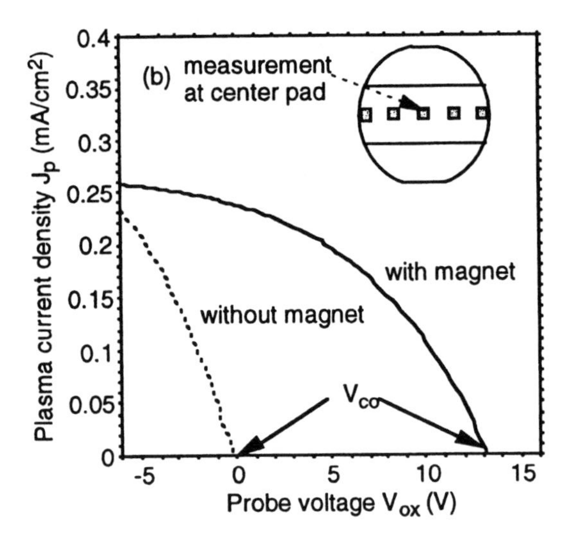

Figure 5.15. Using the SPORT probe like a Langmuir probe, the I-V characteristic of the plasma can also be obtained. Taken from [5].

Note that when using Langmuir probes, the measured floating potential is with respect to ground. There is no easy way to obtain the charging voltage at any location of the wafer without resorting to modeling. The only exception is when the wafer is grounded. In Figures 5.14 and 5.15, the experimental condition is that the wafer was grounded. As a result, the plasma potential is anchored by the grounded substrate. When the plasma potential is uniform (no magnet case), the floating potential of the top electrode will be the same as the substrate and therefore shows zero charging voltage as well as zero floating potential. At first glance, it seems that Figure 5.15 suggests that the charging voltage when a magnet is present is simply the floating potential change. Such information can certainly be obtained by using a Langmuir probe. In reality, this is only an unfortunate choice of experimental condition (grounding the wafer). When the wafer is not grounded the floating potential of the substrate changes with the plasma potential distribution. The charging voltage in the presence of the magnet is not simply the floating potential of the probe pad with the magnet minus the floating potential of the probe pad without the magnet. The ability to obtain directly the charging voltage is clearly an advantage of the SPORT probe over the Langmuir probe. To ensure that the measured charging voltage is meaningful, the SPORT probe should leave ample open substrate area exposing to plasma. The open area should distribute evenly on the wafer. If the open areas are only at the edge of the wafer, its potential is pinned by the plasma at the edge of the wafer. When the open area distribution is not even, the interpretation of data may be difficult.

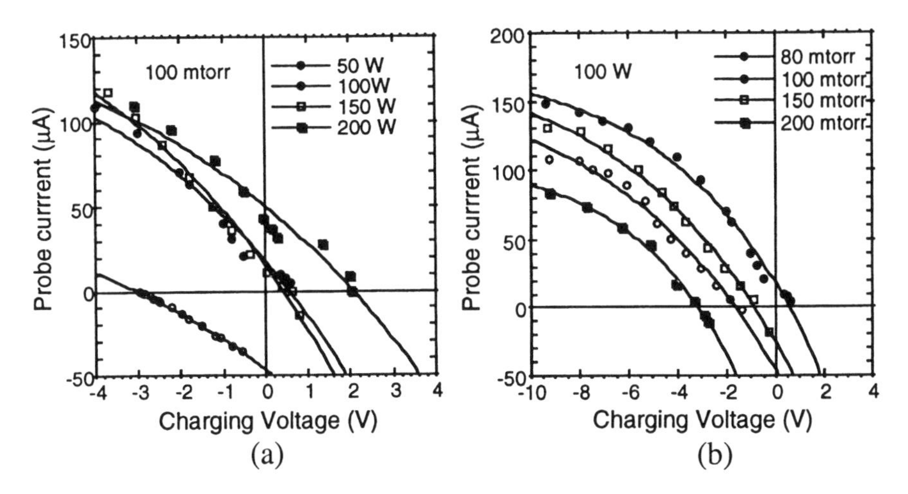

Figure 5.16. (a) Charging I-V as a function of RF power at constant pressure. (b) Charging I-V as a function of pressure under constant RF power. Taken from [5].

The ability to obtain charging voltage as well as I-V characteristics allows one to study the relationship between charging voltage and charging current. Figures 5.16a and 5.16b show examples of charging current as a function of charging voltage can be studied as a function of plasma parameters such as power and pressure [5]. Such study can be useful in understanding how changing the processing recipe can affect the charging damage behavior.

Figure 5.17a and 5.17b show how transient charging voltage during plasma turn-on and turn-off is affected by parameters such as pressure [5]. Transient charging has been predicted theoretically [6] but was never clearly demonstrated experimentally until these results.

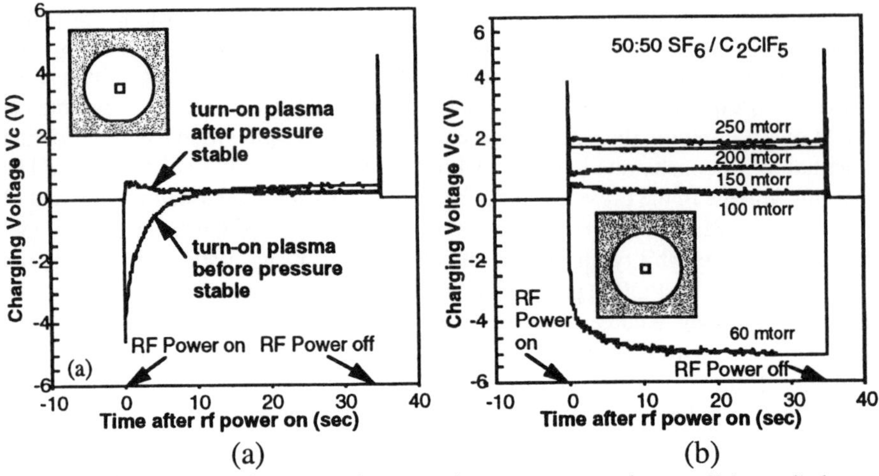

Figure 5.17. Transient charging voltage at plasma turn-on and turn-off is studied as a function of pressure. Taken from [5].

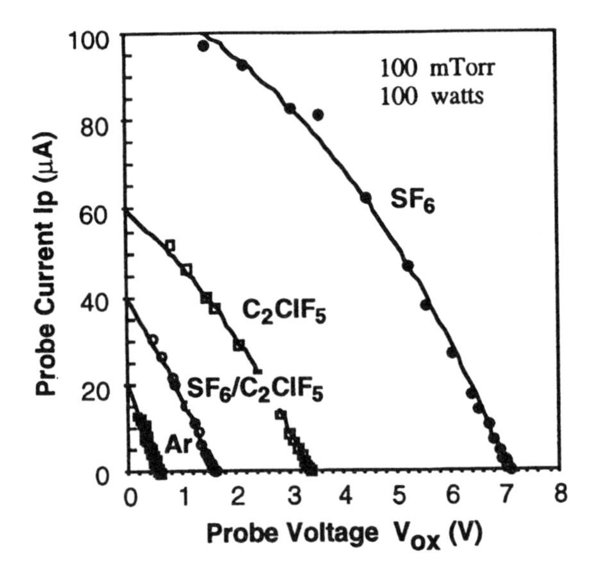

Figure 5.18. Plasma I-V as measured by SPORT probe for different discharge gases under identical discharge power and pressure. Taken from [5].

Obviously, making and setting up the SPORT probe to characterize plasma is not an easy task. Once set up, one would be reluctant to destroy the probe by testing real etching plasmas. However, the charging behavior of real etching plasma can be quite different from a non-etching one. Figure 5.18 is a study of the effect of gaseous species on charging behavior using the SPORT probe [5]. All other conditions being the same, changing the gaseous specie in the discharge changes drastically the plasma I-V characteristics.

As mentioned above, it is near to impossible in most cases to use SPORT wafer in a real processing chamber. In light of that, the shortcoming of not studying the real processing plasma is not so bad. When the goal is to study fundamental plasma charging damage mechanisms, SPORT probes are very useful. For characterizing real processing plasmas, other tools are better choices.

5.3 Using MNOS Device to Measure Plasma Charging Voltage

MNOS capacitors are perhaps the earliest device utilized in plasma charging damage studies to measure the plasma charging voltage. It was discussed in some detail in Section 2.2.1. A typical MNOS capacitor looks like Figure 5.19. It is a capacitor with the composite nitride/oxide layer as dielectric. MNOS transistors are a type of nonvolatile memory device introduced in the 60s [7] and therefore was studied in detail. The device in Figure 5.19 requires more processing steps than the device depicted in Figure 5.20. Needing only one lithography step, the MNOS device of Figure 5.20 is simple to make and was therefore a popular device for use in charging voltage measurement [8].

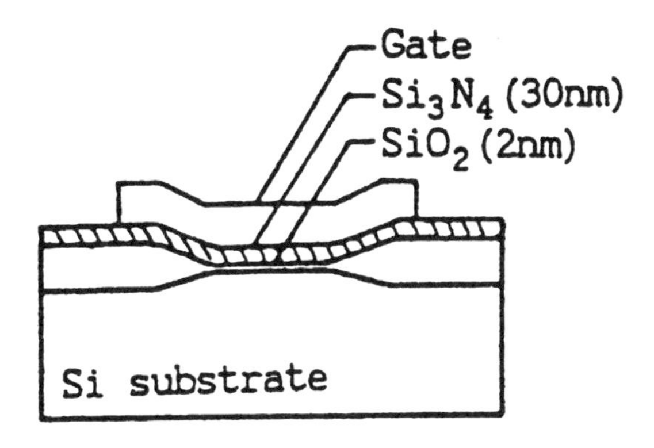

Figure 5.19. A typical MNOS device structure. This one was shown in Figure 2.8.

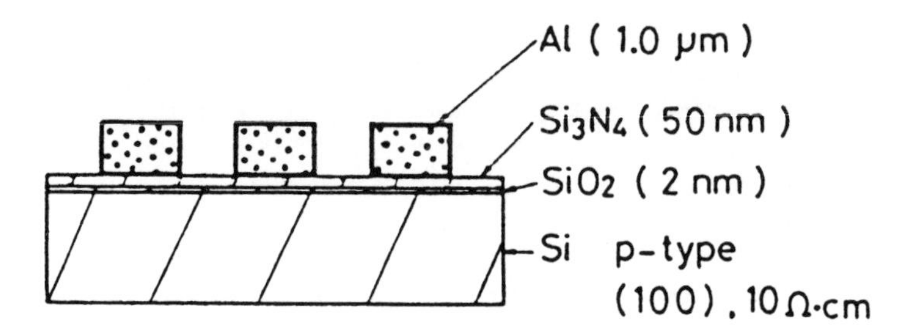

Figure 5.20. An even simpler version of MNOS device that requires only one lithography step. This one was shown in Figure 2.9.

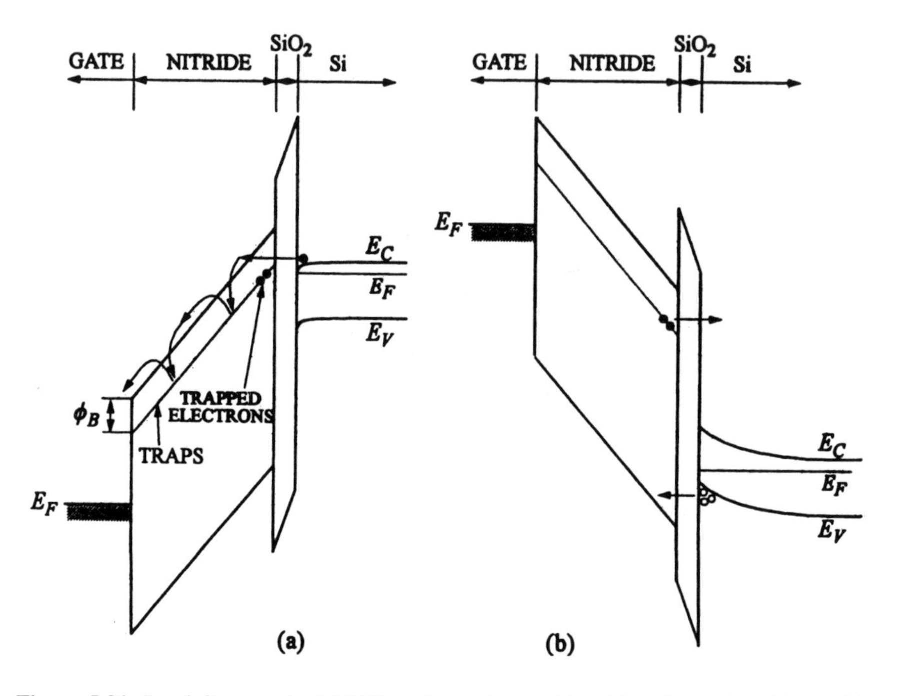

Figure 5.21. Band diagram for MNOS device under positive (a) and negative (b) gate bias. Taken from [7].

Figure 5.21 is the band diagram for the MNOS capacitor under bias [7]. Conduction through the nitride layer is by Frankel-Poole mechanism. It is limiting the over all current flow. Under positive bias (Figure 5.21(a)), electrons tunnel from substrate through the thin oxide and accumulate at the nitride/oxide interface. Under negative bias (Figure 5.21(b)), holes tunnel from substrate through the thin oxide and accumulate at the nitride oxide interface. Figure 5.22 shows the flat-band shift of the MNOS device of Figure 5.20 as a function of bias. When using MNOS capacitors to measure plasma charging voltage, they must be first characterized using known stress to produce the reference plot like those in Figure 5.22. After plasma exposure, the flat-band voltage of MNOS capacitors on the wafer is measured. The equivalent stress voltage experienced by the MNOS capacitors during plasma exposure can then be found from the reference curve. A map of the charging voltage can be constructed. Figure 2.7 in Chapter 2 is a good example of such charging map.

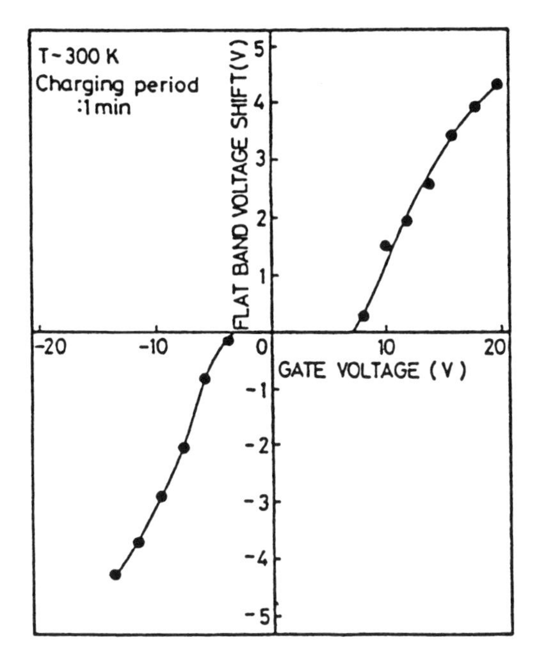

Figure 5.22. Flat-band shift of a MONS device as a function of bias. This plot was shown in Figure 2.9.

In Figure 5.22, there is a range of bias voltage below which no flat band shift is produced. If the plasma charging voltage were within this range, no measurable signal can be detected with the MNOS device. This range is called the dark band. It exist because tunneling current at low voltage is so low that no appreciable net trapped charges were accumulated at the interface of nitride and oxide.

It should be kept in mind that for MNOS devices, after bias is removed, the trapped charges could tunnel back out to the substrate. This detrapping process can introduce uncertainty in the measured results. In fact, poor charge retention is the primary reason why MNOS never did catch on as a memory technology. One can minimize the uncertainty by making the time delay between plasma exposure and measurement the same as the time delay between stress and measurement during the construction of the reference curve.

For the construction of the reference curve, we did not discuss how long the stress should be applied. This is because MNOS capacitors are commonly considered as peak voltage detectors. For a peak voltage detector, the stress duration is of course unimportant. The reason MNOS capacitors are considered a peak detector is that tunneling process through the thin oxide layer controls charge accumulation. For tunneling process, the current level is a very sensitive function of voltage (see Chapter 1). However, for typical MNOS capacitors the oxide is much thinner than the nitride. Most of the applied voltage drops across the nitride layer and the current/voltage relation ship is flattened substantially. So when constructing the reference curve for MNOS capacitors, one must be careful to use a stress time approximately equal the plasma exposure time.

An important consideration to keep in mind when using MNOS capacitors (or any other types of sensors) to measure plasma's charging voltage is the substrate's potential. If the substrate is floating, one must be careful to examine how its potential is affected by any mode of conduction through the dielectric layers. In the presence of UV light, for example, the enhanced conduction through the dielectric in the open area may determine the substrate potential. On the other hand, a broken capacitor due to defects in the dielectric layer may pin the substrate's potential at a value that is completely unpredictable. The interpretation of the result is most straightforward with a grounded substrate. The measured charging voltage distribution is with respect to ground. While it is not what a real wafer would experience, it is possible to predict what a real wafer may experience through modeling.

When use the MNOS capacitors to measure plasma charging voltage, careful planing is required. All the above mentioned problems can render the measurement meaningless. One example of unsuccessful MNOS measurement was reported by Sheu *et al.* [9].

5.4 EEPROM and CHARM®

The first report of using modified EEPROM (Electrically Erasable Programmable Read Only Memory) devices to measure plasma charging voltage was from McCarthy *et al.* [10]. Figure 5.23 shows the structure of the modified device. They call this structure a wafer surface charge monitor (CHARM), which is now a registered trade mark of the plasma charging monitor wafer marketed by Wafer Charging Monitors Inc. Like the EEPROM device, the transistor gate is made of floating-gate/dielectric/control-gate stack. The main difference of the initial CHARM® design from a conventional EEPROM structure is the addition of a large antenna, which they called charge-collecting electrode (CCE), to the control-gate. Instead of using a capacitor structure like MNOS and look for flat band shift in the C-V curve, CHARM® uses transistor and look for threshold voltage shift. Even in its initial form, CHARM® offered improvements over MNOS capacitor. The trade off is that CHARM® requires far more complex processing to make.

The principle of CHARM® is quite similar to MNOS. The main difference is that conduction across the thick dielectric between the control-gate and floating-gate is not by Frankel-Poole mechanism. Rather, it is by tunneling. By keeping the area of the tunnel-oxide small relative to the size of the floating-gate, one can keep tunneling to negligible level everywhere except through the tunnel oxide. Charges tunneled from the substrate through the thin tunnel-oxide accumulate in the floating-gate instead of the nitride/oxide interface in MNOS devices.

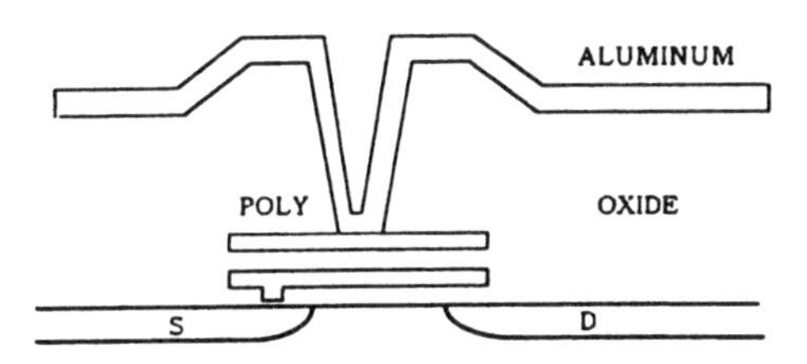

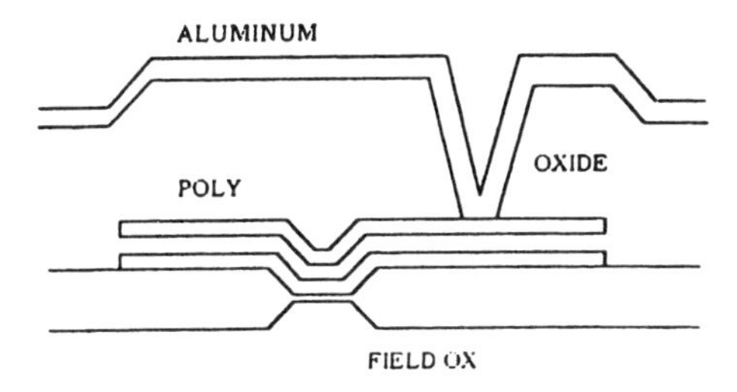

Figure 5.23. Structure of the charge monitor (CHARM) which is a EEPROM with a charge collecting antenna connected to the control gate. Taken from [10].

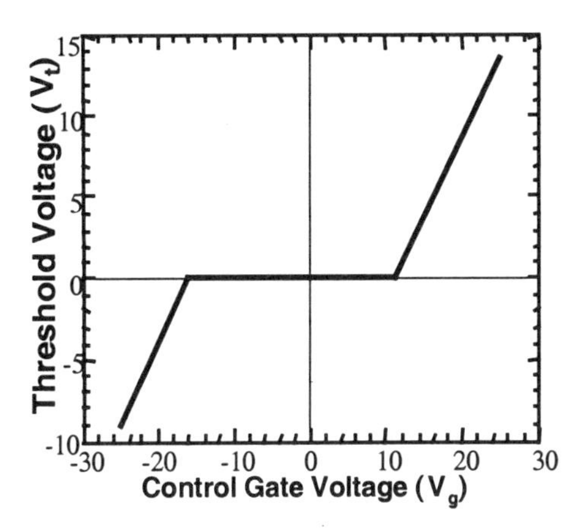

Figure 5.24. Threshold voltage shift in response to control gate voltage stress for a EEPROM or CHARM® device.

A major advantage of the EEPROM design over the MNOS design is charge retention time. With much thicker tunnel-oxide (120Å in McCarthy *et al.*'s original design) than MNOS (20Å seems to be typical thickness), the leakage of trapped charges is much slower. This reduced leakage allows the use of pre-bias technique to improve the voltage detection sensitivity by almost completely eliminating the dark band. Figure 5.24 shows the plot of threshold voltage shift of CHARM® as a function of applied voltage between the control-gate and substrate. Without pre-biasing, there is a rather large dark band that render the device useless except for detecting the extreme cases of charging. However, this dark band can be removed by first inject charges (pre-bias) into the floating gate so that a high field will exist in the tunnel-oxide. When an opposite voltage is applied to the control gate, the field in the oxide will be the sum of the field already there and the field due to the applied voltage. Figure 5.25 illustrates the sequence of positive pre-biasing using the band diagram. Negative pre-biasing is just the opposite. Figure 5.26 shows the full programming characteristic of the CHARM® device. It shows how the device will respond when being biased up and down in full cycle. If the pre-programming (pre-bias) voltage is high enough, a small opposite voltage will cause enough tunneling to produce a threshold voltage shift. If the pre-programming is not high enough, then a small dark band will still exist (dotted curve).

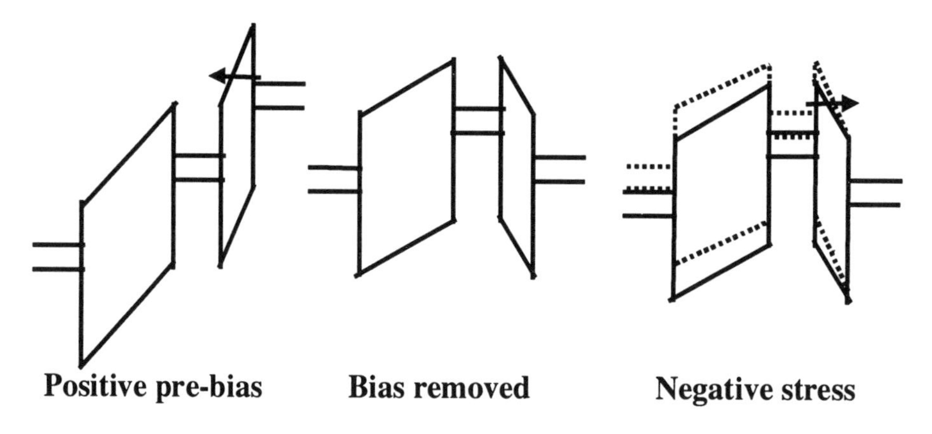

Figure 5.25. Positive pre-bias (pre-program) injects charges into floating gate. After bias is removed, an oxide field exist due to the trapped charges. When a negative voltage is applied, the oxide field is the sum of the applied field and the pre-existing field.

Obviously, to use pre-programming to enhance sensitivity requires two devices to cover the two polarities. One would be positively pre-programmed to sense negative voltage and the other will be negatively pre-programmed to sense positive voltage.

The small tunnel-oxide area offers another advantage, which is that, the field across the tunnel-oxide is very sensitive to the applied voltage at the control gate. Coupled with a thicker tunnel-oxide (Fowler-Nordheim tunneling instead of direct tunneling and therefore tunnel current is much more sensitive to voltage), the

CHARM® device is much closer to being a true peak voltage detector. This characteristic allows the use of CHARM® with less constraint. It is no longer needed to make sure that the programming time for the construction of the reference curve be similar to the plasma charging time. The long retention times also allow a much more lax control on time delays in the programming-plasma-measure sequence. On the other hand, being a peak detector prevents it from sensing the longer-time charging damage events in the presence of a higher voltage transient. It is possible that a longer-time charging event can produce more damage than a very short time but higher stress event.

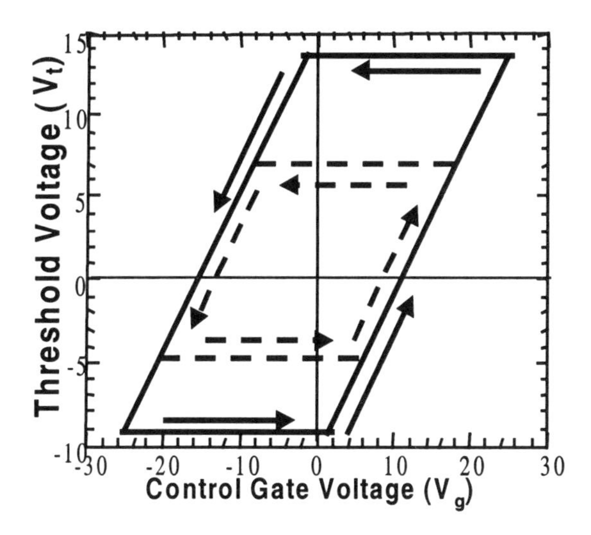

Figure 5.26. Full programming cycle response curve of CHARM®. When fully programmed to saturation at one polarity, maximum sensitivity for sensing voltage of opposite polarity is obtained. When programming is only partway (dotted curve), the remaining dark band is larger.

Figures 5.27 and 5.28 are measured charging voltages of a high-density plasma system using CHARM® wafer. Both maps are produced from the same wafer after exposure to plasma. In each chip, CHARM® devices were pre-programmed in pairs to cover both positive and negative polarities. From these charging voltage maps, one can see that the plasma system being studied is highly non-uniform and therefore highly likely to cause charging damage.

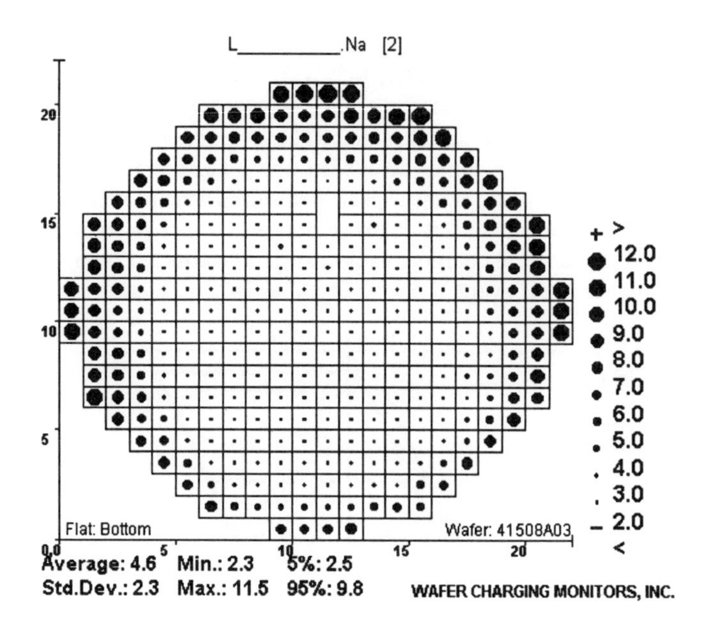

Figure 5.27. Positive charging voltage map of a high-density plasma system as measured by a CHARM® wafer. Courtesy of Wafer Charging Monitors, Inc.

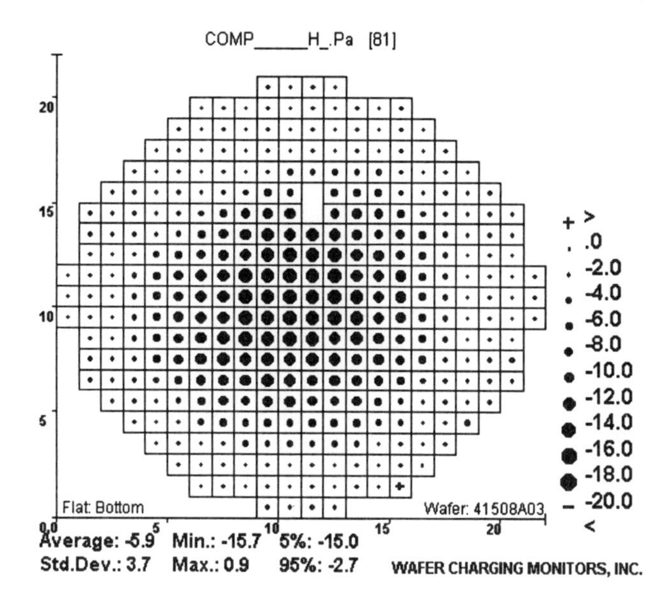

Figure 5.28. Negative charging voltage map of the same high-density plasma system as measured by a CHARM® wafer. Courtesy of Wafer Charging Monitors, Inc.

Knowing only the plasma charging voltage is not enough to determine how serious the charging damage may be. To do that one needs to know what level of charging current can the plasma supply at various charging voltages. By connecting a known resistor between the charge collecting electrode (CCE) and the substrate (Figure 5.29), this question can be answered. The way this method work is that the resistor will clamp the voltage across the CCE and substrate to a level that is limited

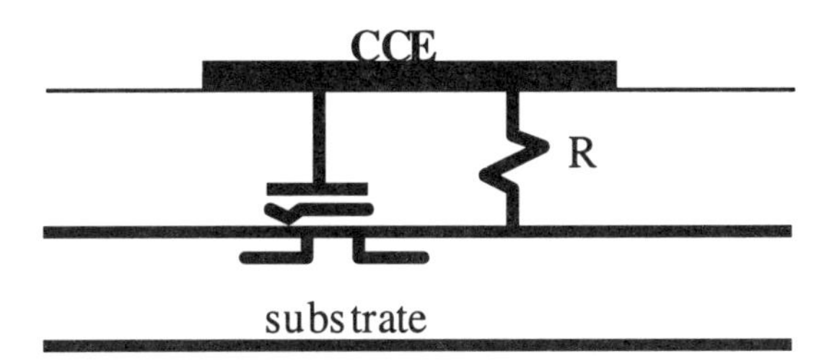

Figure 5.29. CHARM® device with a known resistor connecting the charge collector electrode and the substrate. Courtesy of Wafer Charging Monitors, Inc.

by the current. With a known resistance, the measured voltage can be directly translated into current. With a known CCE size, the current can be converted to current density. This current density is what the plasma would supply at the measured voltage. Using several devices with different resistance, the whole current-voltage (J-V) characteristic of the plasma can be obtained. Since the sensors work in pairs, each chip on the CHARM® wafer thus contains series of paired sensors. Figure 5.30 shows the J-V characteristics of the plasma at peripheral sites in Figure 5.27 and Figure 5.31 shows the J-V characteristics of the plasma at center sites in Figure 5.28.

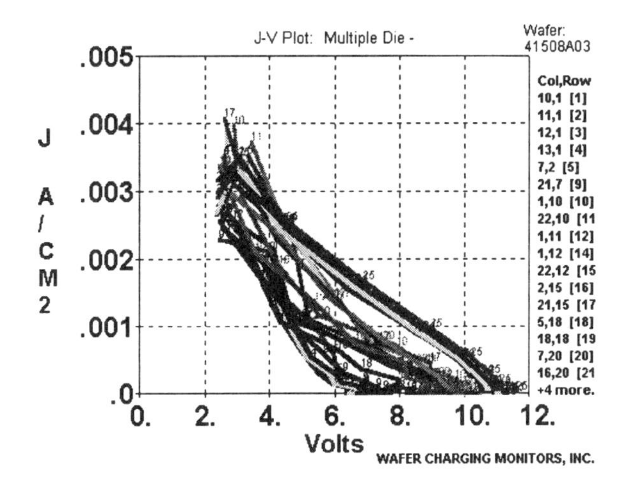

Figure 5.30. Measured current voltage (J-V) characteristics of the plasma at the peripheral sites in Figure 5.27. Courtesy of Wafer Charging Monitors, Inc.

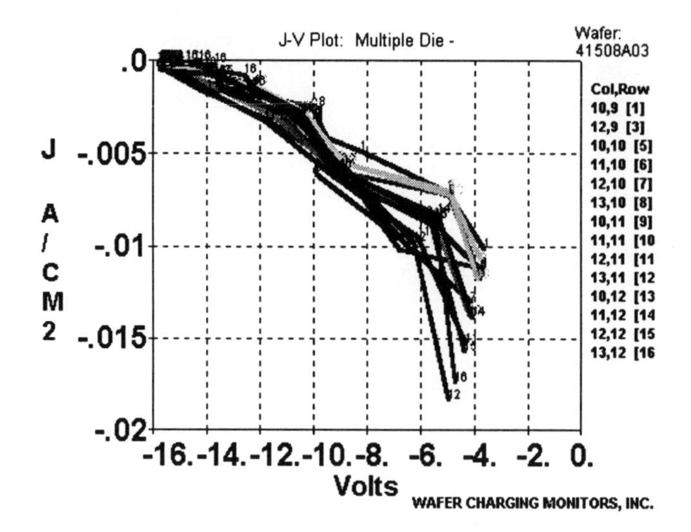

Figure 5.31. Measured current voltage (J-V) characteristics of the plasma at the center sites in Figure 5.28. Courtesy of Wafer Charging Monitors, Inc.

Additional sensor types are also designed in the CHARM® wafer to address more subtle issues relating the plasma charging. For example, plasma charging during a process may have both polarities. In Section 4.4 we have shown one such example (Figure 4.32). This bipolar charging at the same site could confuse the measurement with CHARM® if the bipolar excursion is larger than the hysteresis voltage of Figure 5.26. To deal with this, unipolar sensor design as shown in Figure 5.32 is needed. Again, a pair of sensors one for each polarity is needed.

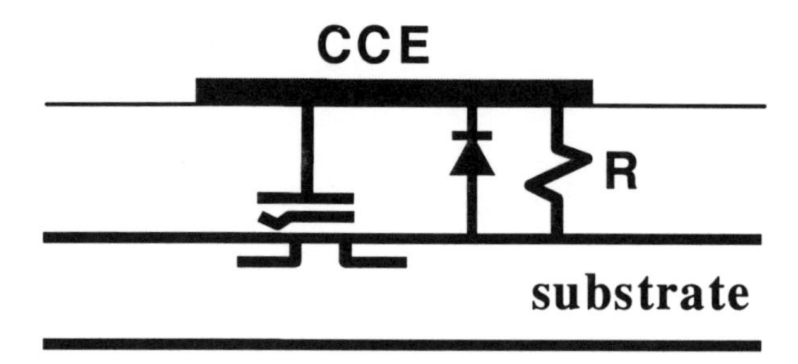

Figure 5.32. Unipolar CHARM® sensor for differentiating bipolar charging. Courtesy of Wafer Charging Monitors, Inc.

UV level in plasma can contribute to damage. This damage is enhanced by the presence of charging (see Section 4.1). Thus it is of interest to measure also the UV intensity distribution. The normal design of CHARM® uses a large CCE not only to serve as charge collector, but also as UV shield. UV irradiation can lead to charge bleeding from the floating gate via mechanisms already discussed in Section 4.1. When not carefully guarded against, this UV induced bleeding of charge will lead to erroneous value for the measured charging voltage. To sense UV, a simple modification of using a small CCE and a small valued resistor (Figure 5.33) will do the trick. The small resistor value assures that the charging voltage will be clamped at very low value. Under such conditions, any change in threshold voltage is due to UV induced charge lost. Although this method does not give an estimate of the absolute UV dose, it does provide a relative dose map. Figure 5.34 is the measured UV intensity map of the high-density plasma of Figures 5.27 and 5.28.

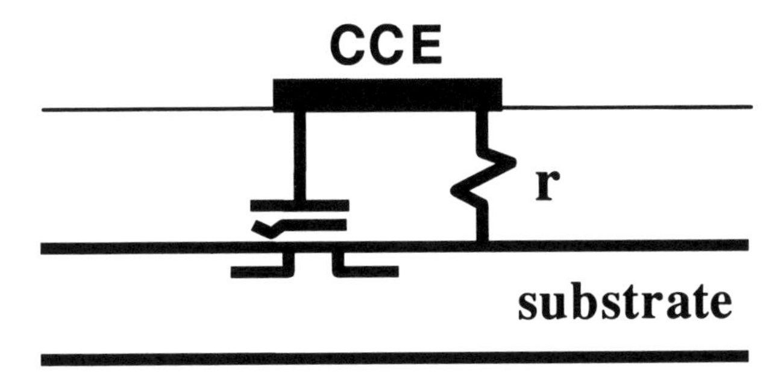

Figure 5.33. CHARM® UV sensor is just a current sensor with particularly small resistance and particularly small CCE size. Courtesy of Wafer Charging Monitors, Inc.

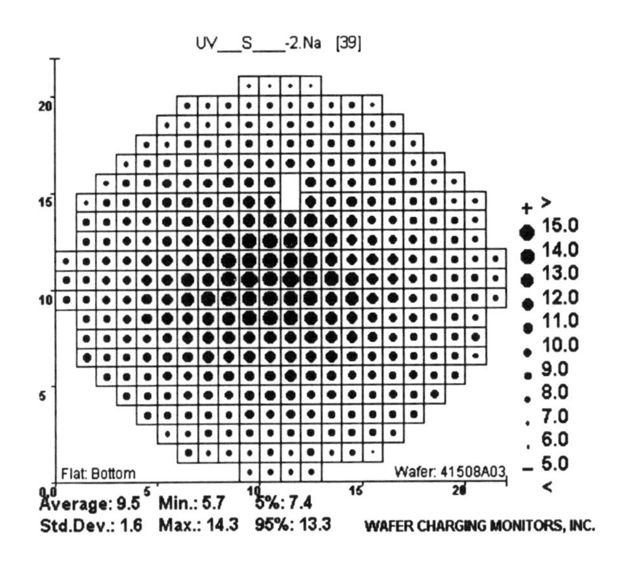

Figure 5.34. Measured UV map for the high-density plasma of Figures 5.27 and 5.28. Courtesy of Wafer Charging Monitors, Inc.

It is clear that CHARM® is a rather extensive set of test devices designed to measure various plasma parameters that are relevant to charging damage. The complexity of its processing and testing is compensated by the fact that it is commercially available. Thus fast turn around experiments to check the health of a plasma system in terms of its propensity for charging damage can be performed. Its ability to measure charging voltage directly and to measured the plasma's response to floating potential change (current-voltage characteristic) is a great plus. Although SPORT probes also allow the plasma's current-voltage characteristic be measured, it do so actively and therefore potentially suffer from disturbing a poorly anchored plasma. Comparing to MNOS capacitors, CHARM® is less prone to errors and provides more information.

5.5 Common Problems with Methods that Measure Plasma Properties Directly

The problems with all the methods discussed so far that measure the plasma's propensity to cause charging damage can be sum up in one sentence, they are not the real product wafers and not measuring real processing conditions. Not that it is impossible to use these methods to measure real processing recipes. The problem is that real processing recipes tend to destroy the sensor wafer after a few uses. Even if one were willing to sacrifice the sensing wafer, the result would still be not quite the same as a real product wafer. The pitfalls of not using real device wafer and not measuring real processing conditions are plenty. Great care must be exercised when using these methods and applying the result to real production situation.

The main problem of not using real device wafers is the substrate potential effect. For example, Hook [11] reported that charging damage during via etching depends on whether the backside of the wafer is bare (conductive) or covered with dielectric film. They found that charging damage happens only to wafers that have a bare backside (Figure 5.35). In another experiment, Lukaszek [12] showed recently that the measured charging voltage and current on CHARM® wafers are quite different when comparing one with scribe lane test devices to one without. Figure 5.36(a) shows the charging voltage map for CHARM® wafer with process monitoring test devices in scribe lanes and Figure 5.36(b) shows the charging voltage map for CHARM® wafer without process monitoring test devices in scribe lanes. Not only are the two maps differ drastically, the current level is also much higher for the high charging voltage points in map (a) as well. This behavior is not a rule, however. Lukaszek [12] also showed that an exactly opposite result is found in another plasma system (Figure 5.37 (a) and (b)). Note that the CHARM® wafers do have low resistance current paths on each site to connect the substrate to the plasma. The big impact of the additional current paths provided by the scribe lane testers is a clear indication of the important role of substrate potential.

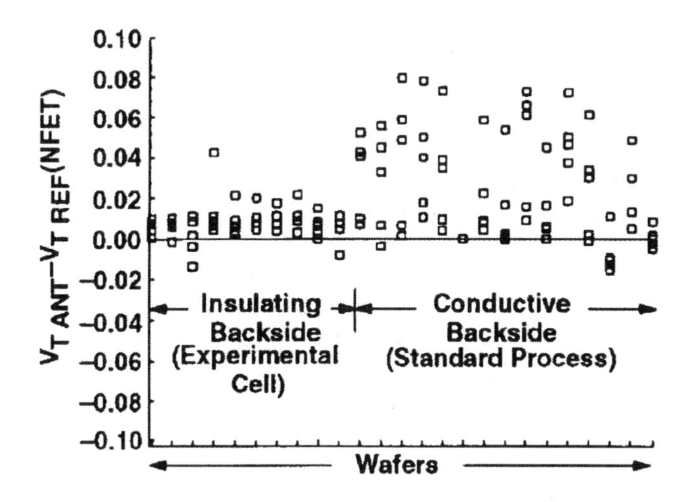

Figure 5.35. Threshold voltage shift due to plasma charging damage in one wafer lot depends on whether the backside of the wafer is conductive or not. Taken from [11].

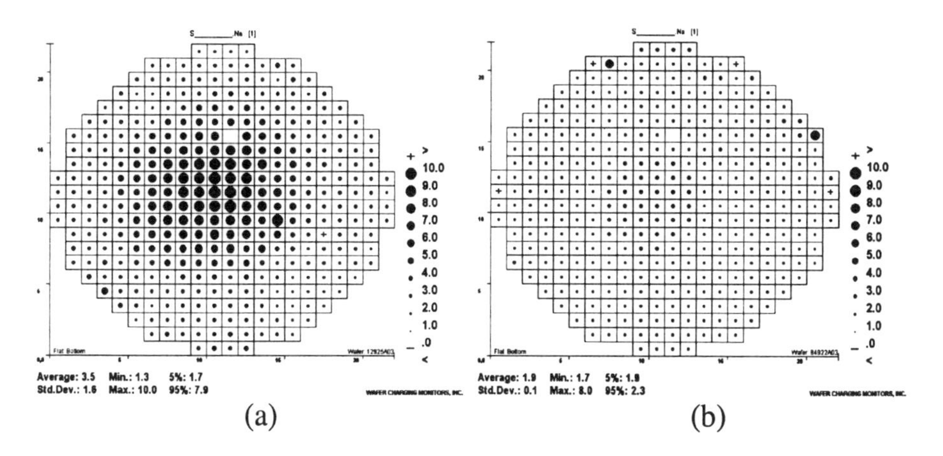

Figure 5.36. (a) Charging voltage map from CHARM® wafer with scribe lane testers. (b) Charging voltage map from CHARM® wafer without testers in scribe lanes. Taken from [12].

Typical CMOS technology wafers have wells and junctions to rectify current flows in the substrate. The effect of each directional current control on the local substrate potential underneath the gate is far more complex than for wafers like MNOS, SPORT or CHARM®. As an example, Lukaszek *et al.* [13] showed how devices in a CMOS structure response differently to CHARM® wafers and how in some cases charging damage may not happen in devices even when CHARM® wafers measure a large charging voltage. Krishnan *et al.* reported [14] that even the subtle difference in well layout can cause major difference in the level of charging damage. Figure 5.38 shows their observed p-MOSFET damage being influenced by the n-Well layout. A simple size change of the n-Well was enough to cause a big change in the damage level.

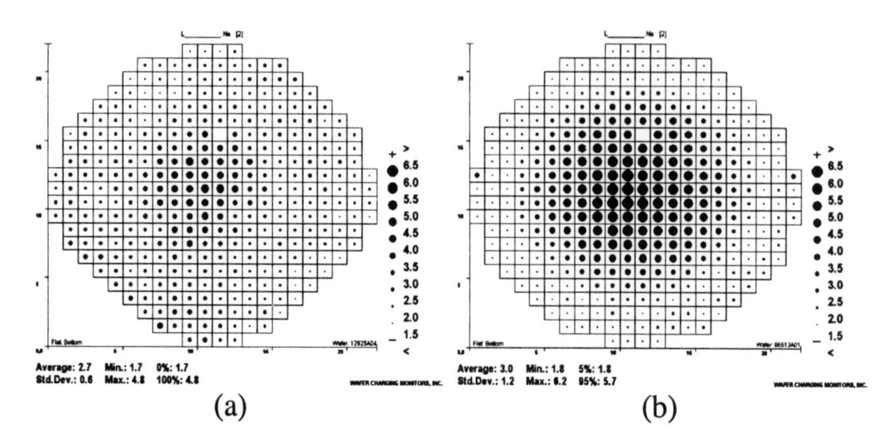

Figure 5.37. (a) Charging voltage map from CHARM® wafer with scribe lane testers. (b) Charging voltage map from CHARM® wafer without testers in scribe lanes. Taken from [12].

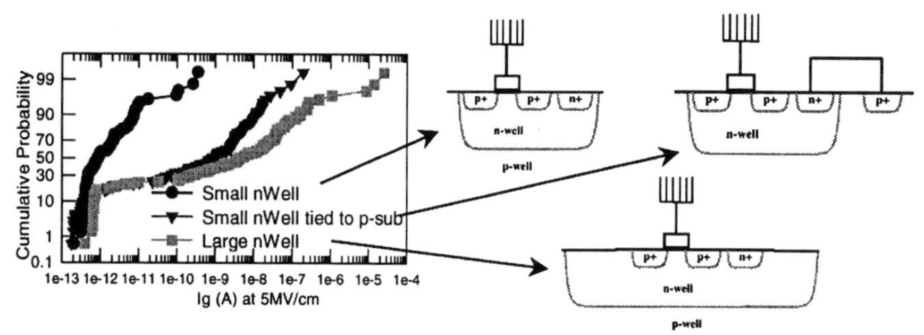

Figure 5.38. Damage level of p-MOS is sensitive to n-Well layout. Small n-Well alone has the least damage. Small n-Well tied to p-substrate has more damage, and large n-Well alone has the most damage. Taken from [14].

The main problem of not measuring under real processing conditions is in the high sensitivity of plasma on processing conditions. In the SPORT probe section of this chapter, we have already seen an example (Figure 5.18) of how sensitive plasma charging behavior is to gases in the plasma. In fact, adjusting these parameters is the main weapon process engineers use to reduce charging damage. For example, Strong *et al.* [15] changes the gas component ratio feeding into plasma to reduce damage in their dielectric deposition process (Figure 5.39).

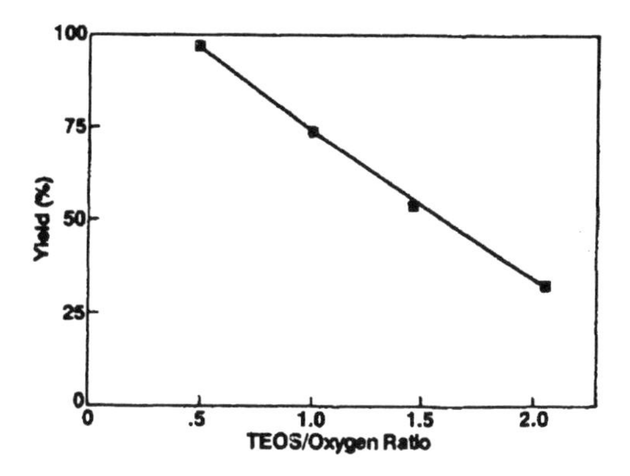

Figure 5.39. Changing the gas ratio in the plasma help reduce antenna test structure yield to improve in a dielectric deposition process. Taken from [15].

A dramatic demonstration of how different the measured charging result can be between measurement done with bare CHARM® wafer and measurement done with resist patterned CHARM® wafers was reported by Lukaszek *et al.* [16]. The resist pattern used in their experiment consisted of various density of large vias (aspect ratio less than 1). When the resist was absent, the charging voltage has a bull's-eye distribution (Figure 5.40 (a)), high charging voltage was recorded at the center only. With the resist, the entire wafer recorded high charging voltage (Figure 5.40 (b)). Since the aspect ratio is low, this high charging voltage was not due to electron-shading effect. It is entirely possible the effect was due to changes in plasma gas composition. Even though the etching recipe was the same, the presence

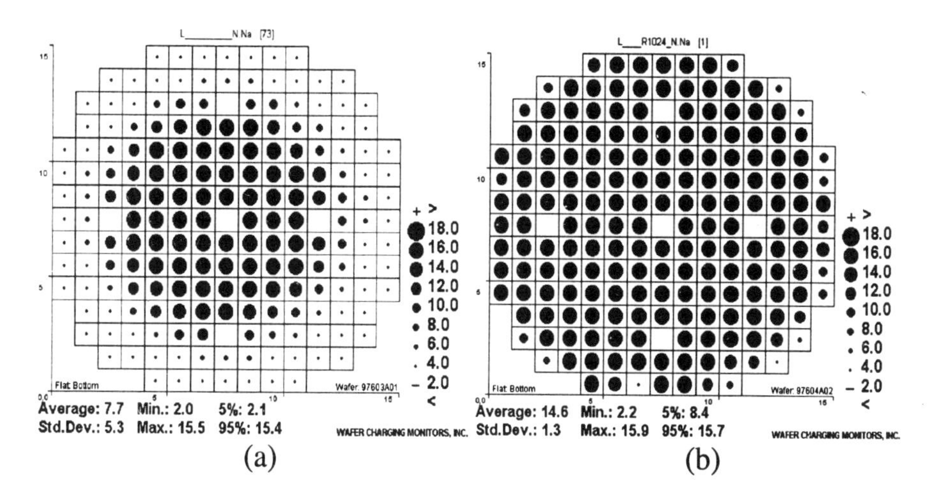

Figure 5.40. Plasma charging voltage recorded from via etching system with (a) bare CHARM® wafer and (b) CHARM® wafer covered with via patterned photoresist. Taken from [16].

of resist allows hydrocarbon compounds in the resist to be introduced into the plasma by the etching process. In manufacturing, the total coverage of resist on top of wafer depends on pattern density, which in turn depends on the product being produced. Different resist coverage means different amount of hydrocarbons being released in the plasma during etching. The charging behavior will also be different. Even if the same etching recipe is used in all products, the amount of damage will not be the same from product to product.

The sensitivity of plasma charging damage to subtle changes in plasma was highlighted in the observation that damage level changes as the number of wafers were processed through the same chamber after a chamber clean. Figure 5.41 shows [17] how in a metal etching process the damage level continued to worsen as the number of wafers were processed after a cleaning.

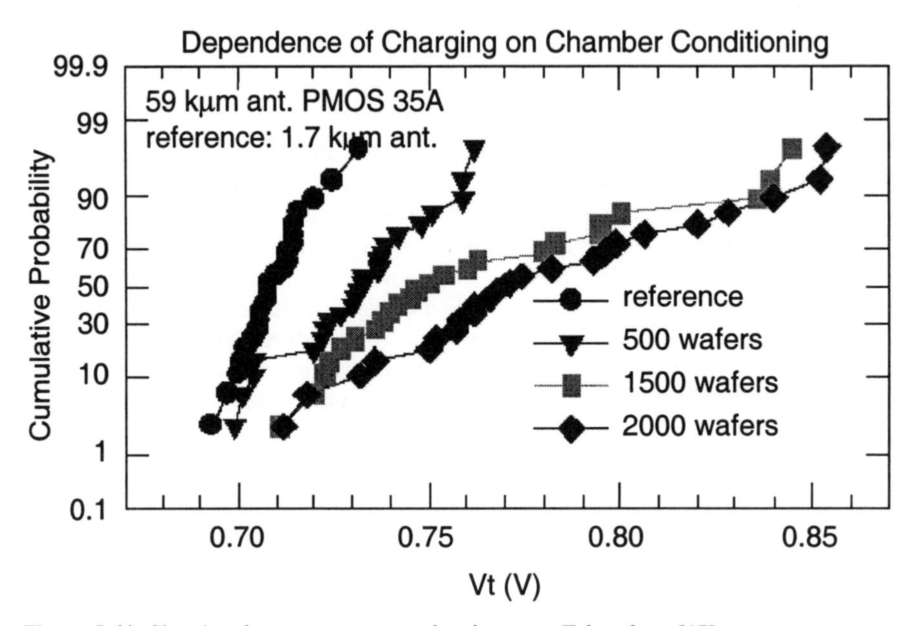

Figure 5.41. Charging damage worsens as chamber ages. Taken from [17].

5.6 Rapid In-line Charge Sensing Methods

In-line testing methods are methods that can be used in the production line for providing rapid feedback during process optimization or monitoring. For plasma charging damage, two methods have been gaining popularity. They are popular because they require only blanket oxide wafers and the testing is simple and quick. These are the Contact-Potential-Difference (CPD) method and the Corona-Oxide-Semiconductor (COS) method. A recent review of these methods was published by

Edelman *et al.* [18]. Although the proponents of these methods like to consider them as charging damage measurement methods, they are included in this chapter instead of next chapter because they are not actually measuring charging damage. At best, they are measuring plasma's propensity for charging damage.

The reason for separating these methods from those discussed in Sections 5.1 to 5.4 is that the measured results of these two methods are not as directly related to charging damage as those covered in Sections 5.1 to 5.4. In fact, the relevancy of most of the measured results to charging damage is highly questionable.

5.6.1 The Contact-Potential-Difference (CPD) Method

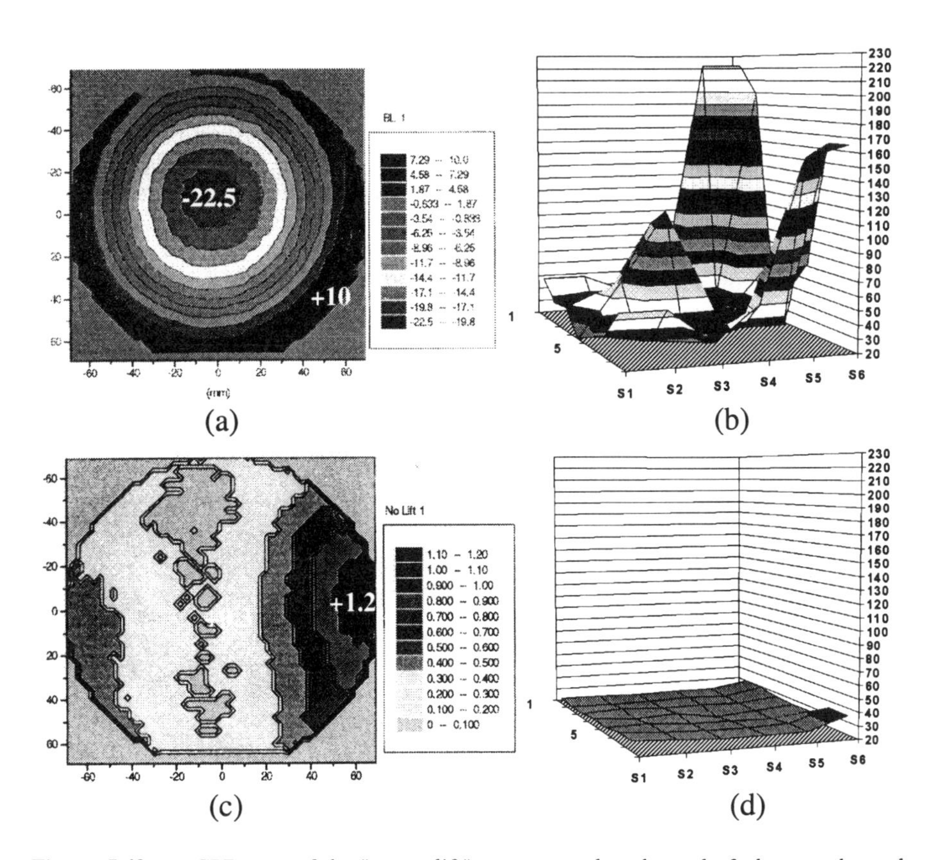

Figure 5.42. (a) CPD map of the "power-lift" process used at the end of plasma enhanced oxide deposition. (b) Charging damage map of the same process measured by the initial electron trapping rate method. (c) CDP map of control. (d) Charging damage map of control.

It has long been known that an oxide-covered wafer would acquire residual charges after exposure to plasma. It is for this reason that some plasma processing tools use "power-lift" method (a short exposure to plasma of different type) to neutralize the residual charge. The CPD method utilizes blanket oxide covered wafer. After exposed to the plasma under study, the surface potential distribution of the wafer due to residual charge is measured using a highly accurate, non-contact Kevin probe. Due to the simplicity of the measurement, highly detailed surface potential map can be generated in minutes.

The proponents of this technique argue that the CPD map is related to the floating potential distribution on the surface of the wafer during plasma exposure [19-22]. The primary support for this belief is from occasional similarity between the CPD map and the charging damage map measured with other well-established measurement methods. One particularly good example is shown in Figure 5.42 [23] for the "power-lift" process at the end of a plasma enhanced oxide deposition process. Here the CPD map (Figure 5.42 (a)) shows a high potential at the center of the wafer. Dropping rapidly outward and becomes highly negative at the edge of the wafer. The corresponding charging damage map is shown in Figure 5.42 (b). The similarity is striking. Figures 5.42 (c) and (d) show the CPD map and the damage map of control wafer. The very low value of surface potential agrees well with the absence of measured damage.

The example shown in Figure 5.42 is probably the best example one can find in the literature, even though there have been numerous reports claiming good correlation between CPD and damage. Careful examination of various reports claiming good correlation would found that it is not clear what parameter is in good correlation with damage. Some would use the magnitude of the surface potential as indicator for serious damage while other would use the gradient of the surface potential as an indicator.

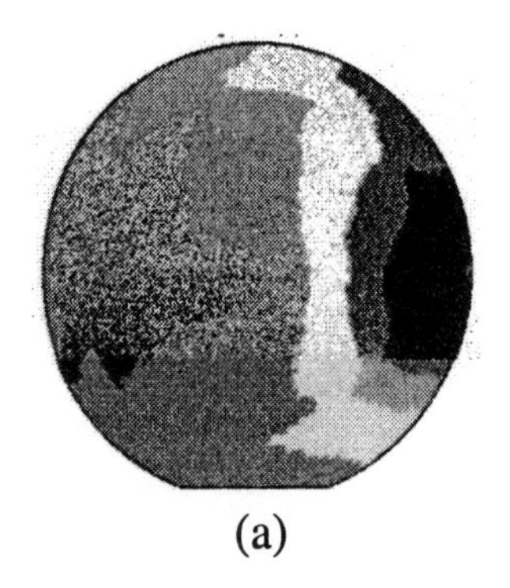

(a)

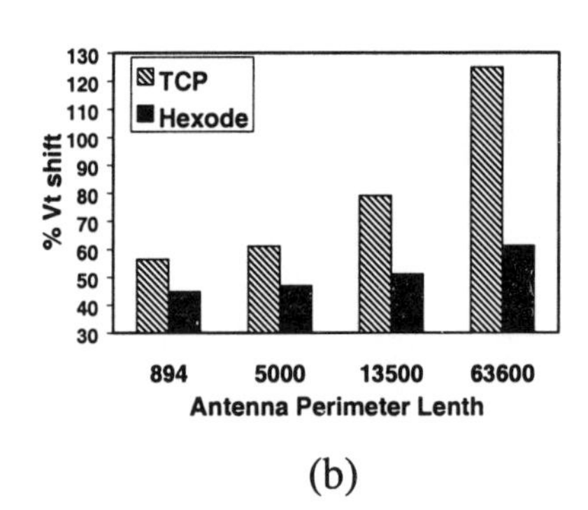

(b)

Figure 5.43. (a) CPD map for main metal etching step in a TCP etcher. (b) Strong antenna dependent threshold voltage shift indicating serious charging damage due to TCP metal etching.

Reports on CPD map that failed to agree with damage measurement are fewer, but they do exist. Figure 5.43 (a) shows the CPD map of the metal etching using a Transformer-Coupled-Plasma (TCP) [23]. The very low and uniform value of surface potential suggests that metal etching in the TCP should be damage free. Figure 5.43 (b) shows the measured charging damage result of the metal etching step. A strong antenna ratio dependent threshold voltage shift is an indication of serious charging damage. Charging damage result from a Hexode etcher is shown also for comparison.

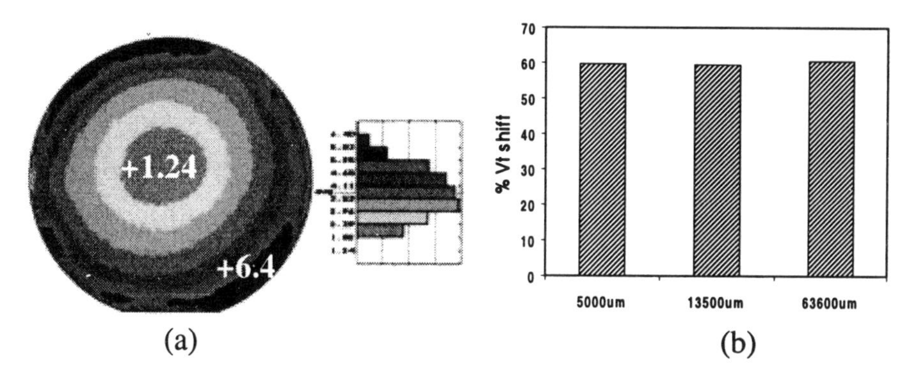

Figure 5.44. (a) CPD map for the resist ashing step. (b) No antenna dependent threshold voltage shift indicating that the resist ashing step caused no charging damage.

Figure 5.44 (a) shows the CPD map for the resist ashing step after metal etching. The higher value and more non-uniform map suggest that the resist ashing step should cause even more charging damage than the metal etching step. Figure 5.44 (b) shows that there is no antenna ratio dependent threshold voltage shift due to the resist ashing step, indicating that charging damage is absent. Figure 5.45 shows the CPD map of the water rinse step after resist ashing. The very large value as well

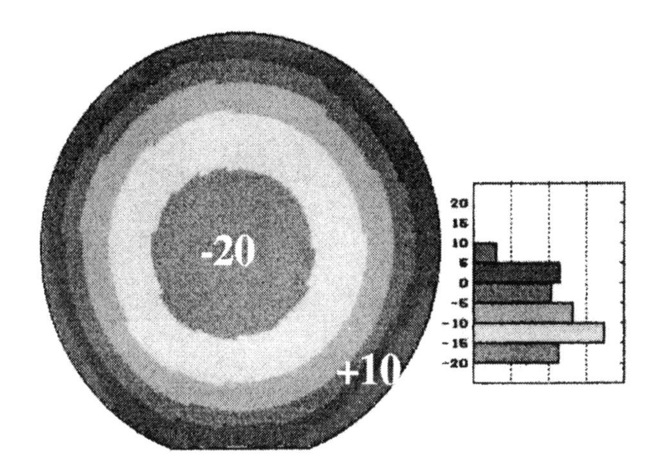

Figure 5.45. CPD map for the water rinse step.

as the very large gradient of change is alarming. It suggests that water rinse should kill most of the devices. This is, of course, not the case. Obviously, CPD map cannot be counted on to determine whether or not charging damage would happen. Recently, Lee *et al.* also reported CPD map leads to wrong conclusion [24].

The CPD method cannot tell where the residual charges are located in the oxide. Its proponents simply assume that the charges are at the surface of the oxide and that these charges are the result of charge imbalance during plasma exposure. It is from this assumption that they argue that the CPD map is a reflection of the floating potential at the surface during plasma exposure. They make this assumption based not on physical model, but on the observation that the map seems to have similar shape as the measured floating potential across the wafer in their experiment [22]. As we discussed in Chapter 2, plasma always adjusts itself to maintain charge balance on everything it contacts. During plasma exposure, all surfaces acquire excess electrons to establish the sheath potential. These excess electrons are neutralized by the in-rush of ions during the extinction of the plasma. The sharp charging voltage transients recorded by SPORT probe during plasma turn-on and turn-off (Figure 5.17) are the results of this in-rush action. During ignition, the plasma supplies excess electrons to establish the sheath. The transient charging potential is due to the difference in electron accumulation rate between probe pad and substrate. This difference arises from the difference in their exposure to plasma (probe pad is fully exposed while substrate is partly exposed). The same difference in exposure also causes the extinction transient. The probe pad collects ions faster than the substrate. However, the fact that the transient fully recovers to zero volt is an indication that the plasma remains active long enough to supply enough ions to fully neutralize the excess electrons. Thus there is no reason to belief that the residual charges measured are due to charge imbalance during plasma exposure. Note that had the plasma fail to neutralize all the charges, the residual charge at the surface should always be negative.

The key to the problem of CPD method is that, no relationship between the CPD map and charging damage has been established. At least nothing based on known physical principle. If CPD map always correlate well with charging damage, then one may be justified to use it for process optimization in the absence of an explanation. However, data indicate that the similarities are exceptions only. The occasional similarity does suggest in some cases that the factors that impact charging damage also contribute to residual charges in the oxide film. A case in point is the example in Figure 5.42. A possible explanation for the similarity has recently been proposed [23]. The proposed explanation, however, cannot be generalized to other charging damage situations. One therefore cannot justify using CPD as a tool for charging damage monitoring or process optimization.

5.6.2 The Corona-Oxide-Semiconductor (COS) Method

The COS method measures, in additional to CPD, the trapped charges in the oxide and the interface state density at the SiO_2/Si interface. It uses a corona source to supply charges to the surface of the wafer. By adjusting the charge concentration at the surface, the surface potential can be controlled without an electrode. The COS method can thus measure the capacitance voltage (CV) curve of the charge-oxide-semiconductor capacitor. From that, the trapped charge and interface state density can be deduced. The methodology is well-established [18]. It is highly useful as an oxide characterization tool.

By providing additional information such as trapped charges and interface state density, proponents of the COS method claim that it really measures damage directly and thus overcome the problem of CPD method. Indeed, some of the CPD method proponents are now admitting that CPD method fails in some situations, and that one must combine CPD and COS to get a true handle on charging damage [18].

In reality, COS method suffers a similar lack of physical explanation problem as CPD method. Proponents of COS method do not address the question of what is causing the trapped charges in the oxide and the interface states at the SiO_2/Si interfaces. Similar to CPD method proponents, they emphasize occasional similarities between their measurement and charging damage to justify the method.

When a blanket oxide wafer is exposed to plasma, there are a few factors that can create trapped charges and interface states. The most important one among them is photons. Plasmas supply a large quantity of photons with energy ranging from UV to VUV. Their efficiency in creating trapped charges and interface states is well-documented [25-31] and should overwhelm all other factors.

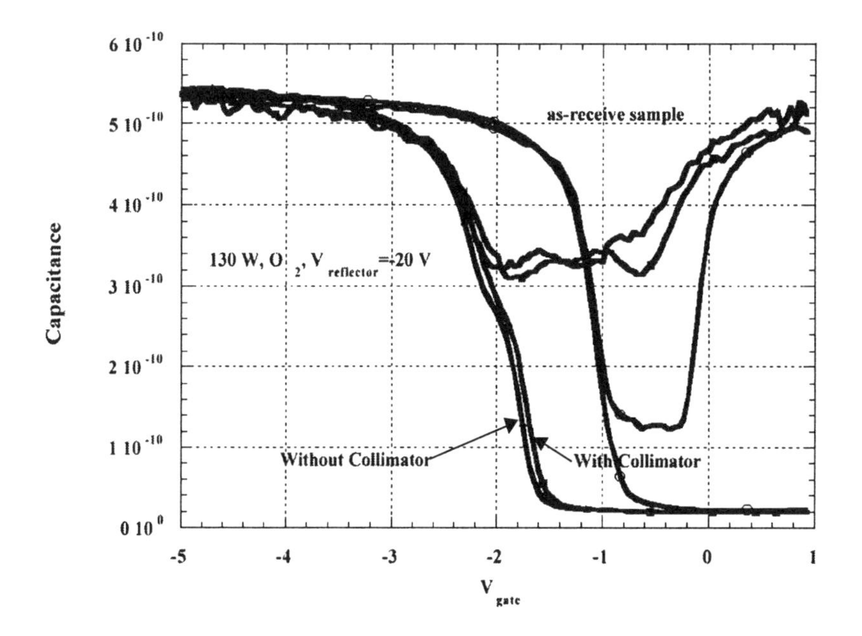

Figure 5.46. CV curves (quasi-static and high frequency) of oxide covered wafer before and after exposure to neutral beam source. Taken from [31].

Figure 5.46 shows the CV curves of oxide covered wafer before and after exposure to neutral beam source [31]. Although the neutral beam source is not completely free of charged particles, serious charging damage is not expected from this kind of source. Yet, the CV curves clearly indicate large amount of trapped positive charges (large negative flat band shift) was created in the oxide. At the same time, very high density of interface state has been created as well. To show that these "damages" are mostly due to photons, the authors inserted a MgF window between the source and the wafer in another experiment under identical condition. They found that the result was unchanged. Since the MgF window block all particles except photons with wavelength longer than 1300Å, thus proved beyond any doubt that the "damage" was due mostly to photons. The authors also performed similar measurement after the wafer was exposed to plasma instead of neutral beam, and found that the amount of trapped charges created was similar to neutral beam.

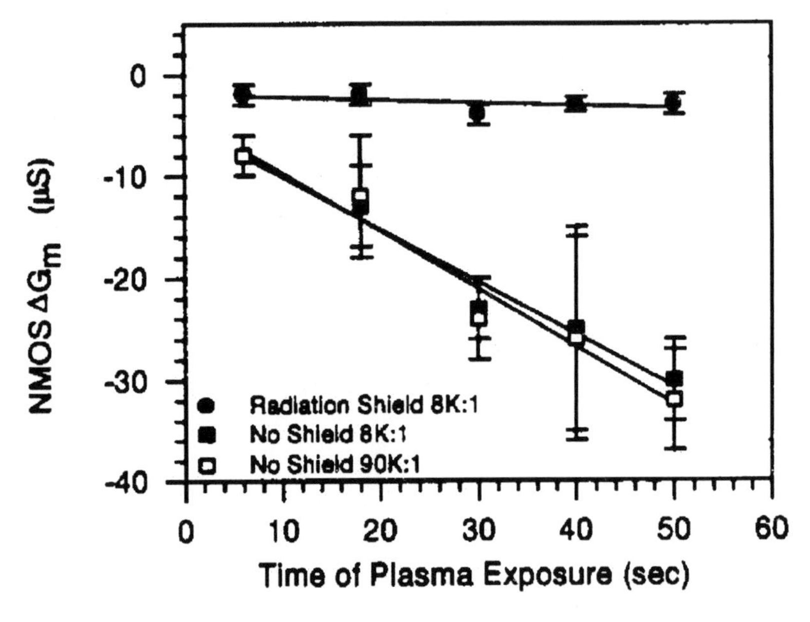

Figure 5.47. Degradation of n-channel transistors is proportional to plasma exposure time, but independent of antenna ratio. UV shielded devices are unaffected. Taken from [32].

It is clear that interface state density of the blanket oxide wafer is related to photon intensity. To optimize plasma processes to achieve a minimum interface state densities in blanket oxide wafer merely results in a recipe that produces the lowest amount of photons. From damage point of view, fewer photons is desirable. For example, Bersuker *et al.* [32] reported transistor degradation due to UV during plasma processing. Figure 5.47 shows degradation of n-channel transistor increases with plasma exposure time, but independent of antenna ratio. Transistors with UV shield (metal shield) are unaffected. In this case, the transistor gate was doped polysilicon that is partly transparent to UV. This mode of damage is not an issue for silicided gate that is more commonly used in advanced technology. This mode of damage is also less important during back end processing. More importantly, the process optimization is supposed to minimize charging damage as well. Simply minimizing UV radiation does not guarantee minimum charging. In fact, there is no physical reason to expect any relationship between charging and UV intensity in plasma. For example, Tao *et al.* [33] reported that while gas chemistry impacts charging damage strongly, UV emission intensity has no relationship with charging damage (Figure 5.48).

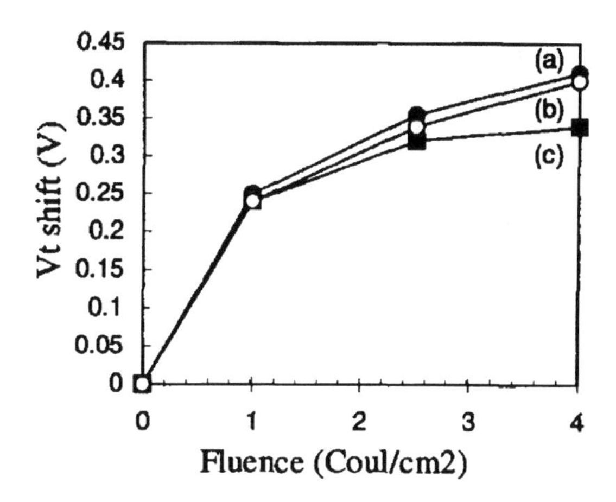

Figure 5.48. Transistor threshold shift as a function of stress fluence for three gas chemistry that have different UV emission intensities. UV emission intensity at 288nm: (a) 80.4, (b) 38.5, (c) 136. Taken from [33].

Another quantity that COS method would optimize is trapped charge. As the main source of trapped charge is UV, minimizing trapped charge also tend to minimize UV. However, the trapped charge is affected by plasma charging to some degree. It is known that UV-created positive charge in oxide can be annealled by charge injection [26]. In fact, when charge injection is large, the trapped charge will change sign (net trapped charge changes from positive to negative). Charge injection can happen during plasma exposure if charging and UV are present at the same time. The mechanism involved was already discussed in Section 4.1. It is the same mechanism proposed [23] to explain the similarity between CPD map and charging damage. This mechanism does not require significant charging to be in effect. For example, the field needed for enhanced internal photoemission is about 10kV/cm, whereas plasma charging damage to gate-oxide requires 10MV/cm. When electrons are injected into the oxide, they neutralize the positive charges that are trapped there. How much neutralization would occur depend on the photon energy distribution as well as polarity of charging voltage. Simply optimizing the processing recipe to obtain a minimum trapped charge is completely meaningless as far as charging damage is concern.

In summary, both CPD method and COS method are not suitable for plasma charging damage monitoring or minimization. It is not to say that they are not useful tools for other applications. They are just not useful for plasma charging damage work.

References

1. Donnelly, V.M., *Plasma Diagnosis Relevant to Damage*, in *Tutorial, International Symp. Plasma Process Induced Damage (P2ID)*. 2000.
2. Malyshev, M.V., V.M. Donnelly, and J.I. Colonell, *Plasma diagnosis and charging damage.* in *International Symp. Plasma Process Induced Damage (P2ID)*. 1999. p. 149.
3. Malyshev, M.V. and V.M. Donnelly, *Trace rare gases optical emission spectroscopy: A non-intrusive method for measuring electron temperatures in low-pressure, low-temperature plasmas.* Phys. Rev. E, 1999. **60**: p. 6016.
4. Malyshev, M.V. and V.M. Donnelly, *Determination of Electron Temperature in Plasmas by Multiple Rare Gas optical Emission and Implications for Advanced Actinometry.* J. Vac. Sci. Technol. A, 1997. **15**: p. 550.
5. Ma, S., *Characterization of plasma processing induced charging damage to MOS devices*, Ph.D thesis, Department of Material Science and Engineering, Stanford University, 1996.
6. Cheung, K.P. and C.P. Chang, *Plasma-charging damage: A physical model.* J. Appl. Phys., 1994. **75**(9): p. 4415.
7. Ng, K.K., *Metal-nitride-oxide-semiconductor transistor*, in *Complete guide to semiconductor devices*. 1995, McGraw-Hill: New York. p. 329.
8. Kawamoto, Y., *MOS Gate Insulator Breakdown Caused by Exposure to Plasma.* in *Dry Process Symp.* 1985. p. 132.
9. Sheu, B.R., *et al.*, *Charging Damage Evaluation in 0.25-0.3um Polycide Gate Etching Processes.* in *International Symp. Plasma Process Induced Damage (P2ID)*. 1998. p. 76.
10. McCarthy, A.M. and W. Lukaszek, *A new wafer surface charge monitor (CHARM).* in *IEEE Int. Conf. Microelectronic Test Structure*. 1989. p. 153.
11. Hook, T.B., *Backside Films and Charging During Via Etch in LOCOS and STI Technologies.* in *International Symp. Plasma Process Induced Damage (P2ID)*. 1998. p. 11.
12. Lukaszek, W., *Influence of Scribe Lane Structure on Wafer Potentials and Charging damage.* in *International Symp. Plasma Process Induced Damage (P2ID)*. 2000. p. 26.
13. Lukasezk, W., M.J. Rendon, and D.E. Dyer, *Device Effects and Charging Damage: Correlations Between SPIDER-MEM and CHARM-2.* in *International Symp. Plasma Process Induced Damage (P2ID)*. 1999. p. 200.

14. Krishnan, S., *et al.*, *Antenna Device Reliability for ULSI Processing*. in *International Electron Device Meeting (IEDM)*. 1998. p. 601.
15. Strong, A.W., *et al.*, *Gate dielectric integrity and reliability in 0.5-um CMOS technology*. in *IEEE International Reliability Physics Symp. (IRPS)*. 1993. p. 18.
16. Lukaszek, W., J. Shields, and A. Birrell, *Quantifying Via Charging Currents*. in *International Symp. Plasma Process Induced Damage (P2ID)*. 1997. p. 123.
17. Krishnan, S., *et al.*, *Inductively coupled plasma (ICP) metal etch dmage to 35-60A gate oxide*. in *International Electron Devices Meeting (IEDM)*. 1996. p. 731.
18. Edelman, P., *et al.*, *Contact Potential Difference Methods for Full Wafer Characterization of Si/SiO2 Interface Defects Induced by Plasma Processing*. in *SPIE Conference on In-line Characterization Techniques for Performanace and Yield Enhancement in Microelectronic Manufacturing II*. 1998. p. 126.
19. Nauka, K., J. Lagowski, and P. Edelman, *Surface Photovoltage and Contact Potential Difference Imaging of Defects Induced by Plasma Processing of IC Devices*. in *DRIP*. 1995. p. 281.
20. Shohet, J.L., K. Nauka, and P. Rissman, *Deposited Charge Measurements on Silicon Wafers After Plasma Treatment*. IEEE Trans. Plasma Sci., 1996. **24**(1): p. 75.
21. Hoff, A.M., T.C. Esry, and K. Nauka, *Monitoring plasma damage: A real-time, noncontact approach*. Solid State Technol., 1996. **July 1996**: p. 139.
22. Cismaru, C., *et al.*, *Relationship between the charging damage of test structures and the deposited charge on unpatterned wafers exposed to an electron cyclotron resonance plasma*. Appl. Phys. Lett., 1998. **72**(10): p. 1143.
23. Cheung, K.P., *et al.*, *Is Surface Potential Measurement (SPM) a Useful Charging Damage Measurement Method?* in *International Symp. Plasma Process Induced Damage (P2ID)*. 1998. p. 18.
24. Lee, M.Y., *et al.*, *Comparison of CHARM-2 and Surface Potential Measurement to Monitor Plasma Induced Gate Oxide Damage*. in *International Symp. Plasma Process Induced Damage (P2ID)*. 1999. p. 104.
25. Buchanan, D.A. and G. Fortuno-Wiltshire, *Vacuum ultraviolet radiation damage in electron cyclotron resonance and reactive ion etch plasmas*. J. Vac. Sci. Technol. A, 1991. **9**(3): p. 804.
26. Druijf, K.G., *et al.*, *Recovery of vacuum ultraviolet irradiated metal-oxide-silicon systems*. J. Appl. Phys., 1995. **78**(1): p. 306.
27. Afanas'ev, V.V., *et al.*, *Degradation of the thermal oxide of the Si/SiO2/Al system due to vacuum ultraviolet irradiation*. J. Appl. Phys., 1995. **78**(11): p. 6481.
28. Druijf, K.G., *et al.*, *Nature of defects in the Si-SiO2 system generated by vacuum-ultraviolet irradiation*. Appl. Phys. Lett., 1994. **65**(3): p. 347.

29. Yunogami, T., *et al.*, *Radiation Damage in SiO2/Si Induced by VUV Photons.* Jap. J. Appl. Phys., 1989. **28**(10): p. 2172.
30. Ling, C.H., *Trap generation at Si/SiO2 interface in submicrometer metal-oxide-semiconductor transistors by 4.9eV ultraviolet irraditation.* J. Appl. Phys., 1994. **76**(1): p. 581.
31. Tang, X., Q. Wang, and D.M. Manos, *Process-induced damage by a low energy neutral beam source.* in *International Symp. Plasma Process Induced Damage (P2ID).* 1999. p. 116.
32. Bersuker, G., *et al.*, *Transistor degradation due to radiation in a high density plasma.* in *International Symp. Plasma Process Induced Damage (P2ID).* 1998. p. 231.
33. Tao, H.J., *et al.*, *Impact of Etcher Chamber Design on Plasma Induced Device Damage for Advanced Oxide Etching.* in *International Symp. Plasma Process Induced Damage (P2ID).* 1998. p. 60.

Chapter 6

Charging Damage Measurement II - Direct Measurement of Damage

Plasma charging damage has the ability to cause a multitude of changes in device and circuit operation. In principle, any of these damage-related changes can be used to develop a plasma charging damage measurement method. For example, plasma charging damage increases the 1/f noise in the transistor drain current. This increase in the 1/f noise would be a useful parameter to indicate damage. In fact, if one were making products that are very sensitive to the low-frequency noise, then the noise itself would be the necessary metric for plasma charging damage and would be the obvious measurement method of choice. In reality, however, the measurement methods in most cases are chosen based on convenience, not necessity.

Most of the plasma charging damage measurements reported in the literature aimed at detecting the existence of charging damage, not quantifying it. Such measurements that monitor the relative changes are useful for process optimization or technology monitoring. However, they generally cannot be used to set up design rules or to determine the impact of damage on product yield and reliability. For this purpose, **absolute** quantitative measurement methods are needed. While most of the measurement methods can be made quantitative, one can find few reports that make the attempt in the literature. Consequently, while the charging damage measurement methods covered in this chapter are in principle better than those in the previous chapter by virtue of actually measuring the damage, they still fall short of being adequate and are sometimes even less reliable. Each method covered here needs careful evaluation of its usefulness. The practical utility of a method often depends on a variety of factors such as measurement speed, oxide thickness and device size.

Many of the device characterization measurement methods discussed here are well established in the semiconductor industry. The details of exactly how to carry out the measurements covered here will not be discussed, for doing so would require a book of its own. A good source for the measurement details on many of the methods covered here is the book by Schroder [1]. This chapter will focus on the application of standard device characterization methods to the evaluation of plasma charging damage.

6.1 Measurement Challenge

The measurement of plasma charging damage can be quite straightforward from the point of view of studying the damage mechanism. Typically, one is studying a particular plasma system and the goal is to detect damage with enough sensitivity to tell whether changing a certain plasma parameter is improving the damage or worsening the damage. The number of devices measured in this type of work is usually limited and the researcher can afford to use a slow measurement method as long as it provides good sensitivity.

The measurement of plasma charging damage is a daunting task from the point of view of monitoring charging damage during the development cycle of a new IC technology generation or during manufacturing. To understand why, we have to look at what is needed to accomplish the task. In modern IC manufacturing, the number of plasma processing steps ranges from 25 to 50 depending on the number of interconnect layers utilized in the technology. If we include those charging steps that do not involve plasma, the number is even larger. Our task is to make sure that all of these steps are at a sufficiently low damage level so that the yield and reliability of the final products are not affected.

To be sure that what we are detecting is indeed due to charging, we need to determine if an antenna effect is present. To do that we need a minimum of 3 antenna ratios per test device type. Since we do not know if the n- and the p-channel transistors will response similarly to charging damage, we need both types of test devices. Charging damage is sometimes sensitive to antenna area, sometimes sensitive to antenna perimeter length and sometimes sensitive to both. We therefore need antenna designs that cover both types. The hidden antenna effect due to RIE-lag (Chapter 4) requires us to include transient-fuse-linked antenna. The electron-shading effect requires us to include varying aspect ratio antennae. Window etching and related damage demand varying the window count. To provide some kind of reference, we also need diode-protected devices. If we were to check for localized charging problems, we would need additional designs that connect the antenna to the source, the drain or the well of the transistor [2].

In principle, we need all these antenna-attached devices for each plasma processing step in order to provide us with the ability to detect all the charging damage as well as to pinpoint where it comes from. This means that ~30 to 50 antenna devices per plasma step, or ~1000 to 2000 antenna test devices per chip are needed, clearly an outrageous number. The usual strategy to limit the testing need is to group the plasma processing steps into zones and then rely on experimental designs to further pinpoint the offending steps. Even then the number of antenna devices tested each time is easily a few hundreds per chip. This may seem like a large number, but it still doesn't cover all the potential problems.

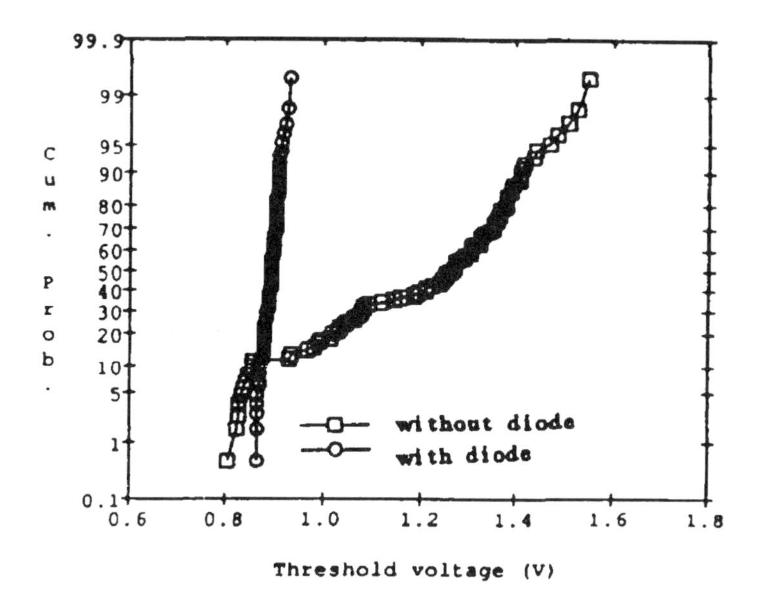

Figure 6.1. The range of n-MOSFET threshold voltage distribution increases in the presence of charging damage. For comparison, the diode-protected n-MOSFET has a tight distribution. Taken from [3].

In addition to having large device count per chip, numerous sites on each wafer need to be measured. This is because plasma charging damage tends to be non-uniform across the wafer. Indeed, one telltale sign of the charging damage is the increase in the standard deviation of the measured transistor parameters. Figure 6.1 shows that the threshold voltage range of n-MOSFETs in the presence of charging damage is very large whereas the diode-protected devices have a tight distribution [3]. Figure 6.2 shows the typical signature of the MERIE (Chapter 4) charging damage [4]. A ring shape region of very large yield loss is observed at the edge of the wafer for the high magnetic field (50G) MERIE case while small yield loss is observed uniformly across the wafer when the magnetic field is lowered to 30G. Figure 6.3 shows the measured threshold voltage (V_t) shift for p-MOSFETs after a FN stress. The highly non-uniform distribution of V_t shift indicates a highly non-uniform damage across the wafer. All of these examples illustrate the need to map out the damage distribution across the wafer. Even if we don't measure every chip on the wafer, we need to measure enough of them to produce a reasonably precise map. The damage map is a very important tool for characterizing and diagnosing the charging damage step. Thus we must measure on the order of 40 to 60 chips per wafer.

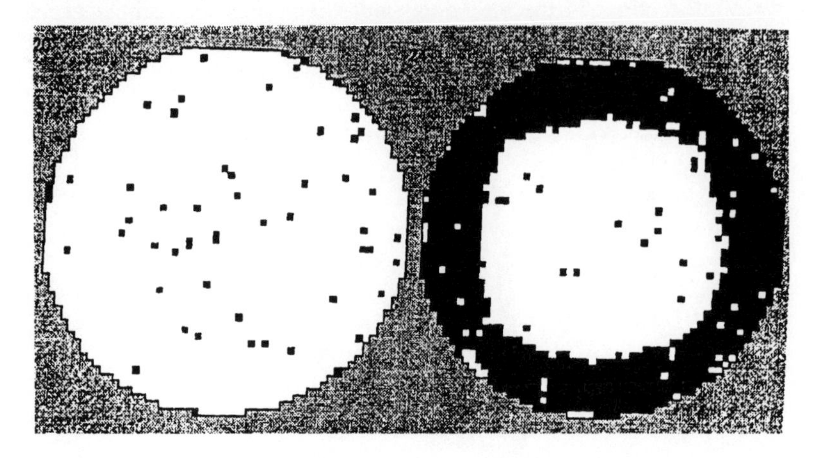

Figure 6.2. Metal-oxide-Metal capacitor yield map shows a characteristic ring shape high damage region at the edge (right side) of MERIE. The ring disappears when the magnetic field is reduced from 50G to 30G. Taken from [4].

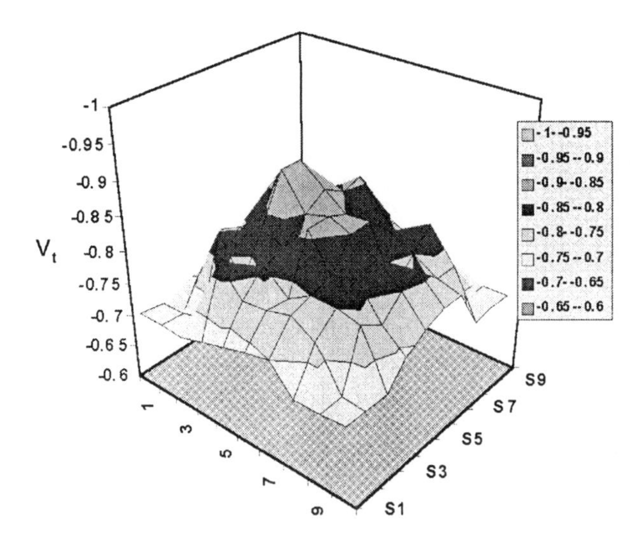

Figure 6.3. Measured p-MOSFET threshold voltage map after FN stress to reveal latent damage. The sharp peak represents high damage region. The damage is clearly highly non-uniform across the wafer.

Another issues further compounds the large number of measurements required for the charging damage assessment: statistical variation. Plasma charging damage is a high-field stressing of the gate-oxide during plasma exposure. Defect generation and oxide breakdown is a statistical process. In order to accurately assess the extent of the damage, repeated sampling of identically processed devices is needed to provide good statistics. Since each site on the wafer needs to be treated generally as having different damage level, we must rely on measuring the same site on many wafers to obtain significant statistics for each site. If the interest is merely to confirm that the damage signal is real and that the relative level of damage is assessed with some degree of confident, 3 to 5 wafers per split is sufficient. In the next chapter, we will be discussing quantitative yield and reliability projection of products from damage measurement data. We will see, for that purpose, much more accurate assessment of the damage level is required. Thus we must measure every wafer on each lot.

We now can see the full scale of the problem. If we test 300 antenna devices (a rather moderate number) per chip and 60 chips per wafer, we would be testing 450,000 devices per 25-wafer lot. If each test requires 2 seconds to complete, we need 10.4 days just to test one lot on one automated test system. In order to have practical measurements, two things are immediately apparent: we need to be selective on what antenna devices to test and we need very fast test methods. It is from this consideration that test speed weights heavily in the discussion of testing methodology.

6.2 Test Devices

6.2.1 Capacitors

A simple capacitor is the most commonly used test device for charging damage detection. It is also the earliest used [5]. For the simple reason that it is easy to make and easy to test. The major drawback is that it is not the main devices used in circuits and hence does not go through the same processing as the main devices. By main device we mean the CMOS devices that make up the building blocks of the integrated circuit.

For the two-terminal device, one of the terminals is usually the silicon substrate while the other is the gate. Antenna structures, either the area intensive type or perimeter intensive type are usually constructed as part of the gate structure. An important complication to watch out for is the series resistance, particularly on the substrate side during an inversion test. Insufficient minority carrier levels may cause a large potential drop across the substrate. The series resistance issue is more severe when the gate oxide is thin. Sometimes even the gate side can be a problem if it is not metallized.

There are two basic types of capacitors, those that terminate on field (thick) oxide (FOX) and those that terminate on gate (thin) oxide (TOX) (Figure 6.4). In general, a capacitor terminated mostly on TOX is preferred because it resembles the situation in a transistor. The exposed gate-oxide edge is vulnerable to ion bombardment and UV attack during plasma processing, particularly when the front-end plasma processing steps are being evaluated. On the other hand, one may want to use capacitors terminated on FOX when evaluating the back-end plasma processing steps to avoid the complication of ion bombardment that is not relevant to the back-end process.

Figure 6.4. (a) Capacitor with gate terminating on FOX. (b) Capacitor with gate terminating on TOX.

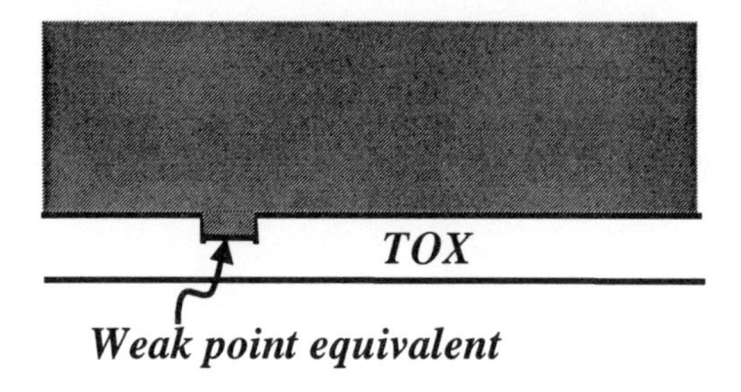

Figure 6.5. A weak point in the gate-oxide is equivalent to a highly localized thinner section and thus turns the capacitor into an antenna structure with a giant antenna ratio.

In general, one should be careful using a very large (TOX area) capacitor. Any weak points in the gate-oxide will be amplified by the "giant" antenna effect (Figure 6.5). This "giant" antenna effect was demonstrated clearly by Shin *et al.* [6]. Figure 6.6 shows the capacitor structures used in their experiment. All three structures have the same overall TOX area of $3x10^5\mu^2$. Structure (a) is one large continuous capacitor while structure (b) and (c) are capacitor arrays. The difference between (b) and (c) is that the capacitors in (b) remain separated during plasma etching of aluminum and plasma ashing of photoresist. Both plasma processes were proven to cause charging damage. All the capacitors in (b) were connected together with another layer of aluminum that was defined by wet etching. Figure 6.7 shows the breakdown voltage distribution of all three structures. Clearly, the capacitors that were connected together during plasma processing have much poorer results due to the "giant" antenna effect. Weak points that cause the "giant" antenna can be from either the TOX or from the isolation edge. In this example, the capacitors were terminated on FOX. The "Giant" antenna effect will distort the real charging damage signal by overwhelming the antenna ratio dependent signal.

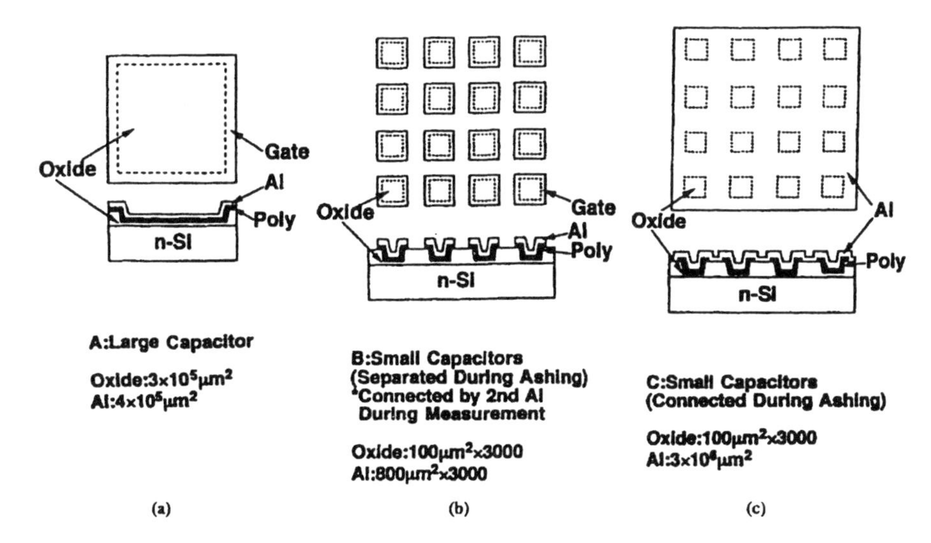

Figure 6.6. Capacitor structures with equal TOX area. (a) is one large continuous oxide area, (b) is an array of small capacitors that are separated during plasma processing, (c) is an array of capacitors that are connected together during plasma processing. Taken from [6].

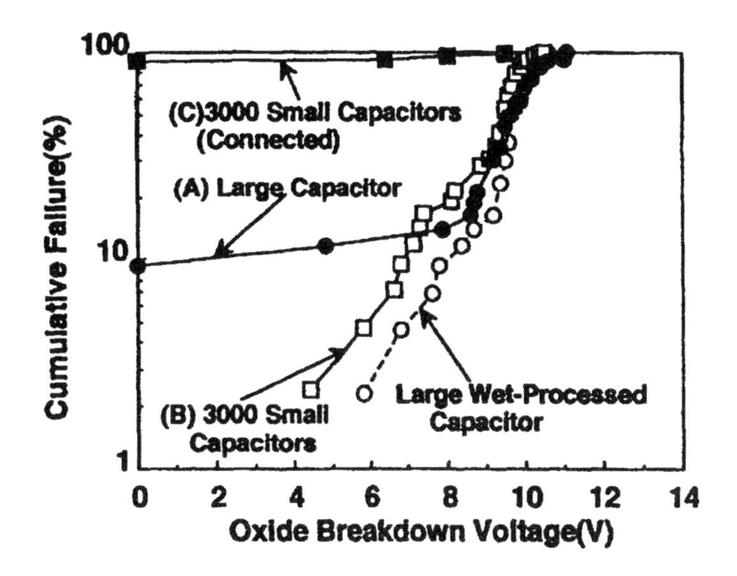

Figure 6.7. Owing to the "giant" antenna effect, capacitors that were connected together during plasma processing have a much higher failure rate. Taken from [6].

Due to the "giant" antenna effect, one should avoid designing the various antenna ratio capacitors by changing the TOX area. Instead, one should keep the TOX area constant and change the antenna area. In addition to the "giant" antenna effect, area-scaling effect will also complicate the measured result when the TOX area is not constant. If it is desired to incorporate an oxide area dependent charging damage measurement, a very careful intrinsic oxide reliability characterization should be performed prior to investigating the plasma charging damage.

6.2.2 Transistors

A transistor is more appropriate than a capacitor to use as charging damage tester by virtue of being the "real thing", particularly when it is built exactly the same way as the IC technology. Using the transistor, there are many different ways to detect damage. One has the option of choosing the method that is important for the technology at hand. The most common design for a transistor antenna test device is to have the antenna attached to the gate. An example of such an antenna-attached device was shown in Figure 2.29. However, as mentioned in Section 6.1, a more complete set would have included antennae attached to source, drain and well, with each antenna situated at a location that may either cancel each other or maximize any local potential non-uniformity [2].

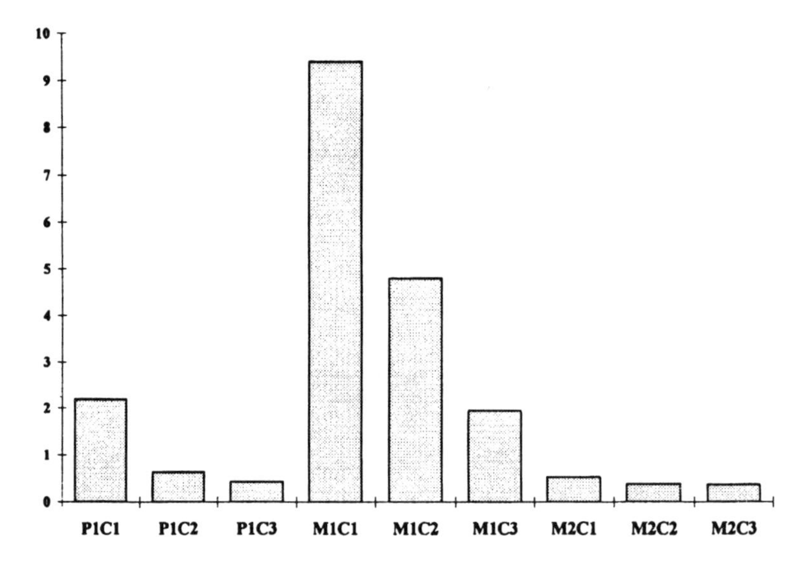

Figure 6.8. Level of damage as measured by electron trapping slope method for various antenna devices on wafers finished metal 2. Vertical scale is relative magnitude of IETS. P1, M1 and M2 are polysilicon, metal 1 and metal 2 respectively. C1, C2 and C3 are large, medium and small antenna ratio respectively.

Usually, antenna attached transistors are designed as a set that contains antennae from various conductor levels. The set needs to provide the ability to pinpoint the damaging step as well as the extent of damage. In other words, the set needs to provide a hierarchy for damage detection. Figure 6.8 shows an example of how such hierarchy of test structures can help pinpoint the damaging step. In this case, all the wafers were tested after metal 2 level processing and annealing in forming gas were completed. Since the metal 1 and metal 2 stacks are identically processed, a negligible antenna ratio dependent signal on the metal 2 devices indicates the damage was not due to metal etching. Both the polysilicon antennae and metal 1 antennae were subjected to plasma enhanced dielectric deposition and both showed an antenna effect suggesting damage was caused by this process. However, the effect is much smaller on the polysilicon antennae, consistent with the fact that there was a 800ºC re-flow step after the dielectric layer was deposited onto the polysilicon antennae. The combined response of the different antenna types plus the relevant processing knowledge allows the damage step to be pinpointed to the plasma-enhance dielectric deposition process.

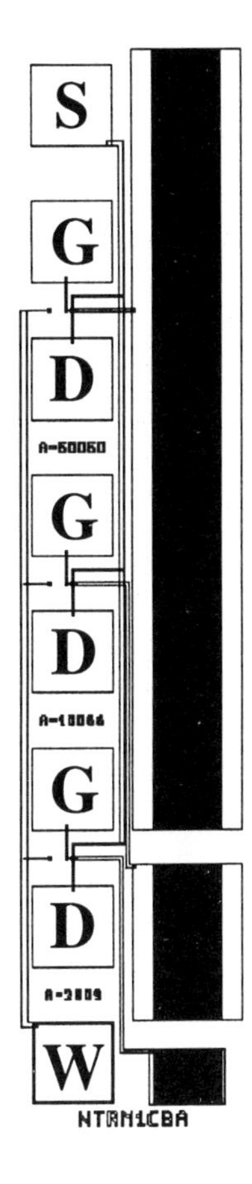

Figure 6.9. Antenna transistor test structure design. Dark area is high-density line and space that represent a large perimeter length. S, G, D and W stand for source, gate, drain and well contact pads. Each of the three transistors requires its own gate contact.

Paul Aum introduced the hierarchy antenna design approach to SEMATECH in 1994 when he was an assignee from Hewlett-Packard. A set of antenna-attached transistors named SPIDER (SEMATECH Process Induced Damage Effect Revealer) was designed and publicized [7]. As a result, many companies use the SPIDER or similar designs in their plasma charging damage work. Figure 6.9 is an example of a similar design. In this case, three transistors with three different perimeter intense antennae are clumped together in one tester. They share a common source and a common well contact pads. Each has its own gate and drain contact pad. This independent gate contact pad design differs from normal transistor tester design where all the transistors share common source, well and gate contact pads. The independent gate contact is of necessity for antenna structures, but it also consumes more contact pads (and therefore spaces), which is undesirable when so many test devices are needed.

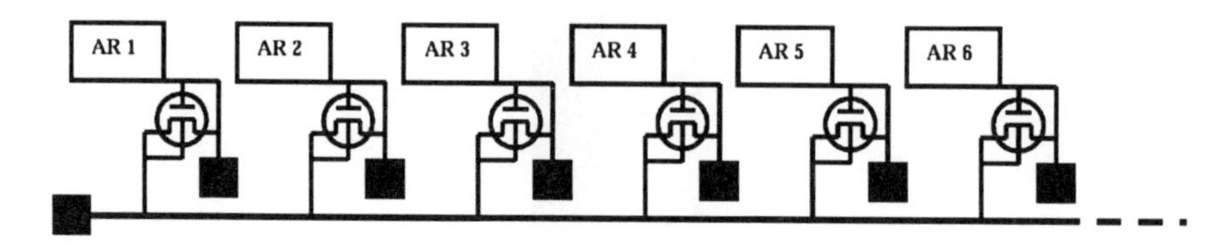

Figure 6.10. Two terminal transistor test structure design. In addition to one common well contact pad, each transistor has one contact pad for both the gate and drain. Taken from [8].

In order to help limit the space required, a two terminal transistor test device design was recently introduced by Brozek [8]. Figure 6.10 shows his design. The source and well contact are all tied together in one contact pad. For each transistor, the gate and the drain are tied together in one contact pad as well. The key to this design is that the gate remains unconnected to the drain during all processing steps except the last layer of metal. This way the drain connection does not become a protection diode during plasma exposure at lower levels. This approach does save contact pads and therefore allows more antenna-devices to be fit in the same space. There is one important drawback to this design; devices cannot be tested until the full process flow is completed. Since the gate and the drain are tied together, the testing method is also different. Most of the normal testing modes is not available for this two terminal design. The two terminal transistor design take advantage of the fact that transistor current in the on-state is large enough that even the smallest transistor can easily be tested (the gate leakage mode of testing is the exception). One can use a very small transistor so that even when a large antenna ratio is needed, the overall size of the test structure can still be small. The main factor that usually limits the number of antenna transistors placed on a chip is the finite probe pad size. Thus Brozek's method to reduce the number of pads per transistor save significant space. Taking that idea to the extreme is to allow a large number of transistors to share the same probe pads by multiplexing using shift registers. Figure 6.11 shows one such approach [9]. In the space of a standard grid tester (2x12 pad), Simon *et al.* managed to cram 1024 transistors consist of 95 different kinds of antennae. Again, the trade-off here is both limited testing modes as well as testing only after the full process flow is completed.

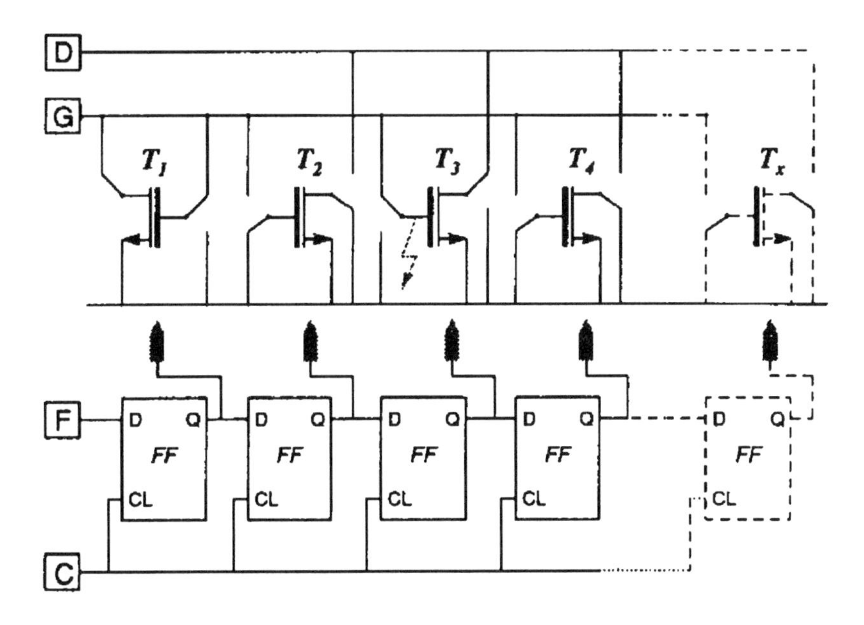

Figure 6.11. Multiplexing using shift registers allows the same probe pad to be shared by large number of transistors. Taken from [9].

6.3 Breakdown Tests

6.3.1 Breakdown Test for Oxide > 50Å

The breakdown test is the earliest test used for plasma charging damage measurement [5]. Figure 6.12 shows the experimental results of Watanabe *et al.* They used simple capacitors with 400Å thick thermal oxide, and the breakdown voltage distribution (presumably measured by V-ramp) method as an indication of damage. They found that for wafers etched in the plasma directly, most of the capacitors have zero breakdown voltage (dead on arrival (DOA)). Such a dramatic signal obviously can convince people that the problem is a serious one.

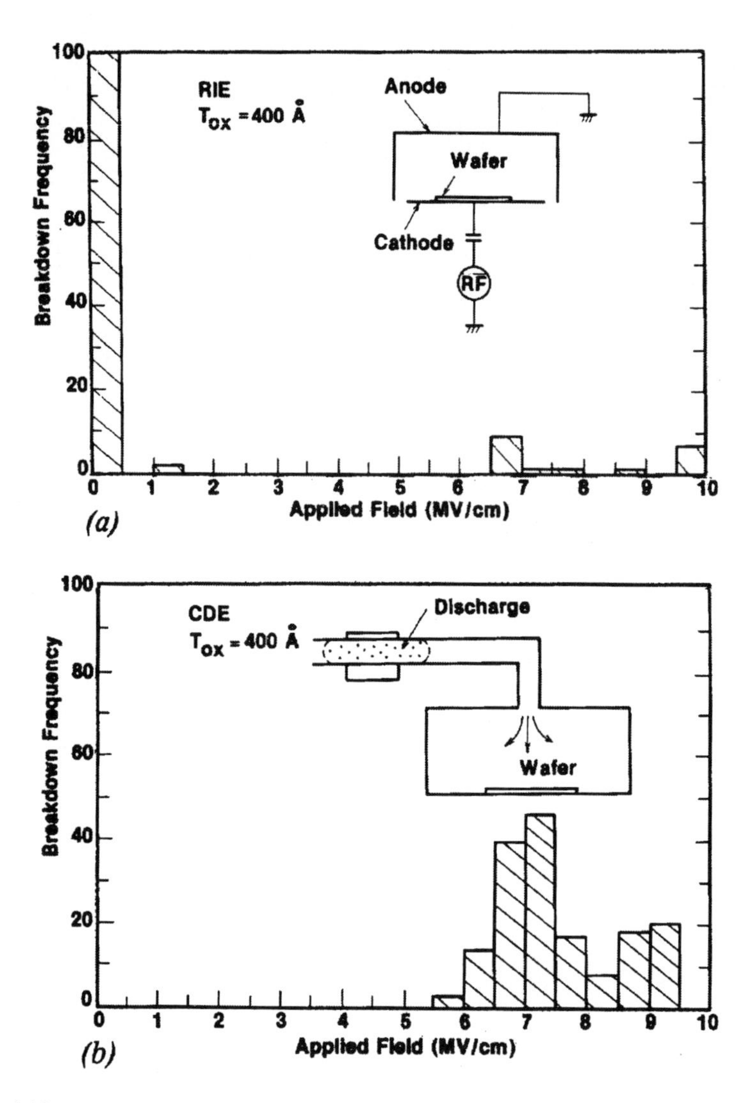

Figure 6.12. (a) The breakdown voltage distribution when wafer is etched in plasma directly. (b) The breakdown distribution when wafer is etched by a remote plasma. Taken from [5].

Breakdown measurement was the popular method in the early history in plasma charging damage. Various breakdown measurement methods were employed. For example, Wu *et al.* [10] showed the T_{BD} distributions (Figure 6.13) and the Q_{BD} distributions (Figure 6.14) measured with the constant voltage TDDB method (see Chapter 1) degraded drastically (compared to the minimally processed control devices) by the time the wafers received one level of metallization. Using the distributions to determined the T_{50} (median time to breakdown T_{BD}), they further

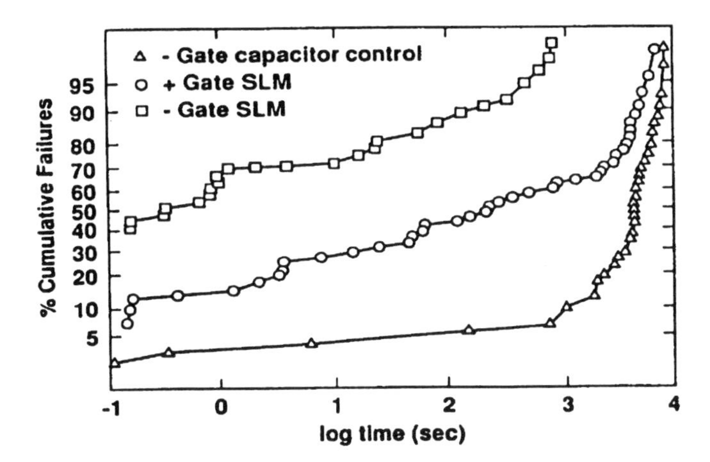

Figure 6.13. The time to breakdown distribution as measured by TDDB method showing gate-oxide degradation by the time wafers were finished at the first level of metal. Taken from [10].

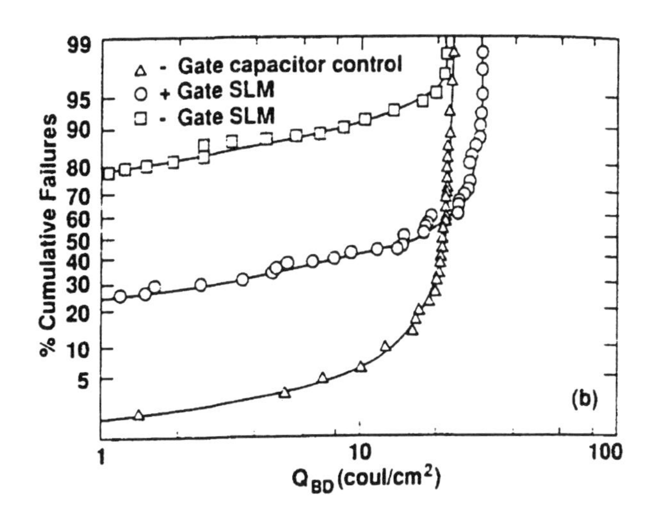

Figure 6.14. The charge to breakdown distribution as measured by TDDB method showing gate-oxide degradation by the time wafers were finished at the first level of metal. Taken from [10].

showed that the situation gets worse by the time the wafers received metal 2 processing (Figure 6.15). They also showed, in the same study using the V-ramp method, some of the earliest examples of the antenna effect. In Figure 6.16, the results from 0.5mm^2 capacitors with various sized metal pads attached to the gate are shown. Compared to the results from 0.01mm^2 capacitors with the same metal pad antennae (Figure 6.17) they found that while the antenna effect can still be seen, the breakdown distributions exhibited very little degradation even though the antenna ratio was 50 times larger. They correctly attributed this observation to the extrinsic defects in the gate-oxide. This is an early example of the "giant" antenna effect discussed above.

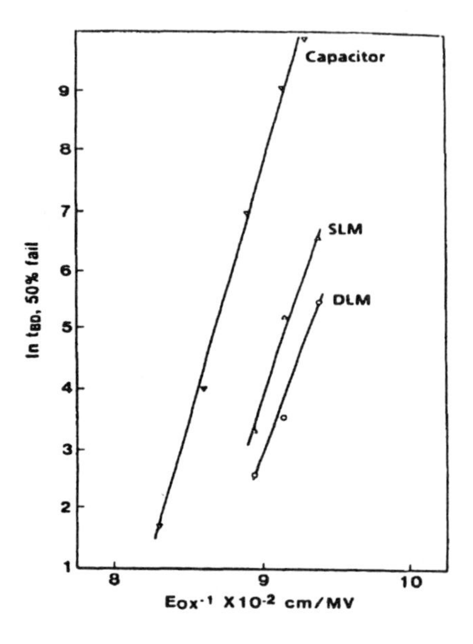

Figure 6.15. The stress field dependent T_{50} clearly indicated degradation is more severe by the time wafers received the second level of metal. Taken from [10].

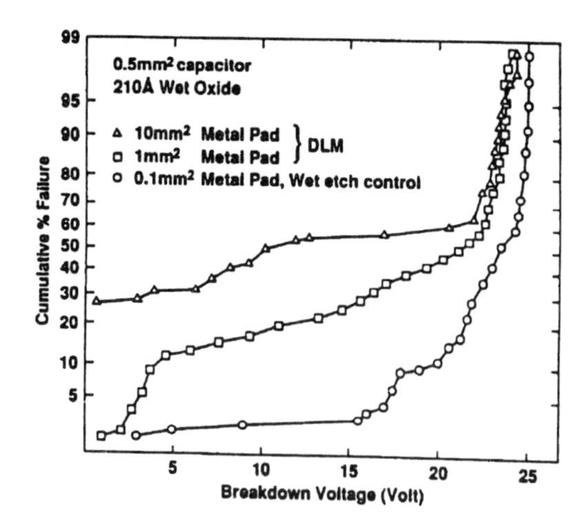

Figure 6.16. Breakdown voltage distributions from V-ramp test show antenna effect. Taken from [10].

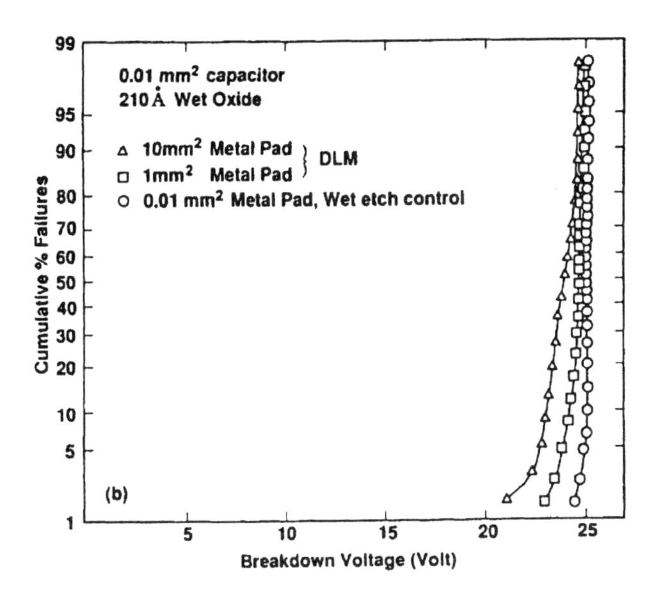

Figure 6.17. Smaller capacitors show much lower degradation even though the antenna ratio is 50 times larger, showing "giant" antenna effect. Taken from [10].

Most of the plasma charging damage studies using the breakdown method is qualitative only. In other words, the relative shift of the breakdown distribution is used as an indication of oxide degradation due to charging damage, but not as a quantitative measure of damage. This need not be the case. Eriguchi *et al.* [11] demonstrated that quantitative measurement of charging damage is possible when data from a range of antenna ratios as well as a control is available. The concept is illustrated in Figure 6.18. For a given capacitor size and gate-oxide thickness, the damage-free T_{BD} of a constant current stress (20mA/cm^2 was used in their experiment) can be determined from a set of control samples. The shorter T_{BD} measured from a set of damaged antenna capacitors are interpreted as the residual lifetime of the damaged oxides. The difference between the damage-free T_{BD} and the residual T_{BD} is the T_{BD} consumed by the plasma charging stress. Since Q_{BD} of a constant-current-stress TDDB measurement is simply the product of current density and T_{BD}, the plasma-injected charge can be obtained from the consumed T_{BD} and the stress current density.

An example of how to use this method is provided in Figure 6.19 [11]. They define the Q_{BD} of the device-under-test (DUT) as residual Q_{BD}, which is calculated from the stress current and T_{BD}. The residual Q_{BD} is the same as initial Q_{BD} for the control (antenna ratio =1). They plot the residual Q_{BD} as a function of antenna ratio for two overetch durations. Both sets of data are fit to straight lines using the least square method. The fact that the longer overetch line has almost twice the slope of the shorter overetch line proves that the damage happens during overetch. They can then use the linear fit to calculate the plasma stress current by realizing that the difference in residual Q_{BD} between a particular antenna ratio and control is simply the product of stress current, antenna ratio and stress time. For example, the residual Q_{BD} for the 3000:1 antenna ratio from the 27s line is about 4 C/cm^2 less than the control. The plasma stress current at the wafer surface is simply $4/(3000x27)=49\mu A/cm^2$.

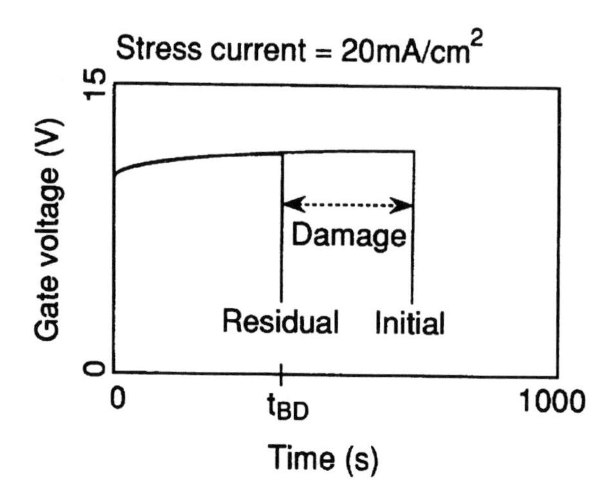

Figure 6.18. Illustration of the concept of quantitative damage measurement using constant current stresses to breakdown the oxide. If the initial (without damage) t_{BD} is known, then the measured t_{BD} is the initial t_{BD} minus the lifetime consumed by damage. Taken from [11].

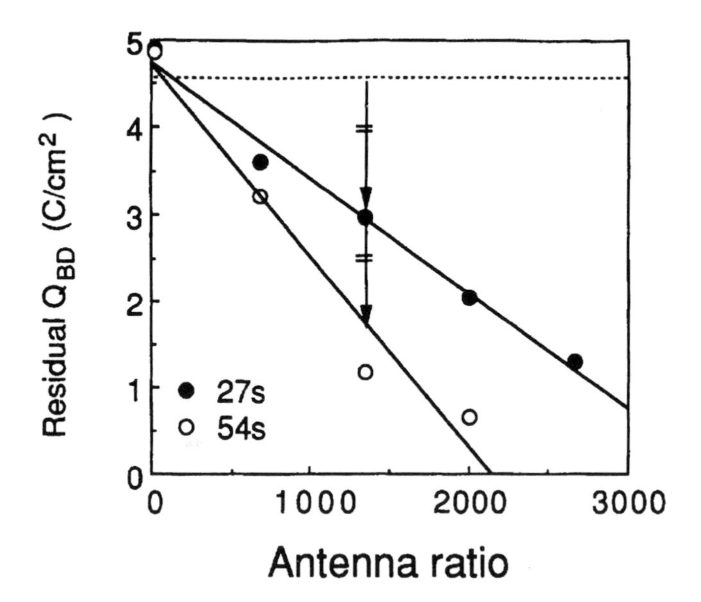

Figure 6.19. The measured residual Q_{BD} as a function of antenna ratio for two overetch durations allow the plasma stress current be determined. Taken from [11].

For a qualitative evaluation of plasma charging damage, any method employed for breakdown measurement is useful as long as it indicates a relative level of damage. However, for quantitative measurements, some methods are far better than others. For example, Kosaka *et al.* [12] showed that constant current TDDB is much better than constant voltage TDDB method. Figure 6.20 shows the Q_{BD} from constant current TDDB as a function of antenna ratio for both positive (substrate injection) and negative gate (gate injection) stress. A monotonic decrease in Q_{BD} as antenna ratio increases is observed as expected.

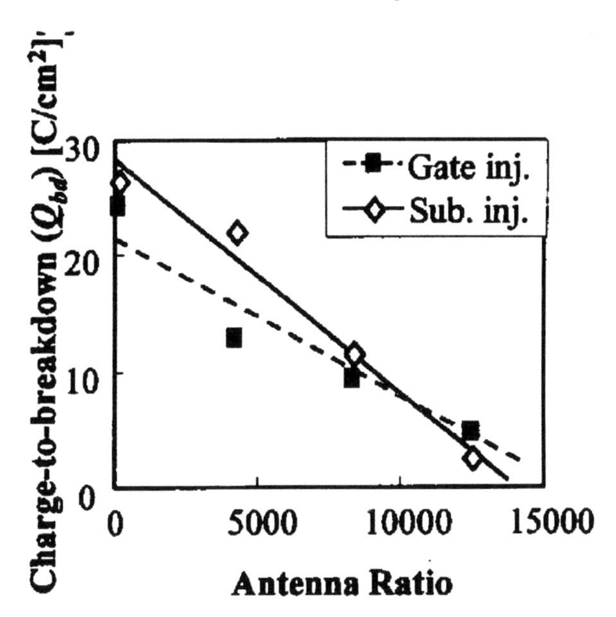

Figure 6.20. The Q_{BD} as a function of AR behave monotonically. Taken from [12].

Figure 6.21 shows the T_{BD} from constant voltage TDDB as a function of antenna ratio for the same plasma process. For the substrate injection stress, the T_{BD} is clearly non-monotonic as a function of antenna ratio. Even for the gate injection case, the trend is not very clear. They explain their result as a consequence of charge trapping in the oxide, which causes the stress current to change during the constant voltage stress. This effect was discussed in Chapter 1 and the current as a function of time during constant voltage stress was shown in Figure 1.9. The reason for the gate injection case being different from substrate injection, they argued, was that the distribution of traps created by plasma charging damage is not uniform in the oxide. Note that the voltage as a function of time during constant current TDDB is also changing due to charge trapping. However, the total injected charge to breakdown remains unaffected by this voltage variation because the time needed to breakdown the oxide will be affected as well. The bottom line is, the percolation model requires a critical trap density for breakdown to occur. For thick oxide

(>50Å) at high stress field, the trap creation per injected charge is not sensitive to the stress voltage.

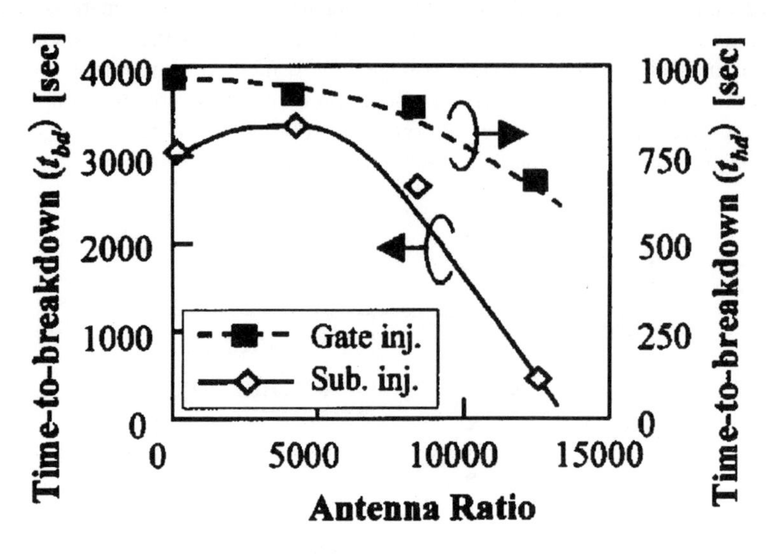

Figure 6.21. The T_{BD} as a function AR is not monotonic. Taken from [12].

In Kosaka *et al.*'s [12] constant voltage TDDB experiment, if they had monitored the current during stress and used it to calculate the Q_{BD}, they would have obtained the similar result as the constant current TDDB. So it should not be concluded that constant current TDDB is better than constant voltage TDDB. Rather, the physics of gate-oxide wear-out dictates that plotting Q_{BD} as a function of antenna ratio is better than plotting T_{BD}.

TDDB tests tend to take a long time. It is not suitable for routine plasma charging damage monitoring in technology development or in manufacturing. One can speed up the test by using very high stress levels. However, since breakdown is a statistical process, some devices will still take a long time to breakdown. For a stress level that is high enough for all devices to breakdown in a short time, the majority of the devices will breakdown in an extremely short time. This causes resolution problems. V-ramp and J-ramp tests help ease this problem by increasing the stress level continuously until breakdown occurs. These tests guarantee a fast breakdown for all devices. Nevertheless, fast-ramp type of tests usually give a noisier distribution than the slower TDDB type tests. Hook *et al.* [13] recently did a direct comparison of the rapid-ramp test and TDDB (on thin oxide) and found that they give qualitatively similar results, but the TDDB test gives a much cleaner signal. Figure 6.22 shows the comparison between the fast V-ramp test result and the slow TDDB test result for the same plasma process. The fast-ramp test result is clearly noisier. Such a noisy distribution will cause the extracted median V_{BD} value to have a relatively large uncertainty.

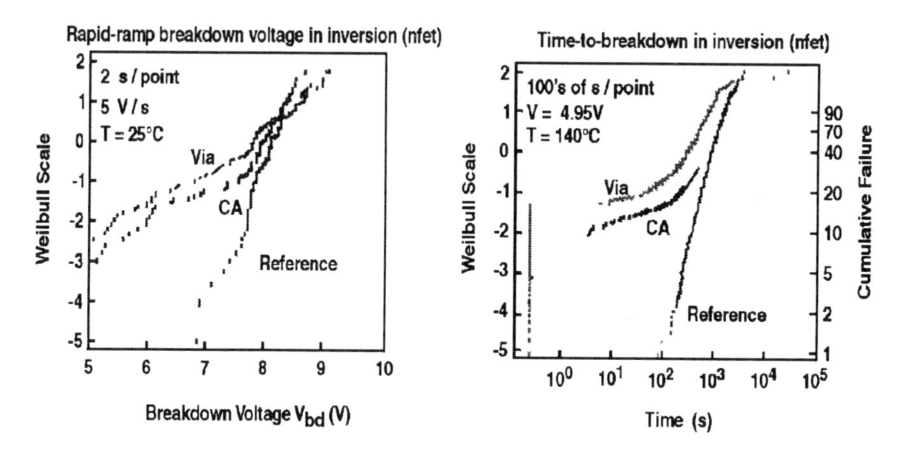

Figure 6.22. Fast ramp test (left-hand-side) produces qualitatively similar result as slow TDDB test (right-hand-side). The TDDB signal is much cleaner. Taken from [13].

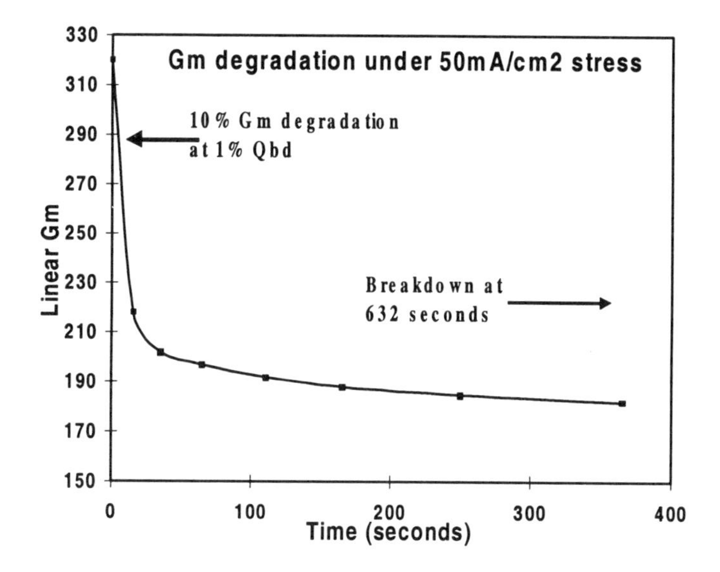

Figure 6.23. The transistor transconductance degraded by 10% in less than 1% of the total stress time to breakdown. The gate-oxide was 90Å thick.

Even with the slow TDDB method, breakdown measurements in thick oxides are really only good for disaster detection. In other words, only very serious charging damage can be detected by breakdown test. This fact can be seen from all the breakdown distributions shown so far in this chapter. It is not practical to rely on breakdown test to detect plasma charging damage that has consumed only 10% or less of the intrinsic lifetime of the oxide. On the other hand, plasma charging damage can cause many of the transistor characteristics to degrade beyond a tolerable level long before it consume 10% of the oxide lifetime. Figure 6.23 shows how transistor transconductance (G_m) degrades as a function of time under constant current (50mA/cm^2, gate injection) stress. In less than 1% of the total stress time to breakdown, 10% degradation was reached. Since 10% G_m degradation is a common limit for transistor parameter variation in circuits, plasma charging damage will exceed this limit long before breakdown measurement will detect any signal.

6.3.2 Breakdown Test for Oxide < 50Å

For thinner gate-oxides, breakdown tests are very important. When gate-oxide is thicker than 50Å, the intrinsic breakdown lifetime tends to be hundreds to thousands times longer than the expected service time of an integrated circuit. Even if plasma charging damage consumed 90% of the intrinsic lifetime, the gate-oxide will still be reliable enough from breakdown point of view. When gate-oxide is significantly thinner than 50Å, the intrinsic breakdown lifetime is comparable to the expected service time of integrated circuits. If plasma charging damage were to consume 10% of the intrinsic lifetime, the gate-oxide may fail reliability requirements.

In addition to breakdown reliability being an important factor, breakdown measurements become important for thin oxide because most other measurement methods lose sensitivity. We shall discuss these other methods separately. We concentrate on breakdown measurements in this section.

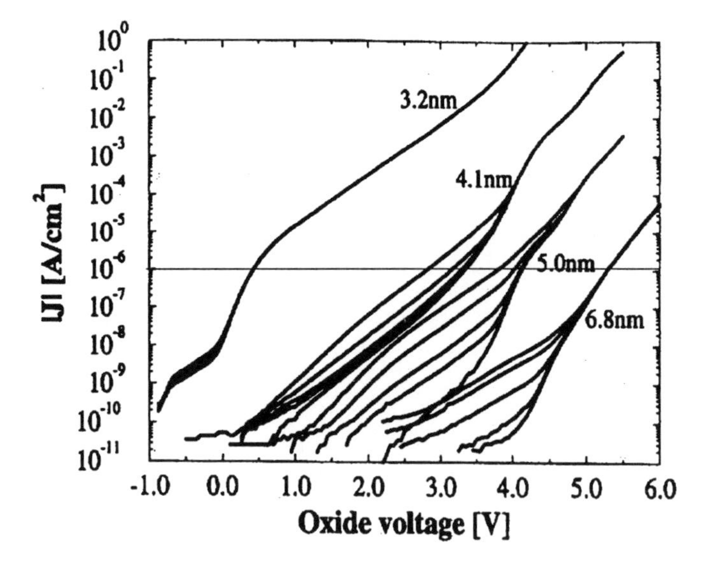

Figure 6.24. SILC increases with decreasing oxide thickness, but so is direct tunneling current. At 3nm or below, direct tunneling current overwhelms SILC. For most oxide voltages, SILC is less than 1μA/cm^2. Taken from [14].

An important characteristic of thin oxide is that when stressed, leakage current will increase. This stress-induced leakage current (SILC) is due to trap-assisted-tunneling and is proportional to trap density. When trap density reaches a certain level, breakdown occurs. Figure 6.24 shows how SILC increases with decreasing oxide thickness [14]. Below a certain thickness, direct tunneling current overwhelms SILC. For most measurement voltage, breakdown occurs before SILC reaches 1μA/cm^2.

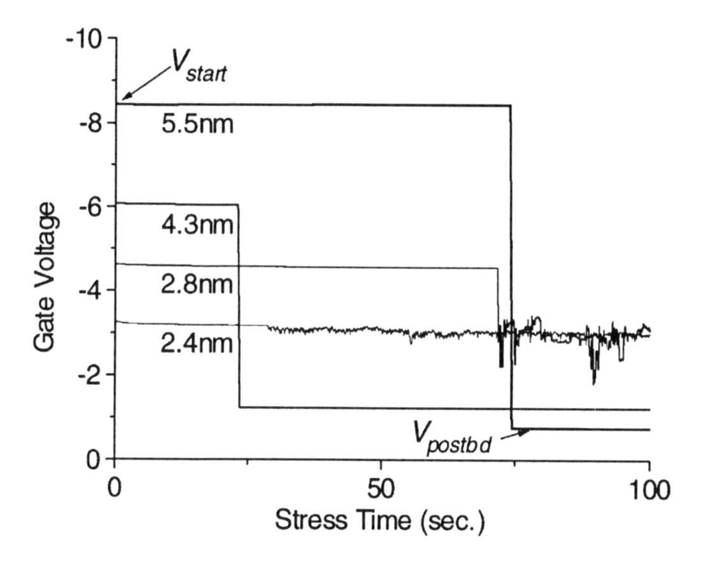

Figure 6.25. Under constant current stress, the pre- and post-breakdown voltage sustaining the current are changing with oxide thickness to opposite direction. The voltage drop at breakdown gets smaller. For the thinnest oxide, the voltage drop is zero and only the noise level of the gate voltage signals a breakdown. Taken from [15].

Unlike the case for thick oxide, breakdown in thin oxides are often very soft (see Section 1.3.4). Soft breakdowns are not as straight forward to detect as hard breakdown. This is because soft breakdowns do not produce an electrical short like hard breakdowns do. The softness of the breakdown increases with decreasing oxide thickness. Figure 6.25 shows the breakdown behavior of various oxide thicknesses under constant current stress [15]. The gate voltage sustaining the constant current drops drastically when breakdown occurs in thicker oxide. The voltage drop decreases for thinner oxide. For the 2.4nm oxide, the voltage drop is non-existent. To detect breakdown in such a case is difficult using the conventional TDDB method. Only the increase in gate voltage noise signals a breakdown has occurred. Thus noise increase is the most promising criterion for breakdown detection. Such a criterion has been implemented in a modified J-Ramp test [16]. Figure 6.26 shows an example of the modified J-Ramp test of a 2.8nm oxide sample [16]. Both the gate voltages during J-Ramp test, as well as the noise level of the gate voltage is shown. When breakdown occurs, the gate voltage change was about 10% while the noise level jumped more than 3 orders of magnitude.

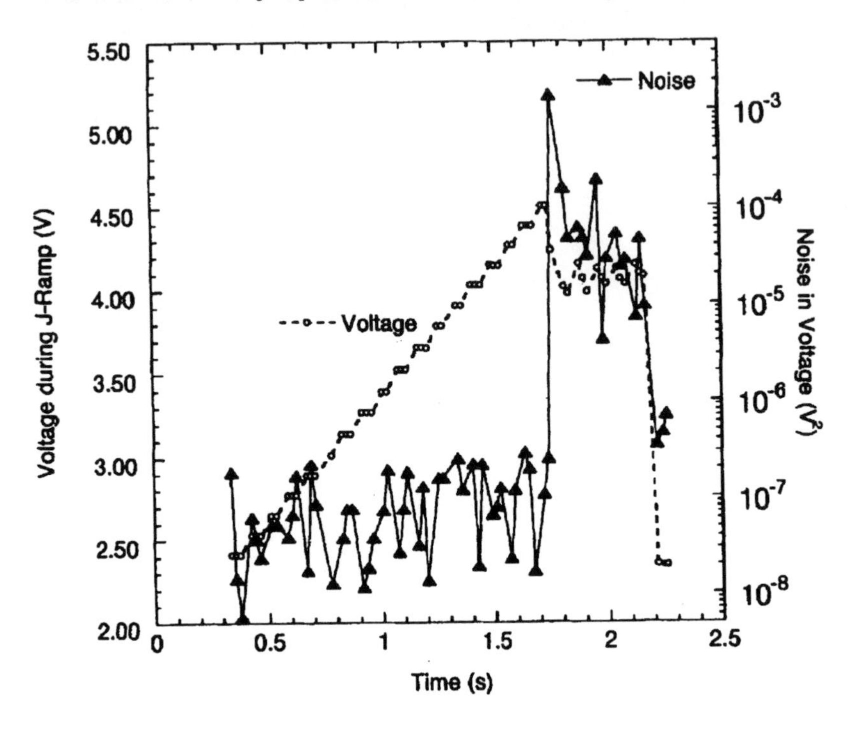

Figure 6.26. Gate voltage and gate voltage noise during a J-Ramp test of a 2.8nm oxide sample. At breakdown, while the gate voltage only drops by about 10%, the gate voltage noise increases by more than 3 orders of magnitude. Taken from [16].

Currently, few reports have appeared in the literature using noise as a breakdown criterion. Most breakdown tests for thin oxides rely on detecting a current jump at low gate voltage. The idea behind the measurement is the belief that when breakdown occurs, even if it is a soft one, there is a minimum current increase corresponding to a minimum breakdown spot size. As long as the direct tunneling current is not significantly larger (small device can limit direct tunneling current), one can always detect the breakdown event.

For measuring plasma charging damage in very thin gate-oxide, the most common method is to measure the initial gate leakage current of the transistor at a medium oxide-field before any stress is applied [17-21]. Typical transistor sizes used range from less than $1\mu^2$ to $\sim 10\mu^2$ in order to allow for a large antenna ratio without occupying too much space. Thus the direct tunneling current component should be small enough for easy detection of soft breakdown even when the gate-oxide is very thin. Figure 6.27 shows an example of using initial gate leakage current I_g to detect plasma charging damage [22].

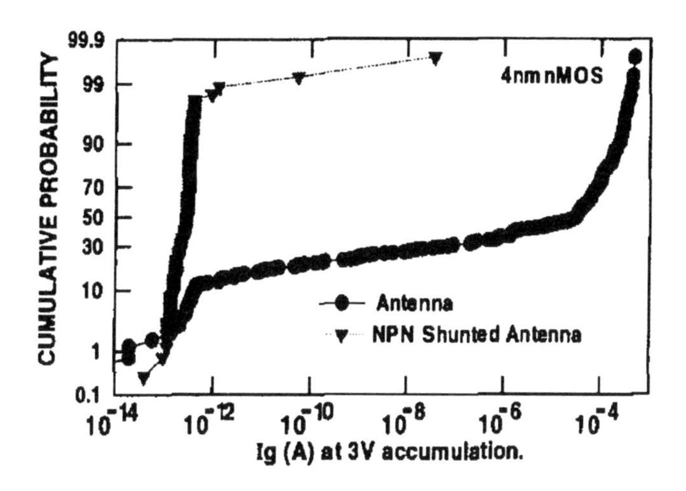

Figure 6.27. An example of using initial gate leakage current I_g to detect plasma charging damage. The antenna device without protection shows far higher percentage of high leakage than with protection. Taken from [22].

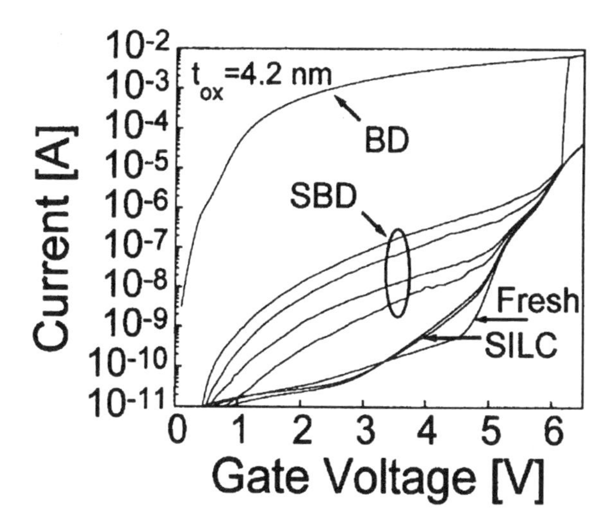

Figure 6.28. Typical progressions of oxide I-V characteristic for thin oxide during stress. First is the appearance of SILC. Then the first soft breakdown, second soft breakdown,.. Etc., until hard breakdown occurs. Taken from [23].

Figure 6.28 shows typical progression of oxide leakage during stressed [23]. At short stress times, leakage increases from the fresh device baseline to show SILC. After further stress, soft breakdown occurs. The first soft breakdown produces a significant current jump. Further stressing produces step-wise current jump. This is interpreted as the creation of additional soft breakdown spots. Since each has a minimum conductivity, the current takes quantized jumps. Eventually, hard breakdown occurs.

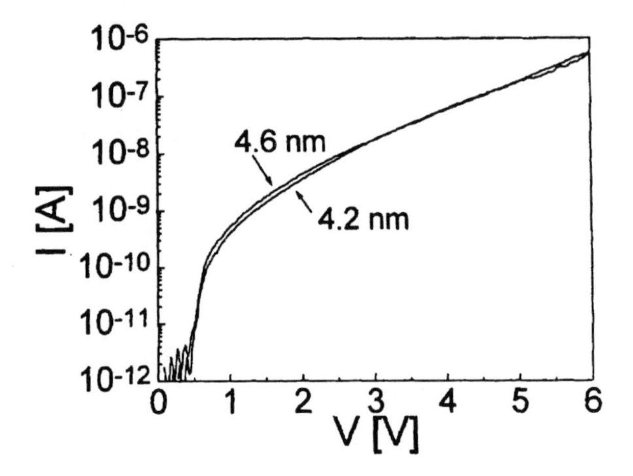

Figure 6.29. Post first soft breakdown I-V characteristic is independent of oxide thickness, supporting the notion of a minimum conductivity associated with a minimum breakdown spot size. Taken from [24].

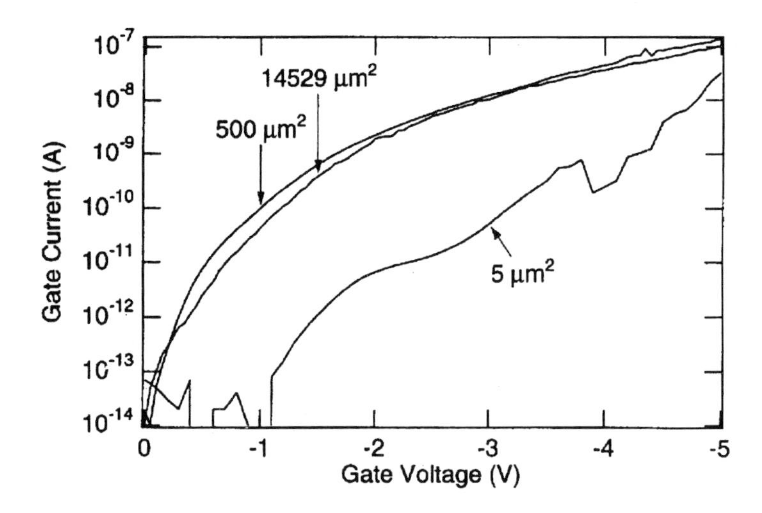

Figure 6.30. The post first soft breakdown I-V characteristic for some case is independent of capacitor size, supporting the notion of a minimum conductivity associated with a minimum breakdown spot size. Taken from [25].

Other experimental evidence that supports this minimum spot size belief is that the I-V curve after the first soft breakdown is independent of oxide thickness

[24] (Figure 6.29), as well as the capacitor size [25] (Figure 6.30). However, in Crupi *et al.*'s work, they also observed a much lower current level when the capacitor is very small. Their explanation is that when the capacitor is too small, there is not enough stored energy to establish a solid breakdown path. This observation suggests that the idea of a minimum breakdown spot and therefore minimum post breakdown conductivity might not be correct.

In Figure 6.27, the I_g distribution for the unprotected antenna devices showed a sharp change from low leakage toward high leakage, seemingly supporting the notion of minimum current jump when breakdown occurs. However, one can also clearly see in Figure 6.27 that there are devices at all leakage current levels, contradicting the minimum current jump idea. Lee *et al.* [26] studied the bias-temperature (BT) stability of plasma charging damaged devices. The measured I_g distribution of their samples is shown in Figure 6.31. They found that devices that have I_g larger than around 4pA have orders of magnitude shorter BT lifetimes than control devices (Figure 6.32). This proves that those devices in the transition region of the I_g distribution are broken devices.

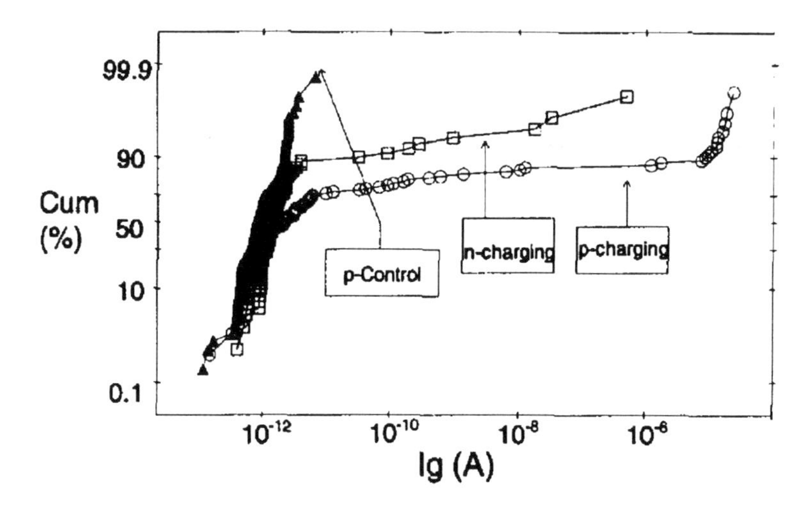

Figure 6.31. The gate leakage distribution of 32Å oxide measured at 2V. Taken from [26].

In Chapter 1 we discussed the modification of the definition of breakdown in the percolation model to account for the measured field dependency of critical trap density for breakdown. According to this modified breakdown definition, a current jump occurs when the last link of the percolation path is formed. This current jump can be any value depending on the straightness of the linkage. The current jump also leads to more traps being formed along the linkage path. It is this feed back process that is responsible for the formation of a finite breakdown spot size. Like any feed back process, the growth of the breakdown spot is a highly non-linear function of the initial current jump. In other words, a threshold exists above which a minimum breakdown spot size will form. This can explain why a minimum spot size like behavior is often observed. Like any threshold behavior, the distribution of conductivity below threshold should be a continuous decay. To observe the sub-threshold behavior, one needs to use large number of measurements. Figure 6.33 shows the transition region between good and broken devices of a 6000 sample distribution [27]. The conductivity of the 2.5μ^2 devices is indeed continuous after breakdown. (Note that the maximum SILC for these devices is ~25fA).

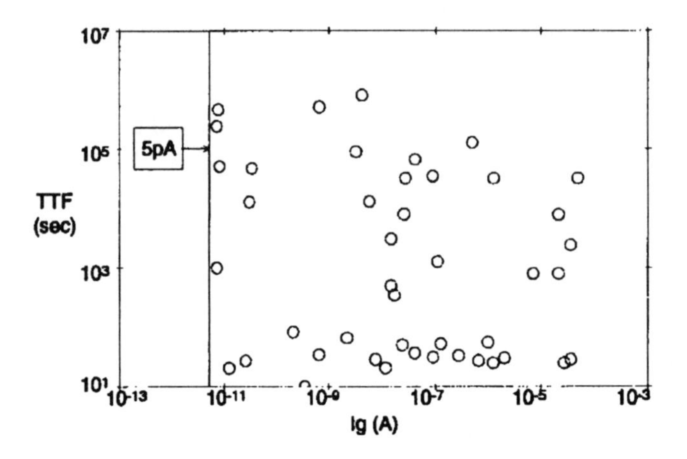

Figure 6.32. The correlation between initial gate-leakage and BT lifetime. BT lifetime for control (off the plot) is larger than 10^9 sec. Taken from [26].

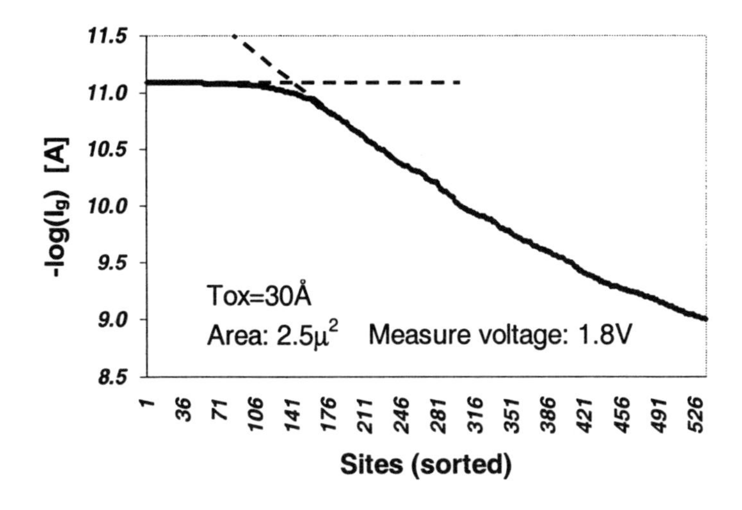

Figure 6.33. Initial gate leakage@1.8V for small devices with large antenna attached. Over 6000 samples were tested. Only the data at the transition region is shown to high-light the smooth transition.

The threshold behavior of soft breakdown conductivity prevents a precise determination of the percentage of broken devices (failure fraction). For a qualitative indication of damage, this is not a big problem. For a quantitative measurement of damage, this is a serious problem. In Chapter 7, we shall discuss how important it is to measure precisely the failure fraction of antenna devices for the purpose of quantitative yield and reliability projection. In here, we end the section by pointing out that the problem is worse when the device size is larger and/or when the oxide is thinner. Figure 6.34 [27] provides an illustration of this point. For the larger devices and/or thinner oxide devices, the direct tunneling current form a back ground level that masks the broken devices that have low conductivity. The higher the back ground level, the bigger the uncertainty in determining the failure fraction. One solution is to use extremely small devices to limit the direct tunneling current (provided the measurement instrument has very low back ground level as well). However, using smaller devices creates a bigger scaling problem (see next chapter).

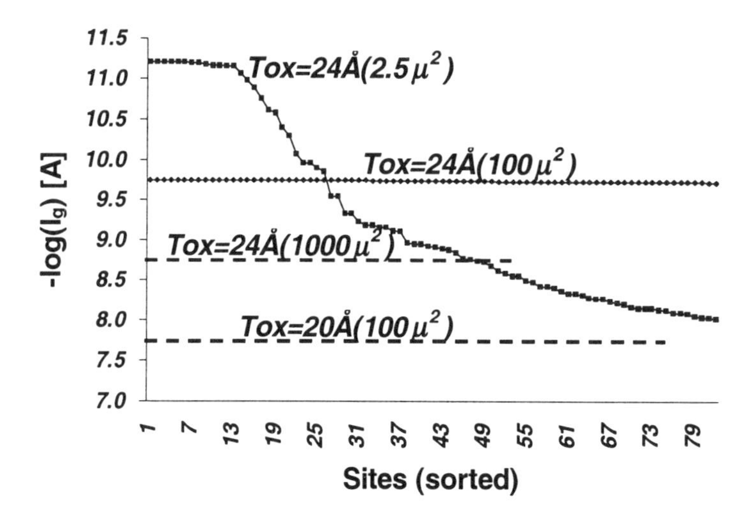

Figure 6.34. Gate leakage data of small antenna devices spans from direct tunneling limit to post soft breakdown conduction (subset of a larger population). Direct tunneling limit for larger area devices and thinner oxide devices are shown for comparison. Dotted line are simulated data and solid points are measured data.

6.4 Wear-out Tests

As mentioned at the end of Section 6.3.1, breakdown tests are not very sensitive. They can detect extreme damage situations, but not moderate damage that may cause devices or circuits to fail reliability specifications. An example of this has already been shown in Figure 6.23. To improve sensitivity, we must use measurement methods that allow us to measure oxide wear-out instead of breakdown. The range of observable effects due to oxide wear-out is large. They include all the transistor or circuit degradations. Although we normally do not call these oxide wear-out phenomena, in truth that is what they really are.

6.4.1 Stress-Induced Leakage Current (SILC) Measurement

Since SILC happens before breakdown, it is naturally more sensitive to damage than breakdown. Using SILC to measure plasma charging damage was probably the original thought behind the use of gate leakage current to detect damage. As we have already discussed, most, if not all, of the gate leakage measurements were actually measuring soft breakdown. However, there are a few reports in the literature that are really using SILC to measure damage [28, 29]. Figure 6.35 shows the antenna ratio dependence of SILC for 160Å oxide [28] samples after plasma etching. The gate area was $25\mu^2$. Note that SILC has been shown to be directly proportional to trap density [14, 30].

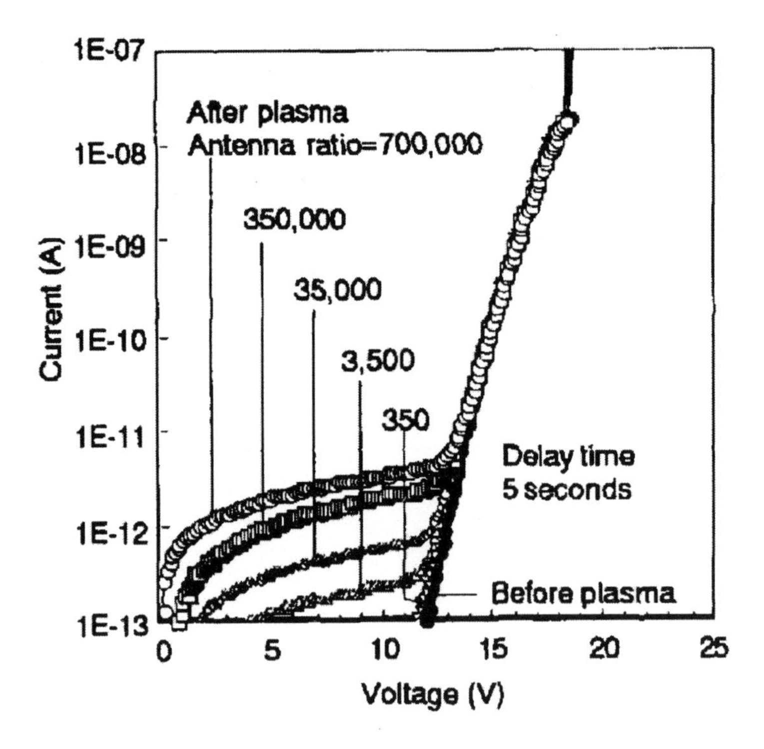

Figure 6.35. SILC as a function of antenna ratio for 160Å oxide after plasma etching. Taken from [28].

Although the antenna ratios of Figure 6.35 cover 3 orders of magnitude, that does not mean that SILC can measure damage down to very low values. As can be seen in Figure 6.35, the SILC change is less than 2 order of magnitude. In fact, the typical SILC measurement has less than two orders of magnitude dynamic range before breakdown occurs. This means that, at best, SILC can be used to detect damage level that has consumed at least 1% of the oxide lifetime. In other word, SILC is only better than breakdown measurements by 1 order of magnitude in sensitivity. This advantage is largely offset by disadvantages such as needing a large device and long measurement times.

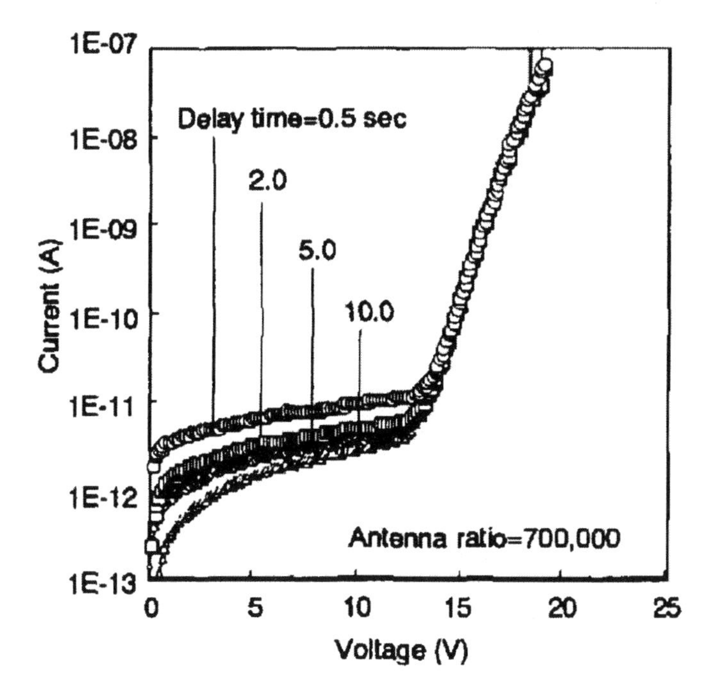

Figure 6.36. The SILC level depends on the time delay, which is the time between the voltage application and measurement at every step of the voltage scan. Taken from [28].

Figure 6.36 [28] shows why the measurement is slow. When voltage is applied to the oxide to measure SILC, there is a transient trap filling component of current that one must avoid to measure the real level of SILC [31]. The measured SILC level depends on the time delays between the voltage change and current measurement. Thus one cannot do a high-speed measurement of SILC. For thinner oxides, the transient is shorter and the measurement can be faster. However, direct tunneling current will be higher and the device needs to be smaller to keep SILC from being overwhelmed. This will require the measurement be done at a lower current level. Since instrument resolution is a function of time constant, the very low current measurements requires long measurement times. Thus, in general, SILC is not a good damage measurement method for development and manufacturing with perhaps one exception – the damage measurement in flash memory. However, as a research tool, it is much better than breakdown measurements. In a breakdown measurement, a large number of devices need to be stressed until all are broken to determine the median Q_{BD}. For a SILC measurement, only a few devices per antenna ratio are needed to determine the SILC level.

6.4.2 Flash Memory Retention Time Measurement

The flash memory is a non-volatile memory. It is very similar to the EEPROM device utilized in the CHARM wafer. Indeed, the full name of flash memory is flash EEPROM memory. It uses a floating gate to retain charges for the memory function. Charges are injected into the floating gate during the programming step. For non-volatile memory application, the injected charges need to stay in the floating gate for at least 10 years. The SILC due to traps created by plasma charging damage in the tunnel-oxide can cause a premature discharge of the floating gate. In the previous section, we mentioned that most SILC measurements have only two orders of magnitude dynamic range, however, that does not imply that SILC does not exist in very low level. For the high-density flash memory application, the maximum allowed SILC is less than 10^{-24}A/cm^2 [32]. This level of current is much too low for any instrument to measure. The only way to detect the low-level SILC is by measuring the memory's data retention statistics. Thus the retention time measurement is the flash memory's own special measurement method. Since SILC is directly proportional to the trap density, the flash memory retention statistics is an ultra-sensitive method for detecting plasma charging damage. At the moment, however, this method has not been utilized as a general damage measurement tool.

An example of using the flash memory retention statistics to detect plasma charging damage is shown in Figures 6.37 (a) and (b) [32]. The experiment used 1Mbit flash memory cell arrays. The arrays were first programmed and the as-programmed cell threshold voltages are shown in Figure 6.37(a). The arrays were measured again 24 hours later and were found to have 10 to 100 failure-bits (Figure 6.37(b)). The cross sectional view of the memory array used in the experiment is shown in Figure 6.38 [33]. 256 cells were connected to one word line that was connected to the upper level metal by an edge via. The authors concluded that it was the via-etching process that caused the damage. They found two ways to reduce the failure-bit count. One was to use thicker tunnel-oxide (Figure 6.39). Either the charging voltage was not high enough to damage the thicker oxide or the thicker oxide suppressed SILC [33]. The other way was to connect the word line to the upper level metal through a diffusion area instead of connecting directly (Figure 6.40). The diffusion area provided a shunt for the charging current during via-etching and therefore protected the tunnel-oxide.

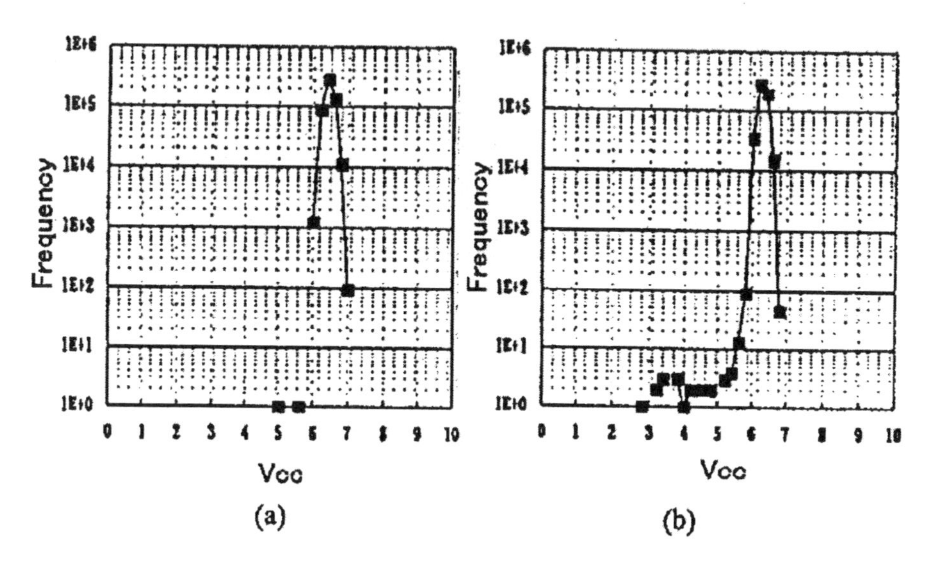

Figure 6.37. Memory cell threshold voltage distribution of a 1Mbit flash cell array after programming (a) and 24 hours at room temperature after programming (b). 10 to 100 tail-bits appeared in (b) but the center of the distribution has not changed. Taken from [32].

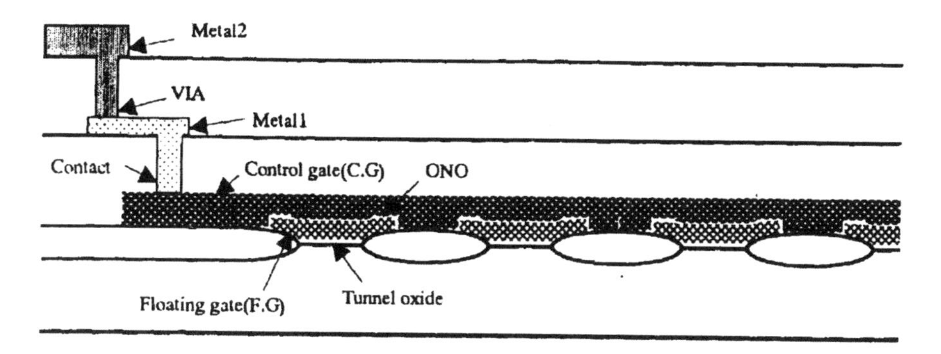

Figure 6.38. The cross sectional view of the flash cell array in the experiment. 256 cells are connected by one word line at the edge to metal 2 by a via. Taken from [33].

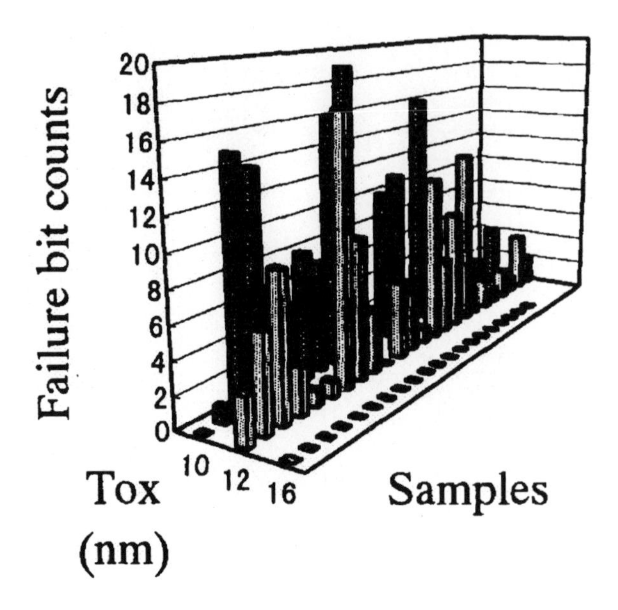

Figure 6.39. Increasing the tunnel oxide thickness can reduce or eliminate the failure bit counts. Taken from [33].

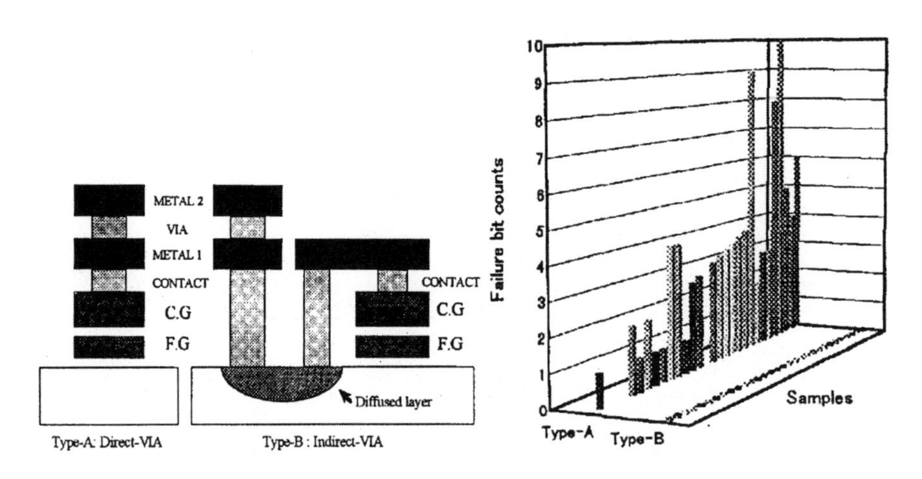

Figure 6.40. Original structure (type A) and modified structure (type B) where the via is connect first to a metal pad that is connected to diffusion. Right-hand-side shows the measured failure bit count of the two structure types. Taken from [32].

The tunnel oxide in flash memory is thick (80Å or thicker) by the standard of gate-oxide in regular MOSFET of similar technology generation. As a result, they are sensitive to the charging conditions that support a larger charging voltage only. In general, the thicker oxides require a higher voltage but not much current to damage while the thinner oxides requires a lower voltage but higher current to damage. Thus the processes that damage the thin gate-oxides may not damage the tunnel oxides and vise versa. Of course, there are plasma processes that can damage both. Nevertheless, we cannot rely on the flash memory retention statistics alone as a general-purpose ultra-sensitive damage measurement tool.

6.4.3 Initial Electron Trapping Slope Measurement

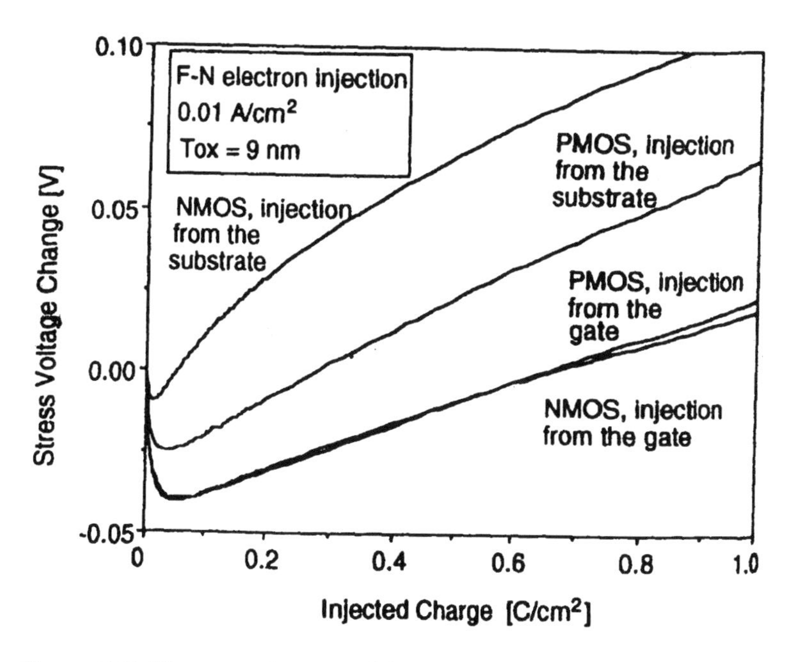

Figure 6.41. The gate voltage sustaining constant current stress for various polarities and channel types. Taken from [34].

In Figure 1.9 of Chapter 1, we saw that the stress current in a constant voltage TDDB test varies with stress time due to charge trapping. When doing the constant current TDDB test, we expect the gate voltage that sustains the current to vary for similar reason. Figure 6.41 shows the gate voltage as a function of the total injected charge (equivalent to time) during a constant current stress for various stressing polarities and channel types [34]. The gate voltage change due to charge trapping can be expressed as [35]:

$$\Delta V_I(+) = -\frac{1}{\varepsilon} Q_{OX}\, x \qquad (6.1)$$

$$\Delta V_I(-) = -\frac{1}{\varepsilon} Q_{OX} (T_{OX} - x) \tag{6.2}$$

where Q_{OX} is the integrated charge in the oxide per unit area and x is the distance of the charge-centroid measured from the gate. From Equation 6.1, we see that the charges at the SiO_2/Si interfaces influence the gate voltage strongly when the stress voltage is positive. This explains why the nMOS and the pMOS devices behave so differently during the substrate injection stress in Figure 6.41. On the other hand, Equation 6.2 suggest that the charges trapped at the interface have no impact on the gate voltage. This is why the nMOS and the pMOS devices behave identically during the gate injection stress in Figure 6.41.

Equation 6.2 suggests that monitoring the gate voltage change during a gate-injection constant current stress can provide information on the trapped charges without the interference of the interface-trapped charges. Figure 6.42 shows the breakdown of various contributing factors to the gate voltage curve [36] during a constant current gate-injection stress. The thick curve is the gate voltage curve as measured. The sharp turn-around at the beginning is often not seen in published results such as those shown in Figure 6.41. This is because most of the measurements do not have the time resolution to capture it.

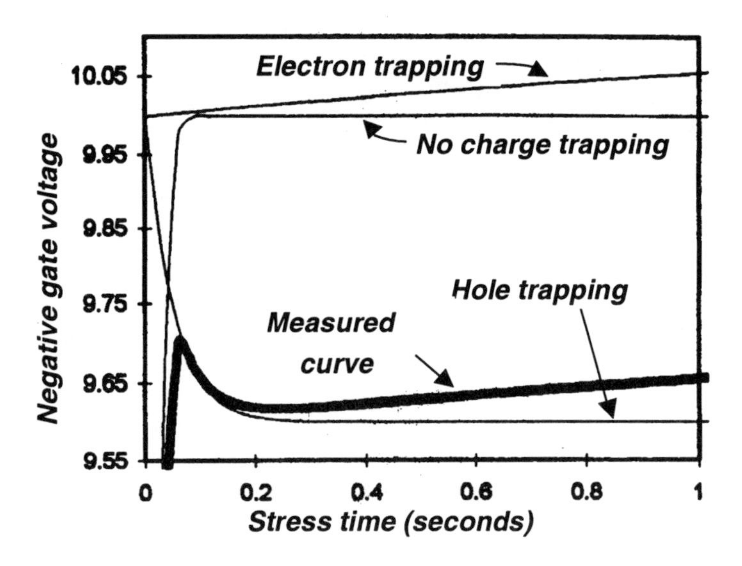

Figure 6.42. Breakdown of various contributing factors to the gate voltage curve during gate injection stress.

Three thin curves are shown in Figure 6.42 representing three contributing factors to the resulting curve. The one labeled "no charge trapping" is the charge up curve of the capacitor, including the capacitance of the measurement setup. This curve should rise up to the final voltage monotonically and level off according to the RC time constant involved. The falling curve is due to the trapping of positive charges. This curve is expected to be monotonic as well and will saturate at some point. The time at which the saturation occurs depends on the stress level. Higher stress levels lead to faster saturation. The rising third curve is due to the trapping of electrons. This curve is also monotonic, but would not saturate.

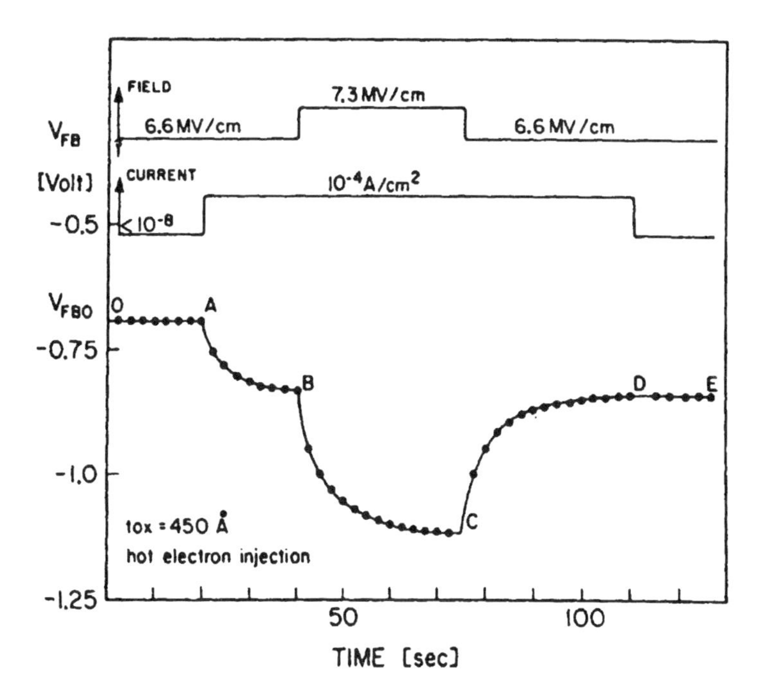

Figure 6.43. The flat-band voltage shift during hot electron injection into oxide with independently controlled oxide-field showing positive charge trapping is determined entirely by the oxide-field. Taken from [37].

The charge-up of the capacitor is determined by the capacitance of the measurement system. The devices under test usually contribute very little to the overall capacitance. So once the stress condition is fixed, the charge-up curve is fixed as well. The falling voltage curve due to the trapping of positive charge is largely fixed by the stress condition. In a series of well-designed experiments, Nissan-Cohen *et al.* [37] demonstrated this point very clearly. Figure 6.43 shows the flat-band voltage shift during a hot-electron injection experiment with an independently controlled oxide field. No trapping of positive charge was observed between point O and point A when only the oxide field was present. Positive charge trapping happened between point A and point B where the injection current was turned-on. More positive charge trapping occurred between point B and point C where the oxide field was increased. Positive trapped charge decayed back to the lower field level between point C and point D where the oxide field was lowered back to the previous level. No further decay occurred between point D and point E where the injection current was turned-off. It is clear from this experiment that the trapping of positive charge is entirely determined by the oxide field. The reason why the oxide field alone (between points O and A) did not cause flat-band shift is because if there is no source of positive charge, nothing will be trapped. The injected current provided a source of positive charge through the anode-hole-injection mechanism. The fact that no further flat-band shift was observed between point D and E is clear evidence that the current does not affect the trapping of positive charges.

As seen in Figure 6.42, the leveling of the capacitor charge-up curve happens very early. This is usually true if the experiment is setup correctly (should be checked to make sure). The gate voltage curve turns upward when the trapping of positive charges is saturated. From this point on, the change of the gate voltage is entirely due to the trapping of electrons. This is the section of the gate voltage curve that allows us to measure the electron trap concentration in the oxide and use it as an indicator for charging damage.

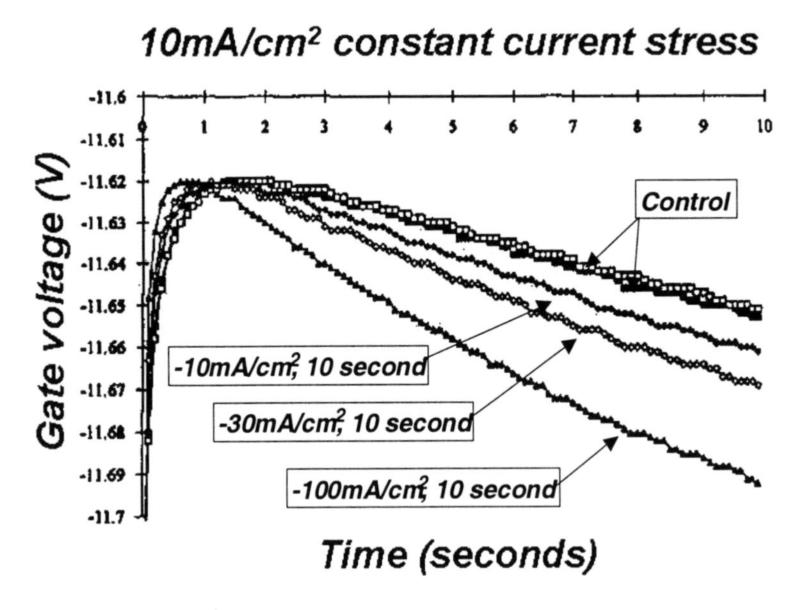

Figure 6.44. Even after a 800°C, 30 minutes anneal in Ar, the pre-anneal stress can be revealed in the electron trapping part of the gate voltage curve.

Figure 6.44 shows an example of using the post turn-around section (after the second turn-around if the first one is captured) of the gate voltage curve to detect charging damage. Note that there is a sign change of the y-axis between Figure 6.44 and Figure 6.43 (the curves are upside down). Here a group of identical capacitors were pre-stressed at various conditions to simulate the plasma charging damage. All of the capacitors were then annealed at 800°C for 30 minutes in an Ar atmosphere to thoroughly detrap all charges and to anneal some of the damages. After annealing, the capacitors were stressed at a constant current of $-10mA/cm^2$. The recorded gate voltage curves are shown in Figure 6.44, and the labels indicate the pre-anneal stress conditions. The two control (no pre-stress) curves are practically on top of each other, showing the reproducibility of the measurement. The post turn-around part of the gate voltage curves clearly displays an increase in slope with the pre-anneal stress level, confirming the method's ability to detect charging damage.

Careful inspection of the post turn-around portion of the gate voltage curves reveals that they are not linear. Only the curves from the controls are relatively linear. This is expected because the trapping of electrons during stress really consists of two processes. One is the filling of the pre-existing empty electron traps and the other is the filling of the traps that are generated by the measurement stress itself. For the short stress duration used, the concentration of newly generated traps should be low, and that it is reasonable to expect the generation rate of new traps to be constant. Thus for the devices that are free of pre-existing empty traps, the gate voltage curve after turn-around should be linear. The filling of the pre-existing empty traps, however, should not be linear. The rate of filling of the pre-existing empty traps can be expressed as:

$$\frac{dN_e}{dt} = \sigma J\left(N_0 - N_e(t)\right) \tag{6.3}$$

where σ is the capture cross section, J is the stress current density, N_0 is the initial concentration of empty electron traps and $N_e(t)$ is the filled electron trap concentration at time t. As time increases, the trap filling rate, and therefore the slope of the gate voltage curve, must decrease due to the decrease in available empty trap concentration.

Equation 6.3 also clearly indicates that when t is small so that $N_0 - N_e(t) \sim N_0$ the measured signal (slope) is the largest and is directly proportional to the pre-existing empty trap density. It is for this reason this measurement method emphasize on the initial portion of the post turn-around gate voltage curve, and that is why it is called the initial-electron-trapping-slope (IETS) method [38]. The overall electron trapping slope as a function of time should be:

$$\frac{dN_e}{dt} = J\left[o\left(N_0 - N_e(t)\right) + \sigma^* N\right] \tag{6.4}$$

where σ^* is the effective trap creation cross section per injected electron and N is the total available sites for new trap creation. Discharging of trapped electron has been neglected in the treatment.

Except for the really extreme cases of damage, N should be independent of N_0. Thus as long as we fixed the stress condition, the fact that the measurement process itself creates new traps does not interfere with the measurement. The background slope owing to new trap creation can always be removed by subtracting the slope of the control from the measurement.

Ideally, we would like to make the measurement of the electron-trapping slope at time zero. However, we must wait for the gate voltage curve to turn around so that the trapping of positive charges does not contribute to the measured slope. In practice, IETS is measured at the steepest part (maximum slope) of the post turn-around gate voltage curve. Theoretically, the steepest part of the curve is also the earliest point at which we can make the measurement. From Nissan-Cohen *et al.*'s work [37], we know that the positive charge trapping level is determined by the stress field while the time to reach saturation is determined by the stress current density. We can shorten the time to reach turn-around by using a higher stress current density. An added bonus of using a higher current density is a larger signal as can be seen from Equation 6.4.

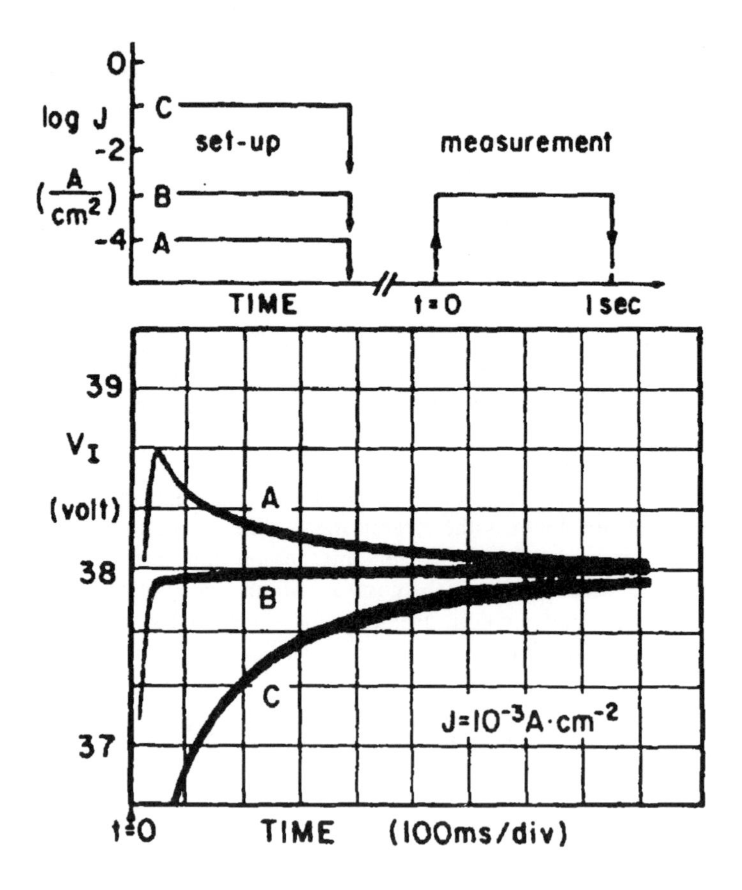

Figure 6.45. Flat-band shift transient is dependent upon the setup stress level but the final value at saturation is the same. The pre-stress and measurement conditions are shown above the plot. Taken from [37].

We mentioned that the positive charge trapping level is entirely determined by the oxide field as is evident in Figure 6.43. The fact that the flat-band shift returned to the same level when the field was lowered in Figure 6.43 suggests that the higher stress field did not create enough new hole-traps to be noticeable. In the same paper, Nissan-Cohen *et al.* did another experiment that involved an even higher stress level and found that the positive charge trapping level was not affected. In this experiment, they pre-stressed the oxide at various levels up to $1A/cm^2$. For each oxide sample, they switched from the pre-stress to measurement and monitored the flat-band voltage shift at the same time (Figure 6.45). Other than the initial transients observed immediately after switching, the saturation level remained unchanged. Thus, even at the $1A/cm^2$ level of stress, the created hole-trap concentration was not significant enough to be noticeable.

Recently, however, there are experimental evidences to show that plasma charging damage would increase the hole trap density in the oxide [39]. If the positive charge trapping level is damage dependent, what impact would that has on the electron-trapping slope measurement? More positive charge trapping means longer stress time is needed at a given stress level before turn-around would occur. As a result, a higher level of damage would cause the IETS to be measured at a later time. On the other hand, a higher damage level also means that the initial empty electron trap density is higher. A higher initial trap-filling rate would push the turn-around to happen at an earlier time. Since the time at which the maximum slope is measured is damage level dependent, there is a systematic error in the measurement method. However, this error is not a large one because the effect of the increase in positive charge trapping is opposing the effect of the increase in empty electron trap density.

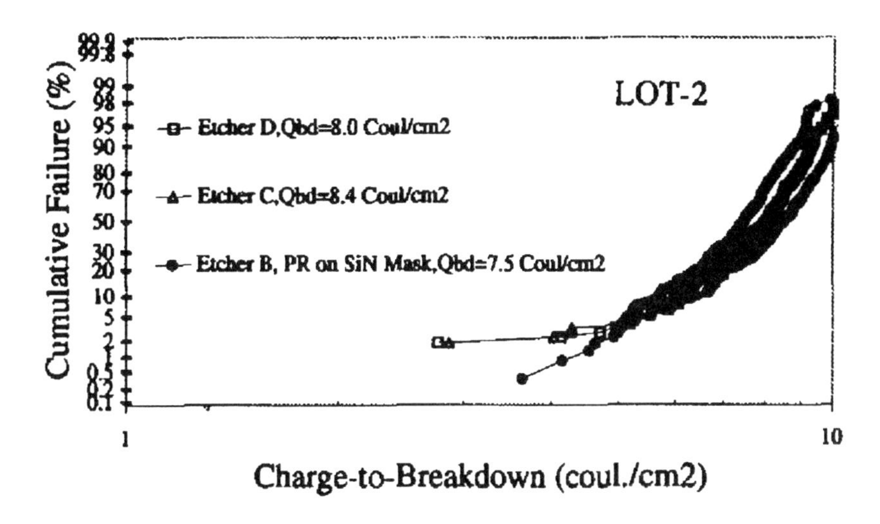

Figure 6.46. The Q_{BD} distribution for three plasma processes as measured by J-Ramp method. It is difficult to distinguish one process from another. Taken from [40].

Although the damage created positive charge traps do not impact the IETS measurement in any significant way, it created an interpretation problem in the transistor V_t shift measurement. We shall discuss that later.

From all the above discussion, it is clear that when using the IETS method for charging damage measurement, a higher stress current gives a better sensitivity and shorter measurement time. Most researchers used a low stress current level when doing a IETS measurement for fear of the errors that may be introduced by the stress-created new traps. Such fear is unwarranted. Nevertheless, even when the stress level is low, the sensitivity of the IETS method is still excellent. Figures 6.46 and 6.47 [40] illustrate a case in point. In Figure 6.46, the Q_{BD} distributions for three plasma processes were measured using the J-Ramp method. It was difficult to distinguish one process from another using the Q_{BD} distributions even though a difference in product reliability was found. Figure 6.47 shows the measured IETS distributions of the same three etching processes measured at 25mA/cm^2. The difference from one process to another was very clear and agreed with product reliability data.

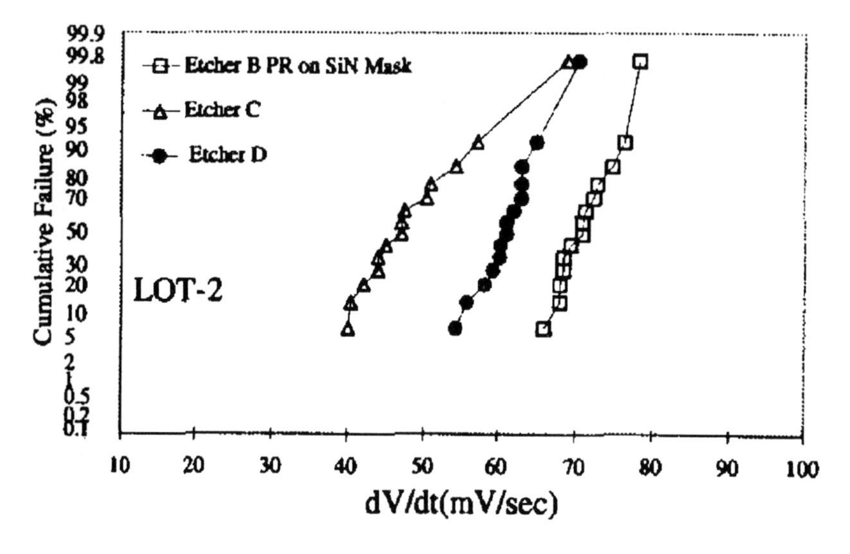

Figure 6.47. The same three plasma processes as measured by IETS method using 25mA/cm^2 stress. The difference between processes is now very clear. Taken from [40].

Figure 6.48 is another example but using a higher stress current. Note that even though the measured slopes for some of the devices are quite high, the low damage plasma system consistently produced very low slopes. Although these low damage devices were not designed to be control devices, they could serve as the control.

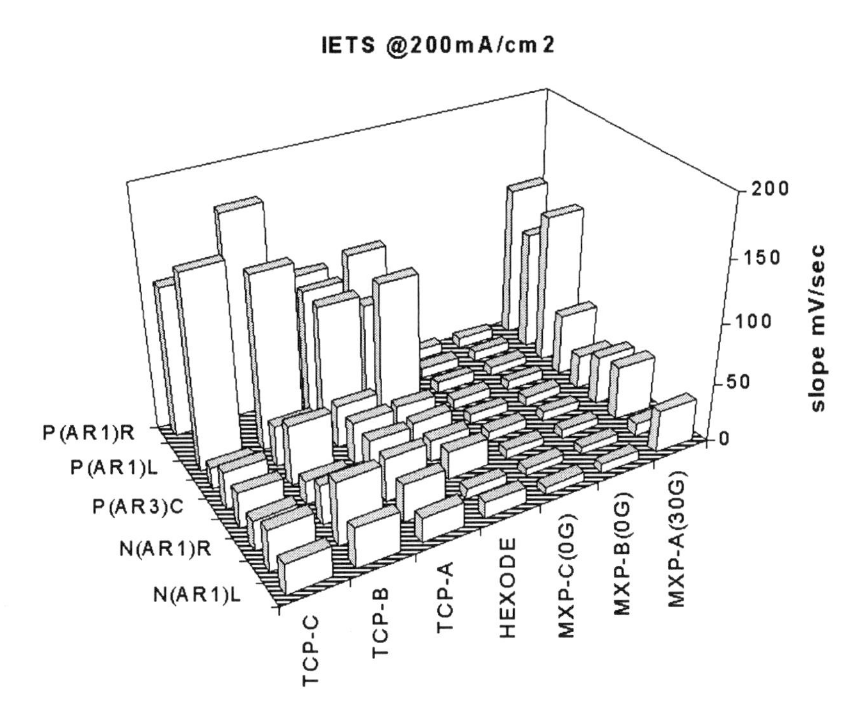

Figure 6.48. IETS values measured at 200mA/cm^2 is used to compare etching systems, die locations and antenna size on the wafer. AR1 = large AR; AR3 = small AR; L = left; R = Right; C = Center.

The values of IETS as measured are not absolute. They cannot be used to project product yield and reliability. To do that they must be calibrated first. Figure 6.49 shows an example of correlating the IETS values to the DC hot-carrier lifetime for the n-MOSFET devices under maximum substrate-current stress condition. If the DC hot-carrier lifetime is the reliability criterion of interest, then this calibration curve would allow the IETS values to be related to reliability.

Since the IETS is a method to measure the rate of filling pre-existing empty electron traps, it is very important that all the electron traps are emptied by an annealing step before the IETS method is performed. The standard forming gas anneal process at the end of wafer processing is perfectly suited for this purpose. One should not use the IETS method on wafers that are directly taken out of the damaging plasma system without first going through an anneal step.

The IETS method is fast, sensitive and unambiguous. Once calibrated, it is an excellent tool for routine plasma charging damage monitoring. When the stress current level is high, the measurement time can be very short. For example, at the 1A/cm^2 stress, the measurement time can be as short as 0.4 second. Thus the IETS method is also well suited for damage monitoring in the development and production environments. The major drawback of the IETS method is that it is not suitable for thinner oxides. For oxides thinner than 50Å, the positive charge detrapping is so efficient that the gate voltage turn-around never occurs. Without the ability t0 separate the positive charge trapping from the electron trapping, we cannot measure the electron-trapping rate.

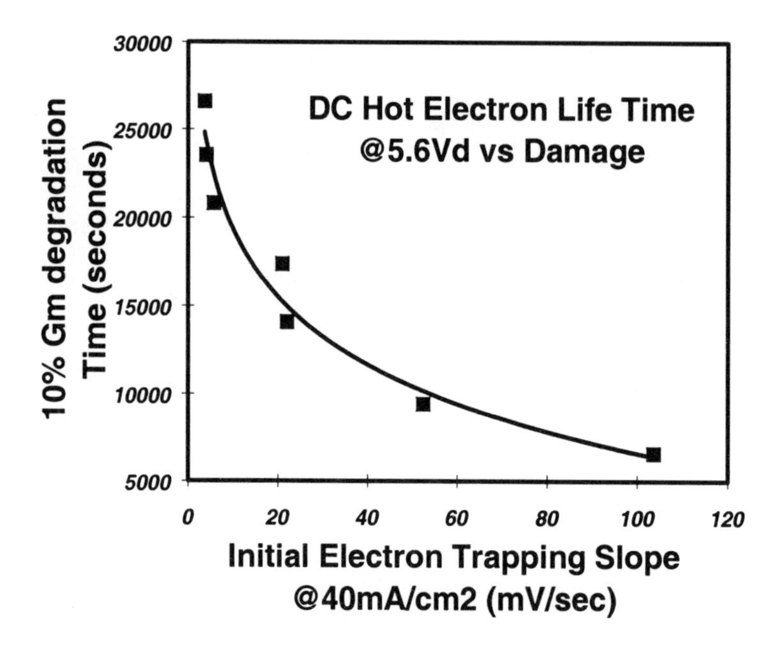

Figure 6.49. IETS values correlate with DC hot carrier lifetime for n-MOSFET under maximum substrate current stress condition.

6.4.4 Transistor Parameter Shift Measurements

Transistor parameter shift, or degradation, is the most direct measure of charging damage. Its importance is therefore well recognized early on [3]. We have already seen in Figure 6.1 that the threshold voltages of transistors spread out due to plasma charging damage. From the same publication, Shone *et al.* [3] showed the transistor transconductance (G_m) degradation due to antenna effect (Figure 6.50) as well as the transistor sub-threshold swing (S) degradation by dielectric deposition charging damage (Figure 6.51).

The measurement of transistor parameter degradation is not straightforward. Early measurements may not have been done correctly. Although a shift in the parameter such as threshold voltage is often observed due to damage, such result may not always be consistent with other damage measurements. The key

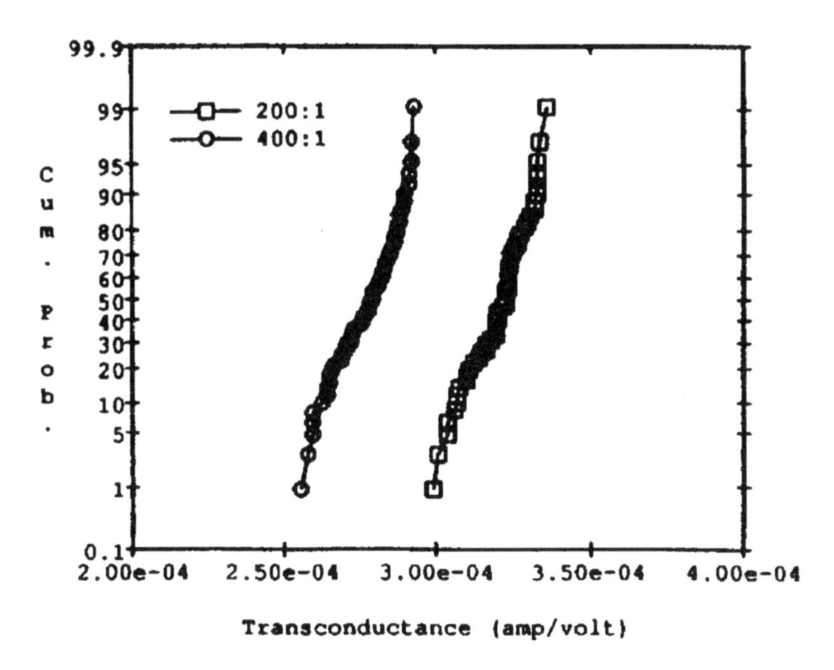

Figure 6.50. The transistor transconductance distribution is a clear function of antenna ratio. Taken from [3].

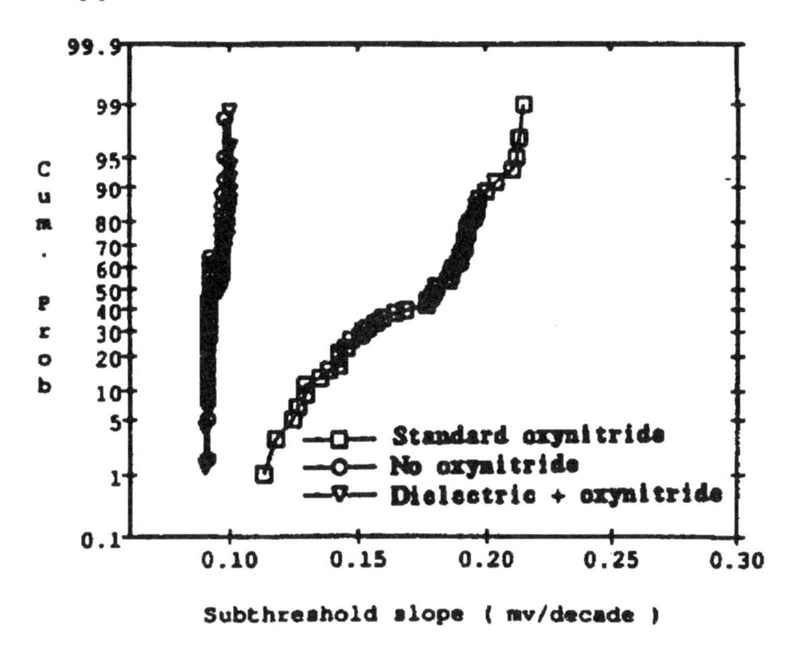

Figure 6.51. The transistor sub-threshold swing distribution is degraded by dielectric deposition charging damage. An oxidenitride linner layer can help prevent the charging damage. Taken from [3].

problem is that most of the transistor parameters are sensitive to charges trapped either in the bulk of the oxide or at the interfaces. Subsequent processing after the damaging step can easily disturb the trapped charges that are introduced during plasma charging damage. The standard forming gas anneal at the end of wafer processing can do a good job in removing all traces of the charging damage signature. The only reason one can still see the charging damage induced transistor parameter shift at the end is that sometimes the forming gas anneal is incomplete (insufficient anneal).

When forming gas anneal is done with sufficient duration and temperature, the transistor parameters should have a tight distribution even if charging damage has occurred during the fabrication process. We have seen an example of this in Chapter 1 (Figure 1.35) where all the devices, regardless of their antenna ratios, have a tight threshold voltage (V_t) distribution when tested as-processed and have a strong antenna ratio dependent V_t distribution after an additional stress was applied. We shall now discuss this latent transistor parameter shift phenomenon in detail.

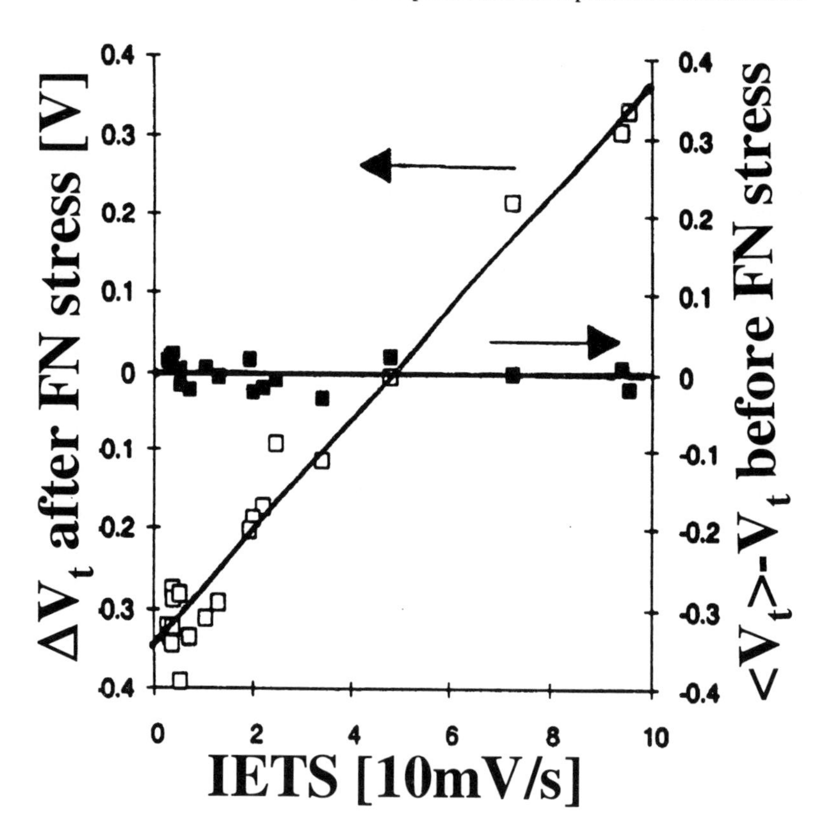

Figure 6.52. The threshold voltage distribution before stress is tight and shows no correlation with IETS values. The distribution spreads out drastically and shows a linear correlation with IETS value after FN stress.

Figure 6.52 shows the distribution of V_t shifts of n-MOSFETs before and after a short FN (40mA/cm^2, 10s) stress. Before the stress, the magnitudes of V_t variation from the mean value are very small and have no correlation with the IETS values measured from similar devices in close proximity. After the stress, the V_t shifts spread out to a very large range and showed a linear relationship with the IETS values. Figure 6.53 shows the corresponding results on the G_m. Again, without the stress, a tight distribution is observed and no correlation with the IETS values is evident. After the stress, a large spread and a linear correlation with the IETS values is seen.

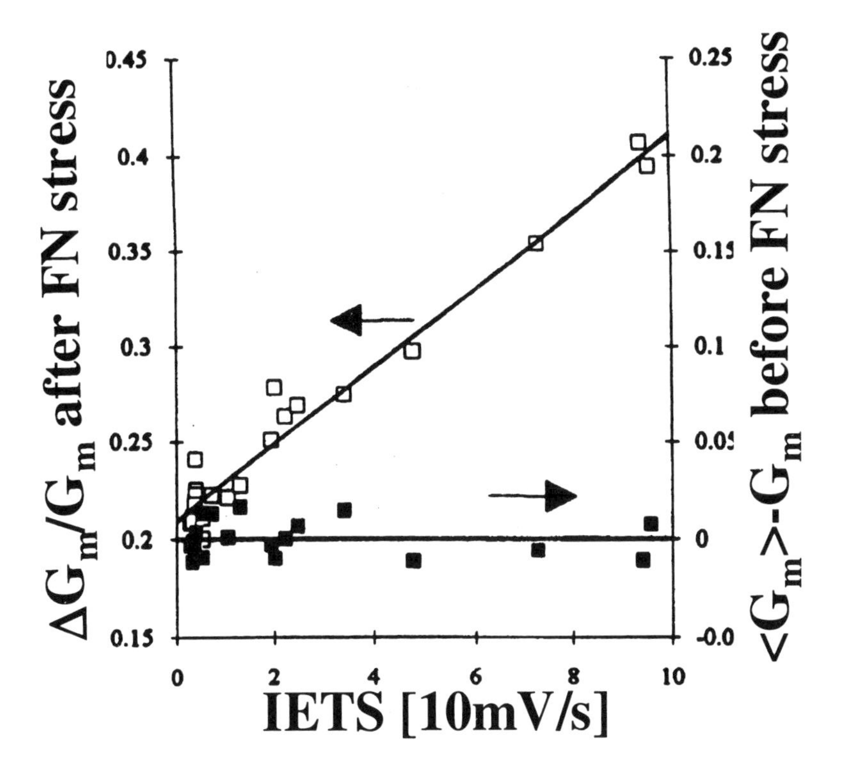

Figure 6.53. The transconductance distribution before stress is tight and shows no correlation with IETS values. The distribution spreads out drastically and shows a linear correlation with IETS values after a FN stress.

In Chapter 1, the example of Figure 1.35 was used to illustrate the latent defect concept. The latent defects that are important to the V_t measurements are in the bulk of the gate-oxide as well as at the interfaces of SiO_2/Si. For the linear peak G_m measurements, only the interface traps are important.

The transistor threshold voltage (V_t) shift is given by [35]:

$$\Delta V_t = -\frac{1}{\varepsilon}\left(Q_{OX}\,x - qN_{it}T_{OX}\right) \tag{6.5}$$

for n-MOSFETs, and by:

$$\Delta V_t = -\frac{1}{\varepsilon}\left[Q_{OX}\left(T_{OX} - x\right) - qN_{it}T_{OX}\right] \tag{6.6}$$

for p-MOSFETs. The N_{it} is the interface state density. The values of qN_{it} are negative for n-MOSFETs and positive for p-MOSFETs at the transistor threshold voltages due to the amphoteric nature of the interface traps. Thus for n-MOSFET, the interface trapped-charges cause the V_t to shift positively, same as the trapped-electrons in the bulk of the gate-oxide. Since charging damage increases both the bulk electron trap density and the interface state density, and since hole trapping is only weakly affected by damage, we can expect, to a good approximation, that the n-MOSFET V_t shift to be an monotonically increasing function with damage. As seen in Figures 6.52 and 6.53, the G_m degradation as well as the V_t shift of n-MOSFET are both linearly proportional to the IETS value, which is a measure of the electron trap density in the bulk. The only way for this to be the case is that they are linearly proportional to each other. Figure 6.54 shows that G_m degradation indeed correlates with V_t shift linearly for a large range of damage.

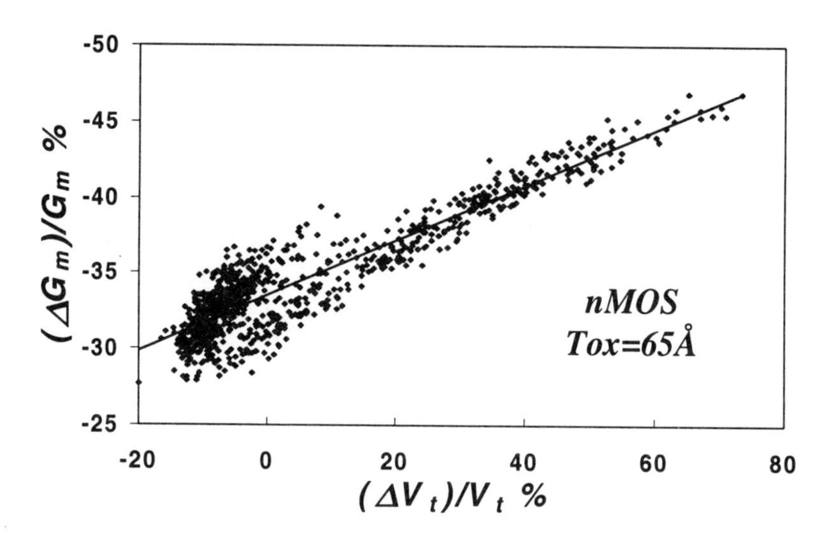

Figure 6.54. The G_m degradation correlateds linaerly with the V_t shift for a wide range of damage.

In Figure 6.52, we see that the V_t shifts for n-MOSFETs at low damage are negative. As damage increases, the V_t shift decreases, passing through zero and eventually becomes positive. This phenomenon can be explained with the help of Figure 6.42, which shows that the hole-trapping process causes a much larger gate voltage shift than the electron-trapping process. A positive net trapped-charges in the oxide is the normal result of a FN stress. The higher empty electron trap density of a damaged oxide causes more electrons be trapped to offset the positive trapped-charges in the oxide after FN stress. The result is a less negative V_t shift. When the damage is very severe, the density of the trapped-electrons can over compensate the positive trapped-charges and leads to a positive V_t shift.

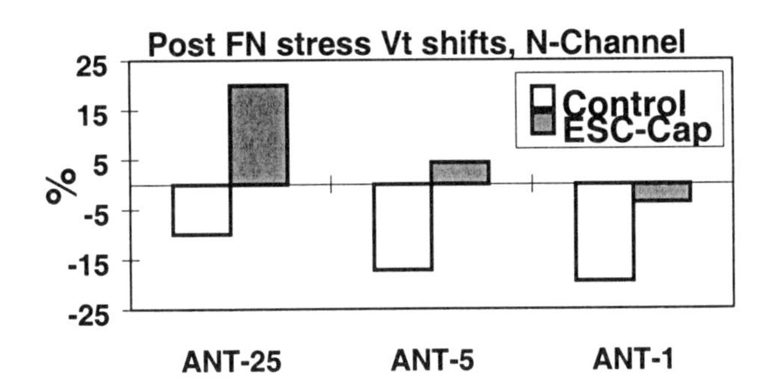

Figure 6.55. Using the n-MOSFET V_t shift after FN stress to detect antenna effect in a plasma etching process. The experiment shows that electrostatic chuck increases charging damage. Control was mechanical chuck.

Figure 6.55 is another example of using the n-MOSFET V_t shifts to detect the antenna effect due to charging damage. In this experiment, the goal was to see if the clamping the wafer on an electrostatic chuck (ESC) would cause more charging damage than on a mechanical chuck (control). In this case, the data shows that all the ESC wafers have positive V_t shifts and that the larger antennae (ANT-25 is 25 time larger AR than ANT-1) have more positive shifts. The controls, on the other hand, have negative shifts and the larger antennae are less negative. From what we have discussed about the V_t shifts in n-MOSFET, we can conclude that the ESC caused much more charging damage than the mechanical chuck, and that even the smallest antenna of the ESC wafers has more damage than the largest antenna of the control wafers.

Figure 6.56 shows the p-MOSFET V_t shift as a function of antenna size for the same experiment. Here, all the V_t shifts are negative. This can be explained with the help of Equation 6.6. Since the interface states are charged positively in the p-MOSFET at threshold, they cause a negative V_t shift, similar to the positive trapped-charges but not the negative trapped-charges in the bulk of the gate-oxide. From Equation 6.6, we see that the interface trapped-charges have bigger impact on the V_t shift than charges trapped in the bulk of the gate-oxide. Coupled with the fact that normally there are more positive charges than negative charges trapped in the oxide after a FN stress, the resulting V_t shifts is very unlikely to be positive.

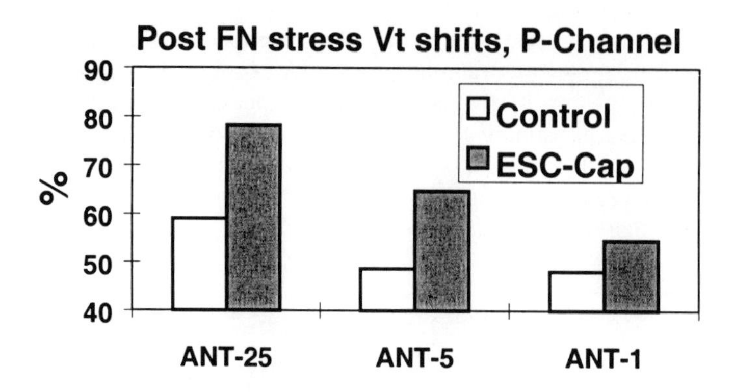

Figure 6.56. The p-MOSFET V_t shift after FN stress show similar result that ESC causes far more charging damage than control.

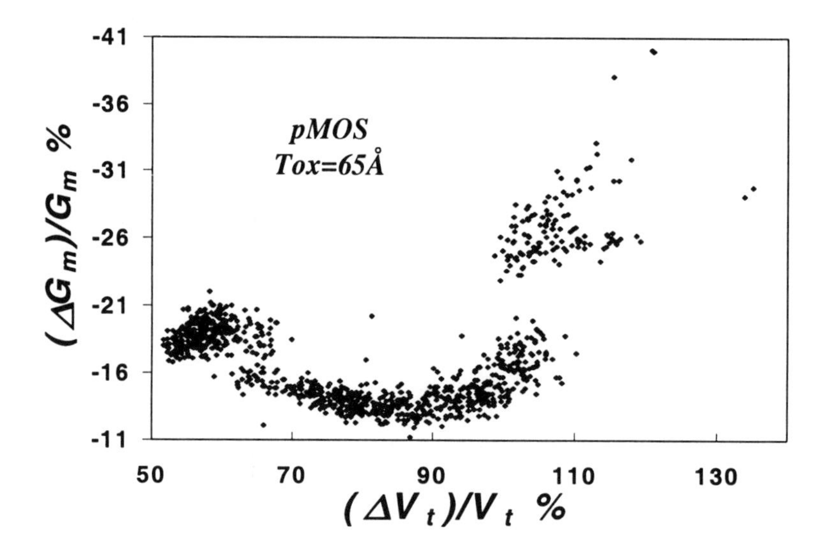

Figure 6.57. The G_m degradation in p_MOSFET does not correlate with the V_t shift.

In Figure 6.56, we see that a larger antenna has a bigger negative V_t shift – an apparent correlation. Such an apparent correlation is actually seen rather often. However, it is not a correlation that has a firm foundation. For the p-MOSFET V_t shift, we have the two types of traps that are most effected by charging damage working against each other. Not only the absolute density of each trap type but also the rate at which each trap type re-appears needs to be factored into the calculation if one were to predict how much a V_t shift would result. It is difficult to predict what the outcome would be under any condition. In Figure 6.57, the V_t shifts of p-MOSFETs are plotted against the G_m degradations. This data were obtained at the same experiment that produced the n-MOSFET data of Figure 6.54. The range of charging damage covered in the data is quite large. Clearly, a more negative V_t shift does not always accompanied by a larger G_m degradation. Thus, one must be very careful when using the p-MOSFET V_t shift as an indicator for charging damage, it may be completely misleading.

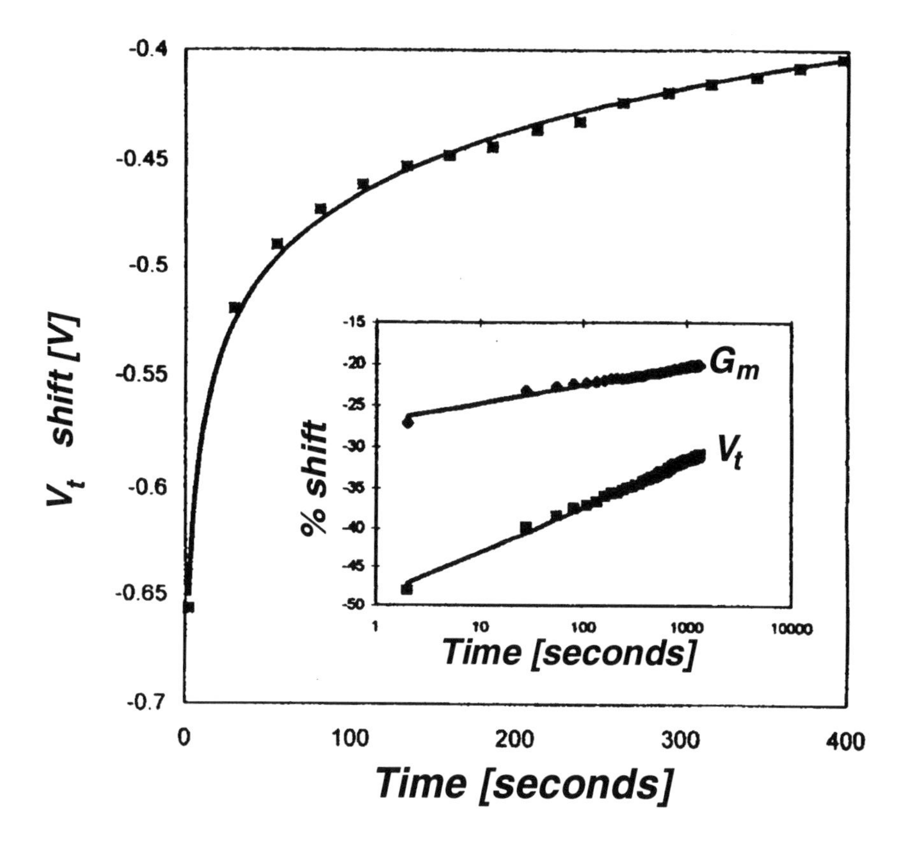

Figure 6.58. Post FN stress relaxation of V_t as a function of time. The inset shows the relaxation of V_t and G_m in log(time) scale. Both are decaying exponentially.

An additional precaution must be observed when using either the n- or the p-MOSFET parameter shift as a damage indicator. After the FN stress, both the bulk trapped positive charges and the interface states relax rapidly. The bulk positive trapped-charges relax by capturing electrons tunneling into the oxide from the electrodes (tunneling front model, see Chapter 1, Section 1.6.1). The interface states relaxation is poorly understood at the moment. Nevertheless, they both relax exponentially with time as shown in the inset of Figure 6.58 [36]. Depending on when does one make the measurement after the FN stress, the measured V_t and G_m can be very different. Thus it is extremely important that the transistor parameter shift measurements be done with a fully automatic measurement setup that can control the time delay between the stress and the measurement accurately. Since the relaxation is exponential in time, it is obvious that the shorter the time between the stress and the measurement, the bigger the error introduced by the delay-time uncertainty. It is therefore highly advisable to build-in a fixed time-delay to minimize the error.

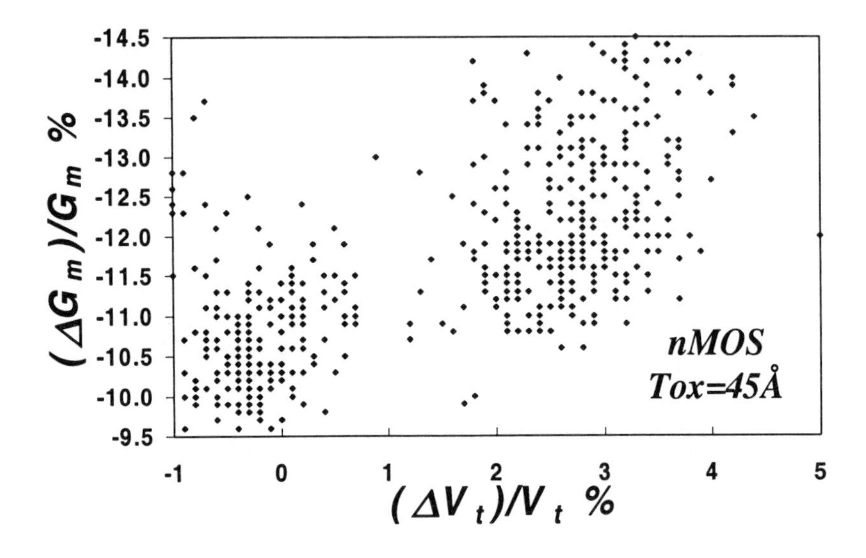

Figure 6.59. For gate-oxide thinner than 50Å, both the measured G_m shift and V_t shift become very small even if the damage level is high.

Finally, there is one more drawback in the transistor parameter shift measurement method, which is not applicable to oxides thinner than ~50Å. The reason is the relaxation phenomenon we just discussed above. For the thinner oxides, the relaxation of the trapped charge is so rapid that by the time one changes from the stressing step to the measurement step, none of the trapped charges are left in the oxide. Figure 6.59 shows the same G_m shift versus V_t shift plot as in Figure 6.54, except that the oxide thickness is 45Å instead of 65Å. Both the maximum G_m shifts and the maximum V_t shifts are so small that they are indistinguishable from the data scatter. The situation is worse for thinner oxides.

In summary, to use the transistor parameter shift measurement method, all the devices should be thoroughly annealed to remove all the trapped charges before the measurement start. The transistor parameters should be measured before and after a stressing step. The stress level does not have to be low. A higher stress gives a better sensitivity [36]. The post stress measurement should be done at a fixed delay. The shift from the before stress value is the damage indicator. Only the n-MOSFET data are reliable.

6.4.5 Hot-carrier and Fast Hot-carrier Stress Methods

An example of hot-carrier lifetime degradation in correlation with IETS was already shown in Figure 6.49. Shone *et al.* [3] was probably the first to report a DC hot-carrier lifetime degradation by plasma charging damage. Since then, large number of studies have been reported [38, 41-49]. Plasma charging damage degrades the p-channel hot-carrier lifetime has been universally observed. Degradation of the n-channel hot-carrier lifetime by plasma charging damage, on the other hand, has many conflicting reports. Some reports [3, 38, 42, 43, 47] observed charging damage induced degradation, while others [44, 46] observed no degradation. The conflicting observations were resolved when it was discovered that the n-MOSFET not-carrier lifetime degradation is dependent on the polarity of plasma charging [50].

The hot-carrier lifetime measurement is normally a very time consuming process. The procedure involves measuring the lifetime on devices stressed at three different elevated acceleration voltages and then extrapolates back to obtain the lifetime at the operational voltage. For fear of the extrapolation model might not be applicable at too high an acceleration voltage, the highest stress conditions are chosen so that the device will take at least many hours to reach the failure criteria such as 10% G_m degradation or 5% I_{dsat} (saturated drain current) degradation. The lowest stress condition thus will take up to a few months to reach the failure criteria.

For plasma charging damage detection, we do not need to obtain the actual hot-carrier lifetime. Instead, we only need to obtain the percent degradation of the lifetime. If there is a direct relationship between the degradation of the lifetime at the highest acceleration stress and the degradation of the lifetime at normal operational voltage, we can simply measure the percent lifetime degradation at the highest acceleration stress. Measuring only the percent degradation at the highest acceleration stress voltage helps to reduce the measurement time. However, as we have already mentioned, even at such condition, an undamaged device will still need several hours of stress to reach the failure criteria. It is still too slow to be used in the routine damage monitoring.

It is possible to further shorten the measuring time by recognizing that the hot-carrier stress-induced degradation is linear in time when plotted in the log-log scale after the initial transient period (Figure 6.60). One does not need to wait until the device has actually reached 10% G_m degradation. Instead, one can extrapolate from the degradation data at earlier time to the 10% degradation time. However, one must be careful not to make the extrapolation before the degradation has reached the linear region. A stress time of a few hundred seconds is still required.

To shorten the measurement time to only a few seconds, Hook *et al.* [47] raised the stress level to ensure that at least a few percent of degradation has occurred by the end of the stress. In other words, they raised the stress condition such that even the undamaged devices will be in the 3% to5% degradation range as shown in Figure 6.60 at the end of their 5-second stress. They then simply use the amount of degradation after the stress as an indicator of charging damage. While the highly accelerated stress condition may be a little bit unorthodox, they appears to be quite successful in applying the method extensively [47, 51-53].

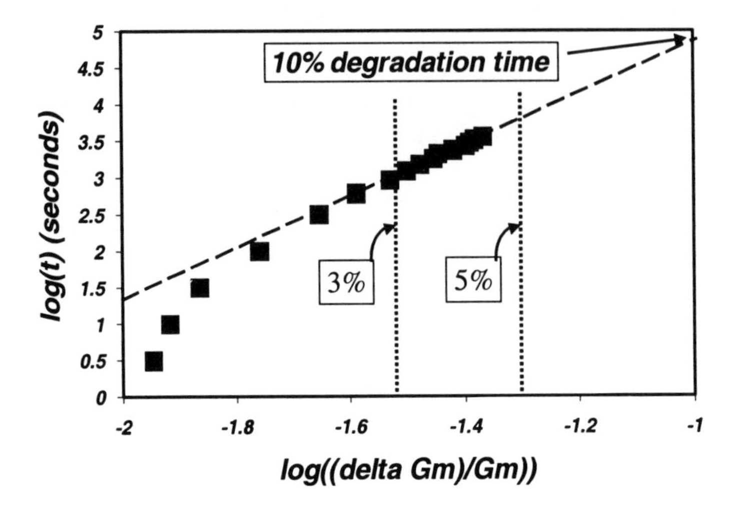

Figure 6.60. The G_m degradation during hot-carrier stress is linear in time when plotted in log-log scale after the initial transient period.

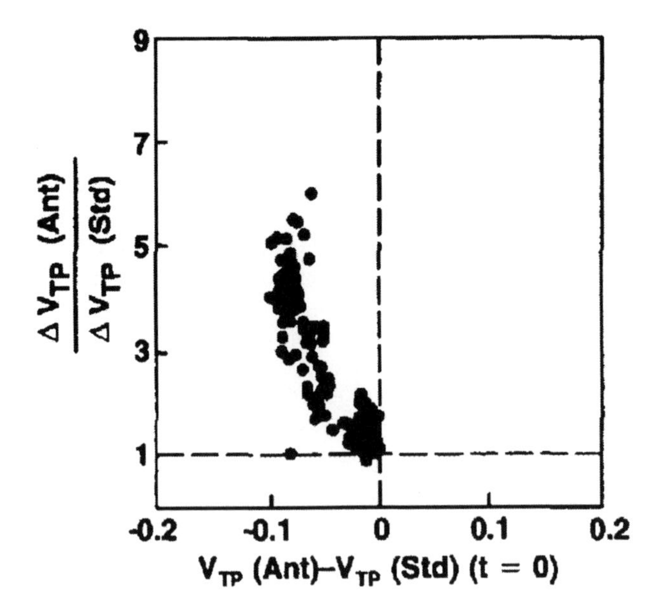

Figure 6.61. The V_t shift ratio between antenna devices and control devices after a p-channel hot-carrier stress for 5 seconds correlates with the pre-stress V_t offset. Taken from [51].

Figure 6.61 is an example of Hook *et al.*'s result [51]. Here they showed that the ratio (the antenna device to the control device) of the p-channel hot-carrier stress-induced V_t shift is related to the pre-stress V_t offset. As we mentioned before in this chapter, as fabricated wafers should not have pre-stress V_t offset if they were annealed properly. However, any residual pre-stress V_t offset is still an indication of damage.

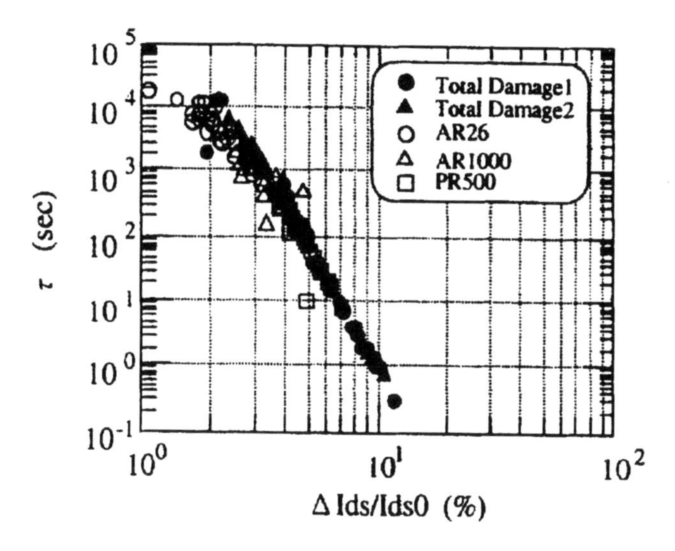

Figure 6.62. The I_{dsat} degradation ratio between antenna and control devices after a 1 second stress correlates well with the 5% degradation time from the same stress. Taken from [48].

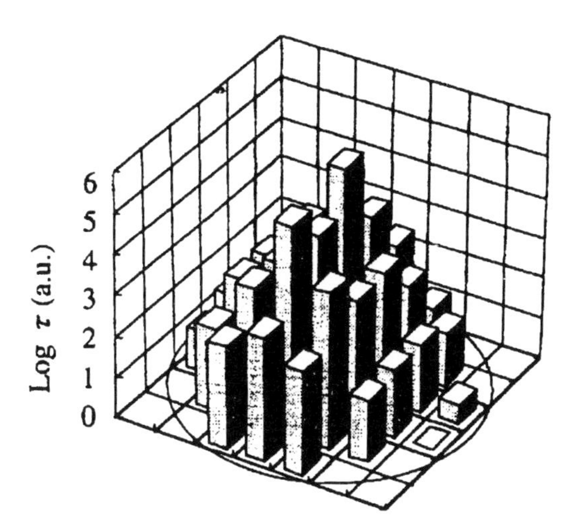

Figure 6.63. The hot-carrier lifetime map of a highly non-uniformly damaged wafer. Taken from [48].

Perhaps a much clearer demonstration of the method is given by Watanabe *et al.* [48]. Figure 6.62 shows their result. In this experiment, they used a high enough stress level so that only a 1-second stress is needed. The measured degradation (I_{dsat} in this case) ratio shows a clear relationship with the lifetime (5% I_{dsat} degradation) under the same stress condition. Note that the undamaged device's lifetime is about 10,000 seconds under this stress condition. Even though the lifetime using I_{dsat} degradation as criterion is usually longer than using the peak G_m or V_t, the acceleration stress used here is not too much higher than the highest level normally used. The 1-second stress-time is fast enough for use in the routine charging damage monitoring. Figure 6.63 shows an example [48] of using this technique to map the highly non-uniform damage level across a wafer.

The transistor parameter degradation induced by the hot-carrier stress is related to the creation of traps either at the interface or in the bulk of SiO_2 depending on the stressing mode. For the plasma damaged oxide, latent traps of both types exist. During the hot-carrier stress, filling of the pre-existing traps is happening along side the creation of new traps. Since filling of the pre-existing traps is an easier process than the creation of new traps, the degradation in the damaged oxide is much faster than in the undamaged oxide under the same stress. To express this quantitatively, we can write:

$$\frac{d^2N_e}{dt^2} = (\kappa_0 + \kappa)\frac{dI_i}{dt} \tag{6.7}$$

where N_e is the trapped charge density, I_i is the impact ionization rate in the channel, K_0 and K are the trap creation efficiency and the latent trap filling efficiency, respectively. Since $K >> K_0$, and since the pre-existing traps have a finite concentration, after a while, all the pre-existing traps are exhausted and the degradation rate becomes the same regardless whether the oxide was damaged or not. This means that the differences in degradation rate due to damage should be observed only at the beginning of the stress and the degradation curves should become parallel at a longer stress time. This is indeed the case as shown in Figure 6.64 [49] where all the differences due to damage showed up in the first second of stress.

Equation 6.7 should be valid regardless of the stress condition or the stress time. Thus, it is not necessary to carry out the hot-carrier stress to the point where the degradation is linear in the log-log scale if we are only interested in the plasma charging damage. Because of that, the unusually high stress level is not needed either. However, one still cannot use an extremely low level of stress because the amount of degradation will be too low to measure. Based on these considerations, a short stress at the highest acceleration condition normally used in the hot-carrier aging test would be sufficient for the purpose of monitoring plasma charging damage. Figure 6.65 shows the relationship of the 1-second stress-induced degradation with the 10% degradation time for the n-channel peak linear G_m [49]. Notice that the damage-free lifetime is around 100,000 seconds, indicating a relatively low acceleration condition. The data here looked noisier because it is in a semi-log plot rather than in a log-log plot. The scatter is mainly due to the quality of the measurement instrument.

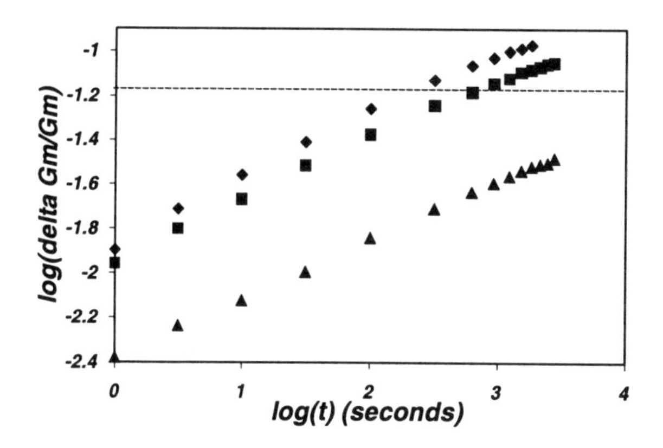

Figure 6.64. The differences in degradation due to damage should show up at the beginning of stress. All degradation curves are parallel at longer time.

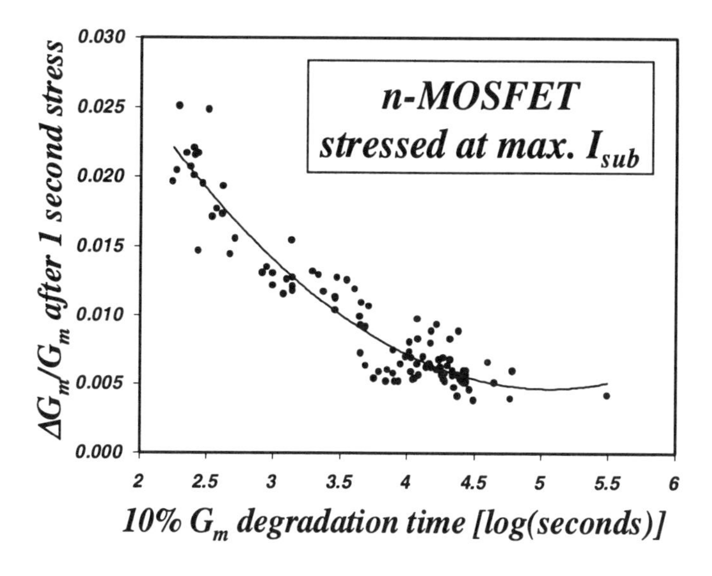

Figure 6.65. The G_m degradation after a 1 second of reduced stress correlates well with the 10% degradation lifetime at the same stress.

The fast hot-carrier stress method described here is sensitive and quick. It is directly measuring the damage to devices and is therefore a most relevant measurement. It can be calibrated directly to the lifetime at operational condition, and is therefore a quantitative method. It works for both the n- and the p-channel devices. It is therefore an excellent method for plasma charging damage monitoring. To use this measurement method, an important requirement must be met. That is, all the devices must have an identical transistor design. Thus one should never use this method in the situation where the charging damage from ion implantation is to be investigated.

6.4.6 Other Methods

Many electrical properties of the transistor are sensitive to the gate-oxide's quality and there are many different ways to measure these properties. All of these measurements can be used as indicators of plasma charging damage. Most of these measurements are suitable for the purpose of plasma charging damage research and have been used for that purpose by many groups. Some of measurements are even suitable for the routine plasma charging damage monitoring if we make the effort. In here we list a number of them but will not discuss in detail for space consideration.

6.4.6.1 Interface State Density Measurement

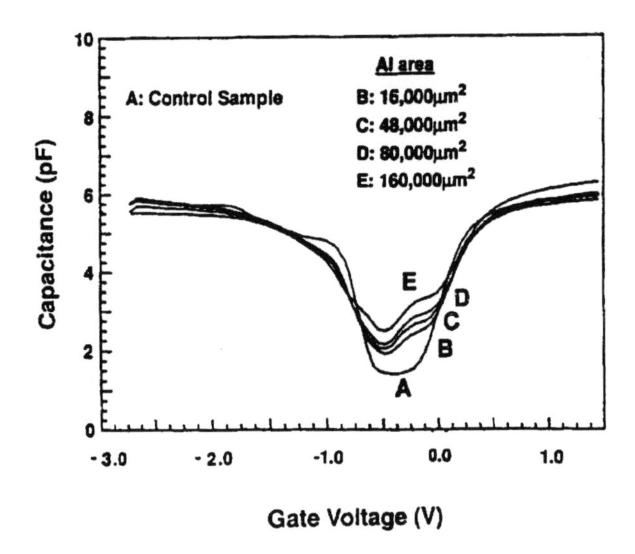

Figure 6.66. Low-frequency CV curve for various antenna size and control. Large antenna clearly has more interface traps. Taken from [54].

Interface state density can be obtained from the capacitance-voltage (CV) curves. Since the interface state density increases with plasma charging damage, the CV curves can be used to quantify plasma charging. This method relies on a set of low-frequency CV curves calibrated at various levels of constant current stress. The measured CV curves from plasma damaged capacitors of the same design are compared to the calibration curves to identify the stress current levels during plasma exposure. Figure 6.66 [54] shows an example of the usage of this method where the CV curves from capacitors with various antenna sizes after plasma processing is shown along with the control. Clearly, the interface state density is the highest for the largest antenna. After comparing to the calibration curves, the plasma stress current for each of the antenna size was found. Using this method, plasma charging current for devices with a given antenna size at various location on the wafer were found for different plasma power (Figure 6.67) [54].

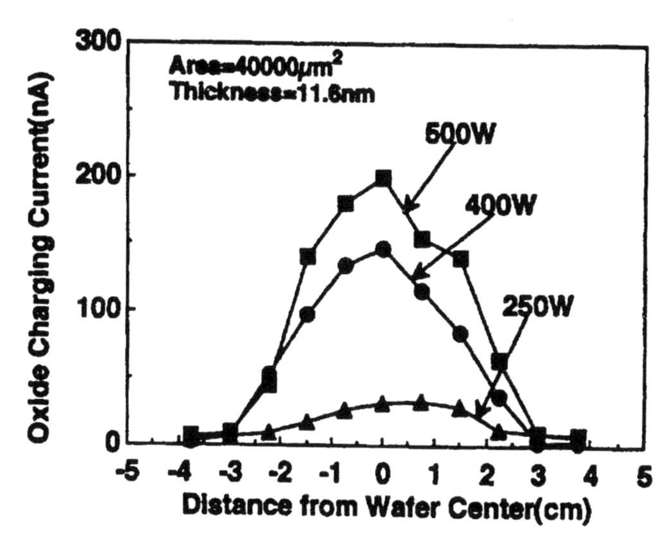

Figure 6.67. Using the low-frequency CV calibration curves, charging current for a given antenna size can be found as a function of chip location and plasma power. Taken from [54].

A quicker way to measure the interface state density is the DCIV method [55]. Guan *et al.* [56] applied this method to the plasma charging damage measurement. Figure 6.68 shows the DCIV current of their device as a function of the constant current stress time. It demonstrates that the peak DCIV current is related to the extent of charging damage.

The relationship between the DCIV peak current and the interface state density is not a simple one. It is not an easy task to make the translation. For a more direct and quantitative interface states density measurement, the charge pumping (CP) method [57] is far better. Figure 6.69 [58] shows an example of using the CP method to study the damage-induced interface state density, as well as the latent interface state effect. The CP method is a powerful method to measure the interface state density. It is very straightforward to deduce the interface state density from the CP current. The measurement can be quick and sensitive. The only reason that it is not a routine plasma charging damage monitoring method is that most of the automated measurement setup that are currently in use cannot handle the short pulses needed for the measurement.

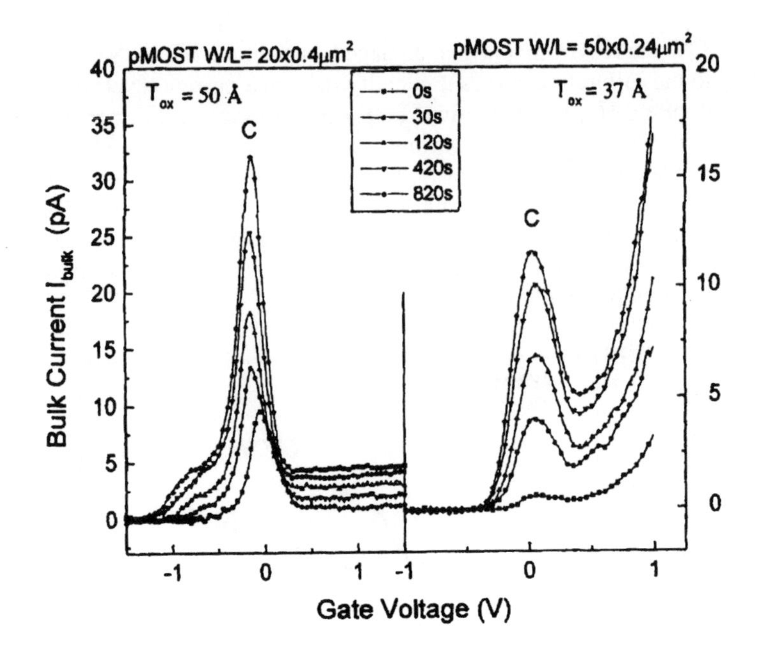

Figure 6.68. The DCIV current as a function of stress time simulating plasma charging damage. Taken from [56].

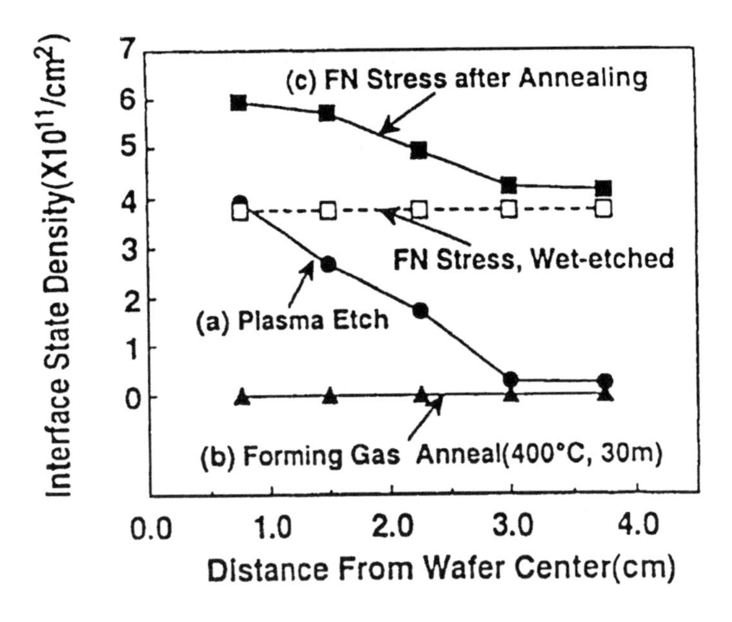

Figure 6.69. Interface state density increase due to plasma charging studied by charge pumping method. Post anneal stress reveal that majority of the interface states become latent upon annealing. Taken from [58].

6.4.6.2 Differential Amplifier Pair Mismatch Measurement

In most circuit design, digital or analog, the differential amplifier is one of the most common circuit elements. Depending on the application, the transistor-pair in the differential amplifier needs to match each other to different degrees of perfection. The signal output of the differential amplifier is extremely sensitive to any mismatch in the transistor-pair. This sensitivity has been utilized to amplify the antenna-induced transistor parameter shift. Figure 6.70 [59] shows the design of the differential amplifier that uses asymmetric antennae to boost the charging damage-induced mismatch in the transistor-pair. Figure 6.71 shows the output-offset voltage of the differential amplifier as a function of the area of the large antenna connected to the transistor-pair.

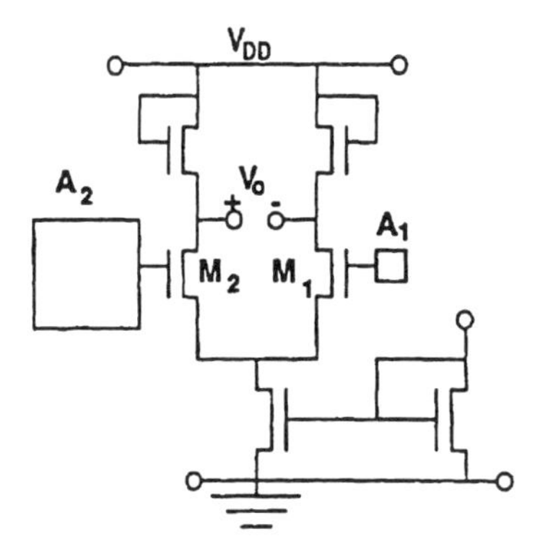

Figure 6.70. Intentionally introduced asymmetric antenna connecting to the differential transistor-pair to make the differential amplifier sensitive to plasma charging damage. Taken from [59].

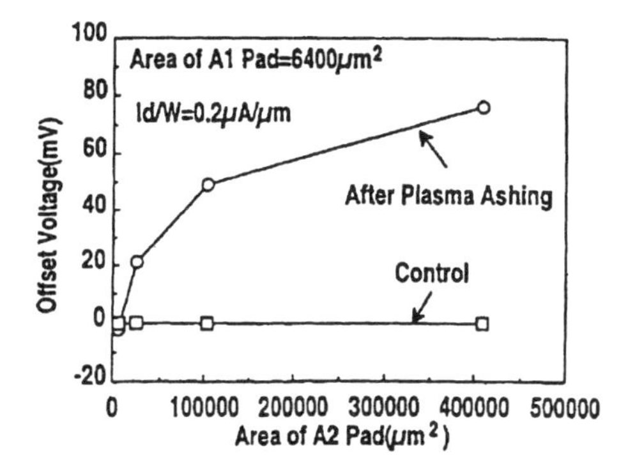

Figure 6.71. Output-offset voltage of the differential amplifier as a function of antenna area of the larger antenna in the differential pair. Taken from [59].

Intentionally introduced asymmetry in the differential amplifier design is fine for the purpose of enhancing the sensitivity to plasma charging damage. Real differential amplifiers are designed to be as symmetric as possible. In addition, the transistor-pair are usually right next to each other so that they would experience the same processing related effects if there are any. It is for this reason, one would not expect the differential amplifier to be sensitive to plasma charging damage. However, Zhao *et al.* [60] reported an antenna ratio dependent V_t mismatch in the transistor-pair in a completely symmetric design (Figure 6.72). A significant mismatch is observed even when the metal antenna ratio is as low as 20:1. This was a surprising result and was very difficult to understand. However, they proved experimentally that the mismatch was indeed due to plasma charging damage.

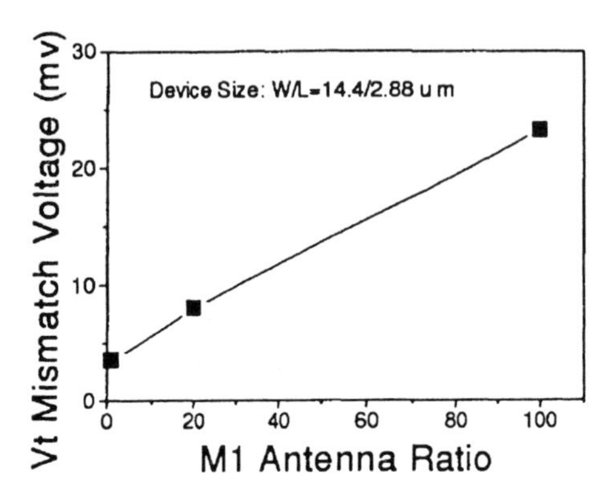

Figure 6.72. The V_t mismatch in transistor-pair in completely symmetric differential amplifier as a function of antenna ratio. Taken from [60].

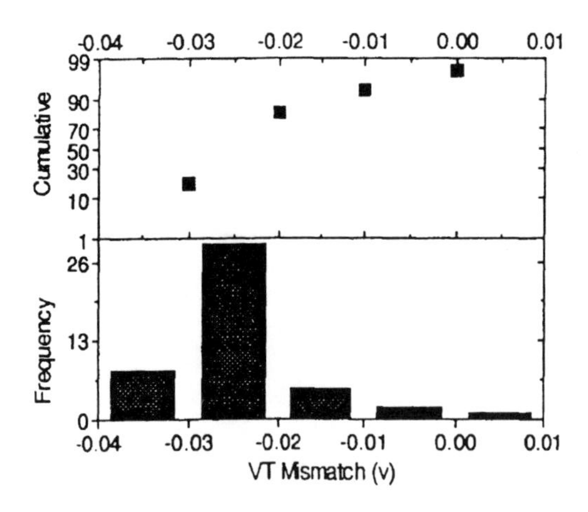

Figure 6.73. The differential-pair V_t mismatch due to plasma damage is not a simple spread out of the distribution. Instead, it is biased distribution. Taken from [60].

Differential amplifier mismatch due to plasma charging damage is a very serious issue in the IC industry. The cause of the problem is currently not understood. The only plausible explanation is that when damage happens, all the transistor parameters spread out due to the statistical nature of trap creation. One can think of the transistor-pair as making two samplings of a distribution. When the distribution is wider, the probability of the two sampling to have widely different value is higher. However, this would predict a wider distribution of the V_t mismatch rather than a biased distribution of the V_t mismatch as shown in Figure 6.73. Thus the true underlying mechanism for the transistor-pair mismatch is probably something else.

6.4.6.3 Drain Current Noise Measurement

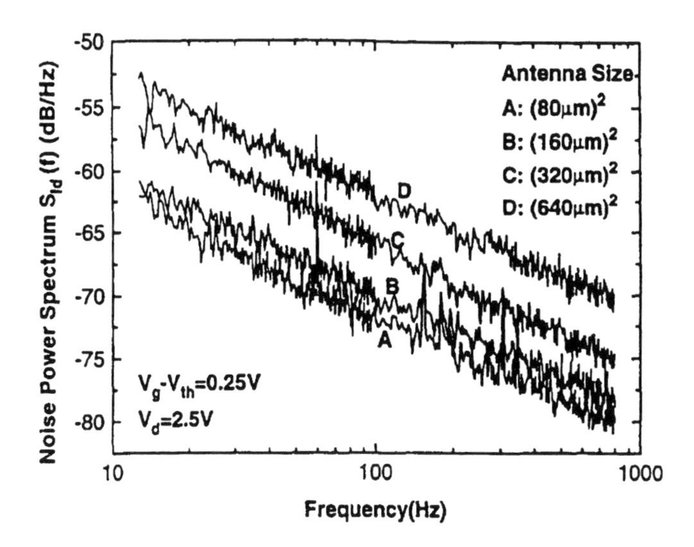

Figure 6.74. The low-frequency drain current noise increases with antenna size after plasma treatment. Taken from [59].

The MOSFET drain current noise in the low-frequency range is a channel-electron number-fluctuation phenomenon. This number-fluctuation comes from the charge exchange between the channel and the traps in the gate-oxide or at the interface. Naturally, we expect the plasma charging damage to cause this fluctuation to increase. This was indeed observed (Figure 6.74, [59]). The low-frequency drain current noise is an important parameter in many analog circuit designs, particularly in the designs for RF applications. Thus this charging damage-induced noise-increase should be an important criterion for minimizing the plasma charging damage. Unfortunately, this is a rather difficult measurement to carry out and is not commonly monitored. The low-frequency drain current noise-increase is dependent on the polarity of the charging damage [61]. The more commonly observed substrate injection damage causes larger noise increase (Figure 6.75).

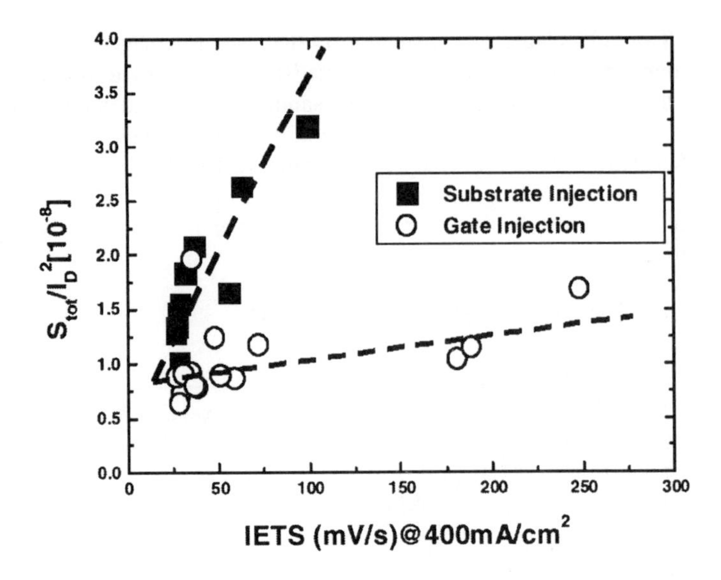

Figure 6.75. The normalized drain current noise spectral density as a function of IETS and charging polarity.

References

1. Schroder, D.K., *Semiconductor Material and device characterization*. 1990, New York: Wiley-Interscience.
2. Aum, P.K., *et al.*, *Controlling plasma charge damage in advanced semiconductor manufacturing - Challenge of small feature size device, large chip size, and large wafer size*. IEEE Trans. Elect. Dev., 1998. **45**(3): p. 722.
3. Shone, F., *et al.*, *Gate oxide charging and its elimination for metal antenna capacitor and transistor in VLSI CMOS double layer metal technology*. in *VLSI Technology Symp*. 1989. p. 73.
4. Harris, E.B., *et al.*, *Charging Damage in Metal-Oxide-Metal Capacitors*. in *International Symp. Plasma Process Induced Damage (P2ID)*. 1998. p. 15.
5. Watanabe, T. and Y. Yoshida, *Dielectric Breakdown of Gate Insulator Due to Reactive Ion Etching*. Solid State Technol., 1984(April): p. 263.
6. Shin, H., M. Je, and C. Hu, *Overestimation of Oxide Defect Density in Large test capacitors Due to Plasma Processing*. IEEE Trans. Electron Dev., 1997. **44**(9): p. 1554.
7. Aum, P., *et al.*, *Use of SPIDER for the Identification and Analysis of Process Induced Damage in 0.35um Transistors*. in *IEEE Conf. Microelectronic test Structure*. 1995. p. 1.

8. Brozek, T., *New Two-Terminal Transistor Test Structures and Test methodology for Assessment of Latent Charging Damage*. in *International Symp. Plasma Process Induced Damage (P2ID)*. 2000. p. 149.
9. Simon, P.L.C., *et al.*, *Multiplexed Antenna Monitoring Test Structure*. in *International Symp. Plasma. Process Induced Damage (P2ID)*. 1998. p. 205.
10. Wu, I.-W., *et al.*, *Damage to gate oxides in reactive ion etching*. in *SPIE conf. Dry Processing for Submicrometer Lithography*. 1989. p. 284.
11. Eriguchi, K., *et al.*, *Quantitative Evaluation of Gate Oxide Damage during plasma processing Using Antenna-Structure Capacitors*. Jpn. J. Appl. Phys., 1994. **33**(Part 1, No. 1A): p. 83.
12. Kosaka, Y., K. Eriguchi, and T. Yamada, *Stress Mode of Gate Oxide Charging during the MERIE and the ICP Processing and Its Effect on the Gate Oxide Reliability*. in *International Symp. Plasma Process Induced Damage (P2ID)*. 1998. p. 209.
13. Hook, T.B., D. Harmon, and C. Lin, *Detection of thin oxide (3.5nm) dielectric degradation due to charging damage by rapid-ramp breakdown*. in *International Reliability Physics Symp. (IRPS)*. 2000. p. 377.
14. Ricco, B., G. Gozzi, and M. Lanzoni, *Modeling and Simulation of Stress-Induced Leakage Current in Ultrathin SiO2 Films*. IEEE Trans. Electron Dev., 1998. **45**(7): p. 1554.
15. Weir, B.E., *et al.*, *Ultra-Thin Gate Dielectrics: They Break Down, But Do They Fail?* in *International Electron Device Meeting (IEDM)*. 1997. p. 73.
16. Alers, G.B., *et al.*, *J-Ramp on sub-3nm Dielectrics: Noise as a Breakdown Criterion*. in *International Reliability Physics Symp. (IRPS)*. 1999. p. 410.
17. Krishnan, S., *et al.*, *High density plasma etch induced damage to thin gate oxide*. in *International Electron Devices Meeting (IEDM)*. 1995. p. 315.
18. Jiang, J., O.O. Awadelkarim, and J. Werking, *Gate leakage current: A sensitive characterization parameter for plasma-induced damage detection in ultrathin oxide submicron transistors*. J. Vac. Sci. Technol. A, 1997. **16**(3): p. 1664.
19. Lin, H.C., C.H. Chien, and T.Y. Huang, *Characterization of Antenna Effect by Nondestructive Gate Current Measurement*. Jpn. J. Appl. Phys., 1996. **35**(Part 2, No. 8B): p. L1044.
20. Sridharan, A., *et al.*, *Leakage Current due to Plasma Induced Damage in Thin Gate Oxide MOS Transistors*. in *International Symp. Plasma Process Induced Damage (P2ID)*. 1997. p. 29.
21. Krishnan, S., *et al.*, *Antenna Device Reliability for ULSI Processing*. in *International Electron Device Meeting (IEDM)*. 1998. p. 601.
22. Amerasekera, A. and S. Krishnan, *ESD Protection Design And Application To Bidirectional Antenna Protection For Sub-5nm Gate Oxides*. in *International Symp. Plasma Process Induced Damage (P2ID)*. 1997. p. 33.
23. Miranda, E., *et al.*, *Soft Breakdown Conduction in Ultrathin (3-5nm) Gate Dielectrics*. IEEE Trans. Electron Dev., 2000. **47**(1): p. 82.

24. Miranda, E., *et al.*, *Switching Behavior of the Soft Breakdown Conduction Characteristic in Ultra-thin (<5nm) Oxide MOS Capacitors.* in *International Reliability Physics Symp. (IRPS).* 1998. p. 42.
25. Crupi, F., *et al.*, *On the Properties of the Gate and Substrate Current after Soft Breakdown in Ultrathin Oxide Layers.* IEEE Trans. Electron Dev., 1998. **45**(11): p. 2329.
26. Lee, Y.H., *et al.*, *Reliability Characterization of Process Charging Impact on Thin Gate Oxide.* in *International Symp. Plasma Process Induced Damage (P2ID).* 1998. p. 38.
27. Cheung, K.P., P. Mason, and D. Hwang, *Plasma charging damage of ultra-thin gate-oxide -- The measurement dilemma.* in *International Symp. Plasma Process Induced Damage (P2ID).* 2000. p. 10.
28. Fukumoto, Y., *et al.*, *Behavior of Plasma-Induced Pre-Tunneling Leakage Current in MOS Capacitor.* in *International Symp. Plasma Process Induced Damage (P2ID).* 1997. p. 99.
29. Park, D., *et al.*, *Stress-Induced Leakage Current Due to Charging Damage: Gate Oxide Thickness and Gate Poly-Si Etching Condition Dependence.* in *International Symp. Plasma Process Induced Damage (P2ID).* 1998. p. 56.
30. Takagi, S., N. Yasuda, and A. Toriumi, *Experimental Evidence of Inelastic Tunneling and New I-V Model for Stress-induced Leakage Current.* in *International Electron Devices Meeting (IEDM).* 1996. p. 323.
31. Dumin, D.J. and J.R. Maddux, *Correlation of Stress-Induced Leakage Current in Thin Oxides with Trap Generation Inside the Oxides.* IEEE Trans. Electron Dev., 1993. **40**(5): p. 986.
32. Takebuchi, M., *et al.*, *New Interconnect Plasma Induced Damage analyzed by Flash Memory Cell Array.* in *Internationa Electron Device Meeting (IEDM).* 1996. p. 185.
33. Takebuchi, M., *et al.*, *Tunnel Oxide Degradation Mechanism in Flash Devices for Via Etching Process.* in *International Symp. Plasma Process Induced damage (P2ID).* 1998. p. 116.
34. Brozek, T., Y.D. Chan, and C.R. Viswanathan, *A Model for Threshold Voltage Shift under Positive and Negative High-Field Electron Injection in Complementary Metal-Oxide-Semiconductor (CMOS) Transistors.* Jpn. J. Appl. Phys., 1995. **34**: p. 969.
35. Nissan-Cohen, Y., J. Shappir, and D. Frohman-Bentchkowsky, *Characterization of simutaneous bulk and interface high-field trapping effects in SiO2.* in *International Electron Device Meeting (IEDM).* 1983. p. 182.
36. Cheung, K.P., *On the use of Fowler-Nordheim stress to reveal plasma-charging damage.* in *International Symp. Plasma Process Induced Damage (P2ID).* 1996. p. 11.
37. Nissan-Cohen, Y., J. Shappir, and D. Frohman-Bentchkowsky, *High-field current induced positive charge transients in SiO2.* J. Appl. Phys., 1983. **54**(10): p. 5793.

38. Cheung, K.P., *An Efficient Method For Plasma-Charging Damage Measurement.* IEEE Electron Dev. Lett., 1994. **15**(11): p. 460.
39. Brozek, T. and C.R. Viswanathan, *Increased hole trapping in gate oxides as latent damage from plasma charging.* Semicond. Sci. Technol., 1997. **12**: p. 1551.
40. Sheu, B.R., *et al.*, *Charging Damage Evaluation in 0.25-0.3um Polycide Gate Etching Processes.* in *International Symp. Plasma Process Induced Damage (P2ID).* 1998. p. 76.
41. Lee, Y.H., *et al.*, *Correlation of plasma process induced charging with Fowler-Nordheim stress in P- and N-channel transistors.* in *International Electron Device Meeting (IEDM).* 1992. p. 65.
42. Li, X., *et al.*, *Plasma-Damaged Oxide Reliability Study Correlating Bot Hot-Carrier Injection and Time-Dependent Dielectric breakdown.* IEEE Electron Dev. Lett., 1993. **14**(2): p. 91.
43. Rakkhit, R., *et al.*, *Process Induced Oxide Damage and its Implications to Device reliability of Submicron transistors.* in *International Reliability Physics Symp. (IRPS).* 1993. p. 294.
44. Mistry, K.R., B.J. Fishbein, and B.S. Doyle, *Effect of Plasma-Induced Charging Damage on n-Channel and p-Channel MOSFET Hot Carrier reliability.* in *International Reliability Physics Symp. (IRPS).* 1994. p. 42.
45. Noguchi, K. and K. Okumura, *The Effect of Plasma-Induced Oxide and Interface degradation on Hot Carrier reliability.* in *International Reliability Physics Symp. (IRPS).* 1994. p. 232.
46. Li, X.Y., *et al.*, *Degraded CMOS Hot Carrier Life Time -- Role of Plasma Etching Induced charging damage and Edge damage.* in *International Reliability Physics Symp. (IRPS).* 1995. p. 260.
47. Hook, T., *et al.*, *A comparison of hot-electron and Fowler-Nordheim characterization of charging events in a 0.5um CMOS technology.* in *International Symp. Plasma Process Induced damage (P2ID).* 1996. p. 164.
48. Watanabe, H., *et al.*, *A Wafer Level Monitoring Method for Plasma-Charging damage Using Antenna PMOSFET Test Structure.* IEEE Trans. Semicond. Manufacturing, 1997. **10**(2): p. 228.
49. Cheung, K.P. and E. Lloyd, *Fast Hot-Carrier Aging Method of Charging Damage Measurement.* in *International Symp. Plasma Process induced Damage (P2ID).* 1999. p. 208.
50. Cheung, K.P., *et al.*, *Is n-MOSFET hot-carrier lifetime degraded by charging damage?* in *International Symp. Plasma Process Induced Damage (P2ID).* 1997. p. 186.
51. Hook, T.B., A. Stamper, and D. Armbrust, *Sporadic charging in interlevel oxide deposition in conventional plasma and HDP deposition systems.* in *International Symp. Plasma Process Induced Damage (P2ID).* 1997. p. 149.
52. Hook, T.B., *Backside Films and Charging During Via Etch in LOCOS and STI Technologies.* in *International Symp. Plasma Process Induced Damage (P2ID).* 1998. p. 11.

53. Stamper, A.K., A. Chou, and T. Hook, *Dual Gate Oxide Charging Damage in Damascene Copper technologies*. in *International Symp. Plasma Process induced Damage (P2ID)*. 2000. p. 109.

54. Shin, H. and C. Hu, *Monitoring plasma-process induced damage in thin oxide.* IEEE Trans. Semicond. Manufacturing, 1993. **6**(2): p. 96.

55. Neugroschel, A., *et al.*, *Direct-Current measurement of oxide and interface traps on oxidized silicon.* IEEE Trans. Electron dev., 1995. **42**(9): p. 1657.

56. Guan, H., *et al.*, *Nondestructive DCIV Method to Evaluate Plasma Charging Damage in Ultrathin Gate Oxides.* IEEE Electron Dev. Lett., 1999. **20**(5): p. 238.

57. Maes, H.E., *et al.*, *Understanding of the hot carrier degradation behavior or MOSFET by means of the charge pumping technique.* Applied Surface Sci., 1989. **39**: p. 523.

58. Shin, H.C. and C. Hu, *Thin gate oxide damage due to plasma processing.* Semicond. Sci. Technol, 1996. **11**: p. 463.

59. Shin, H., Z.J. Ma, and C. Hu, *Impact of Plasma Charging damage and Diode Protection on Scaled Thin Oxide*. in *International Electron device meeting*. 1993. p. 467.

60. Zhao, J., H.S. Chen, and C.S. Teng, *Investigation of Charging Damage Induced Vt Mismatch fo Submicron Mixed-Signal Technology*. in *Internatonal Reliability Physics Symp. (IRPS)*. 1996. p. 33.

61. Cheung, K.P., *et al.*, *Impact of plasma-charging damage polarity on MOSFET noise*. in *International Electron Device Meeting (IEDM)*. 1997. p. 437.

Chapter 7

Coping with Plasma Charging Damage

The impact of plasma charging damage on the yield and reliability of IC products and how to cope with the problem is the subject of this chapter. Since yield and reliability data of real products are sensitive information, they are rarely reported in the literature. As a result, this is the one area of the plasma charging damage field that is the weakest. Plasma charging damage degrades many transistor parameters. It is not difficult to imagine products that are sensitive to these parameters would suffer yield loss and reliability degradation. Due to the lack of published results, quantitative treatment is difficult. The only reported work on quantitative yield and reliability projection from plasma charging damage test result is based on the gate-oxide breakdown failure mode. This is because the gate-oxide breakdown reliability is one of the most pressing issues in deep submicron technologies. Currently, there is widespread belief that plasma charging damage is negligible in thin oxides. Whether this is true or not is discussed in this chapter.

7.1 Impact of Plasma Charging Damage on Yield and Reliability

The impact of plasma charging damage to product yield is rarely easy to identify. Most of the time, yield loss is dominated by defects unrelated to plasma charging. Out of habits, most yield improvement effort concentrate on minimizing defects rather than minimizing charging damage. An additional factor is that severe charging damage usually can be identified early in the development cycle before product yield issue is encountered. When the gate-oxides are broken, it is easy to recognize. This is probably the case in the first report [1] of charging damage. It is the moderate plasma charging damage that escaped detection during development phase that can cause yield problem. Such situation is not supposed to happen if our charging damage monitoring methods are effective. However, as we have seen in the last two chapters, the detection methods are chosen mostly by convenience and may not adequately address the right failure criterion.

Figure 7.1 shows an example of a product involving embedded flash memory. The damaging process involved in this case was a product specific step not in the core CMOS technology. The failure mode of poor retention time is also product (flash) specific. The only available signals were the persistent low yield and the failure map. There were a number of suspected processing steps that can cause the problem and charging damage was not the only possible mechanism of failure. Thus the trouble-shooting process was difficult, slow and expensive. When the damaging step was found and fixed, the difference was drastic as shown.

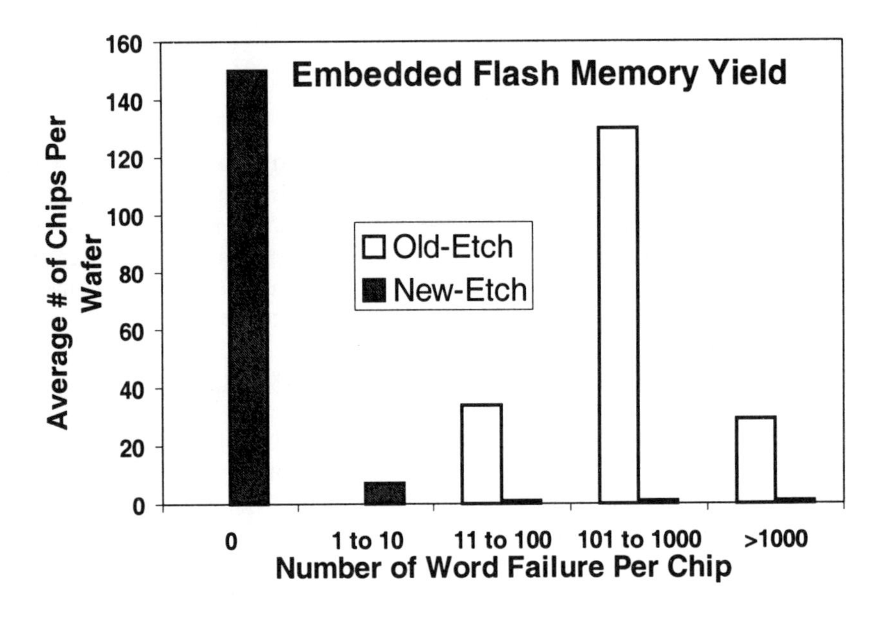

Figure 7.1. The yield of the embedded flash memory improved drastically when the damaging plasma processing step was fixed.

When the charging damage is not severe, the yield impact can be quite subtle. Figure 7.2 (a) and (b) [2] shows an example of very subtle yield loss due to plasma charging damage. In (a) the yield distribution of the charged wafers of product A is compared to the normal wafers. The average V_t shift of a p-MOSFET located in the scribe-lane with 550:1 perimeter intense antenna ratio defines whether a wafer is charged or not. The criterion was a 25mV V_t shift. Above that the wafer is labeled charged and below that the wafer is labeled normal. Without the large number of wafers, one would not have been able to tell the subtle difference in yield distribution. In (b) the yield distribution of product B shows no significant difference between the normal and the charged wafers. This difference in behavior between product A and B was explained by the antenna ratio distribution of the two products (Figure 7.3). Product A has larger number of antenna in almost every ratio bin than product B and is therefore more susceptible to charging damage.

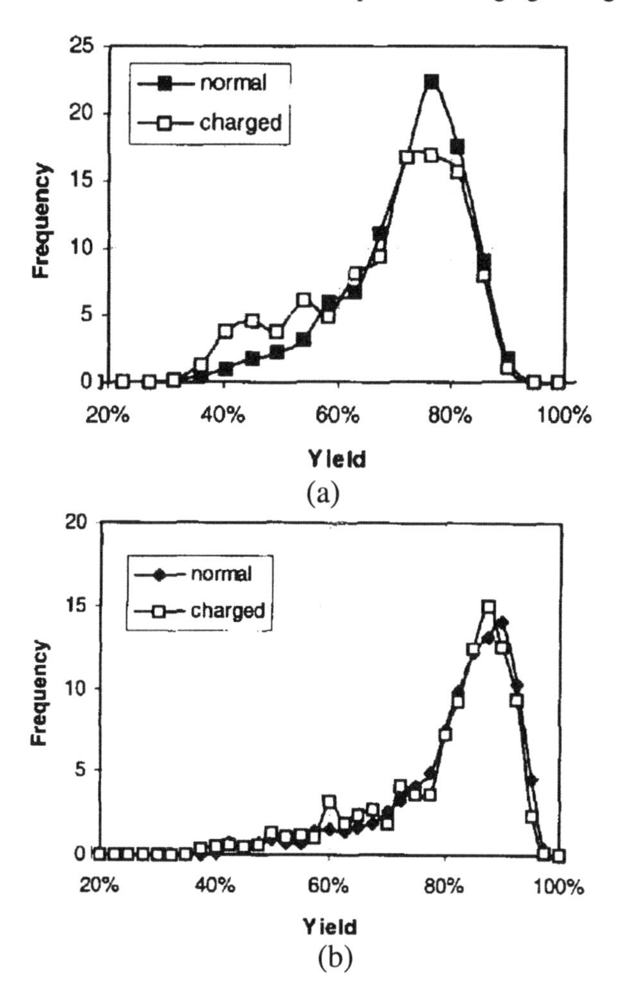

Figure 7.2. (a) The yield distribution of 1484 wafers of the product A for normal and charged wafers. (b) The yield distribution of 527 wafers of the product B for normal and charged wafers. Taken from [2].

Because of the various anneal cycles as well as the final anneal in forming gas, the damaged devices that are not broken would behave normally during tests at

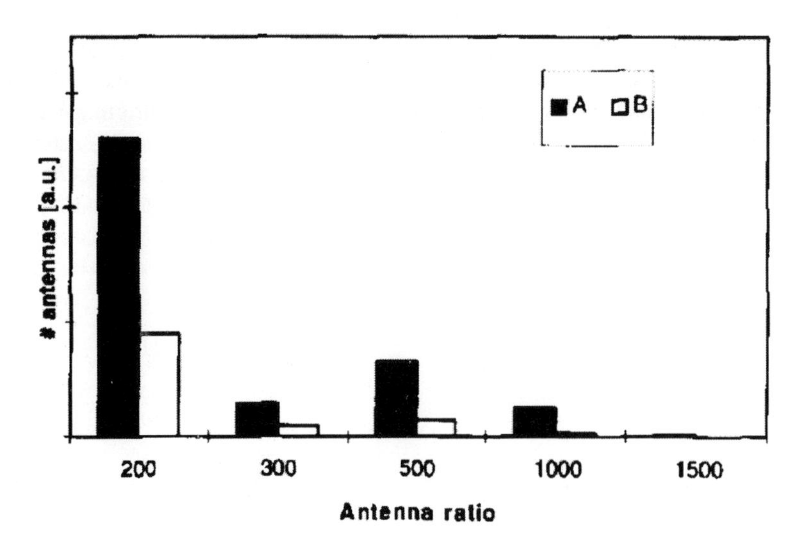

Figure 7.3. The number of antenna for each antenna ratio bin was higher for product A than for product B. Taken from [2].

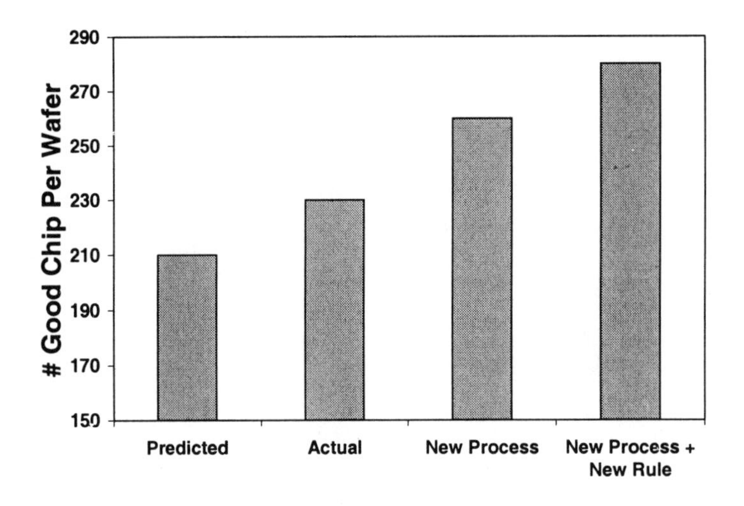

Figure 7.4. The product yield was improved by fixing the plasma process that caused the damage. The improvement was larger when tighter antenna rule was also applied to the product design.

the wafer level. During subsequent stress tests such as burn-in, they may show up as increased failure rate. Figure 7.4 shows a real-life example. The actual yield of a product was found to be higher than the projected yield based on the D_0 (defect density) level of the technology. Other than clearly indicating that the yield projection model is in need of improvement, the yield result suggested that everything else must have been in good shape it to be better than the projection. However, this product failed reliability test. Analysis of the failure chips showed that the failure node was associated with the largest antenna ratio, indicating that charging damage might be the cause. Two courses of action were taken, the circuit layout was redesigned with tighter antenna rule and the offending plasma process was fixed. Process improvement was only able to reduce the charging damage, not eliminate it. Yet, a clear improvement in yield was evident. The combination of the new process and the new tighter design rule achieved the best yield, and the product passed reliability test.

Latent damage related reliability problem could show up in many different ways. Not all of them are within the normal reliability test parameter range. For example, drain current noise due to charging damage is latent after forming gas anneal (Figure 7.5). Only after stress would they show up as increased noise (Figure 6.75). Since only some application would have a stringent noise requirement, normal reliability test would not have captured this particular problem.

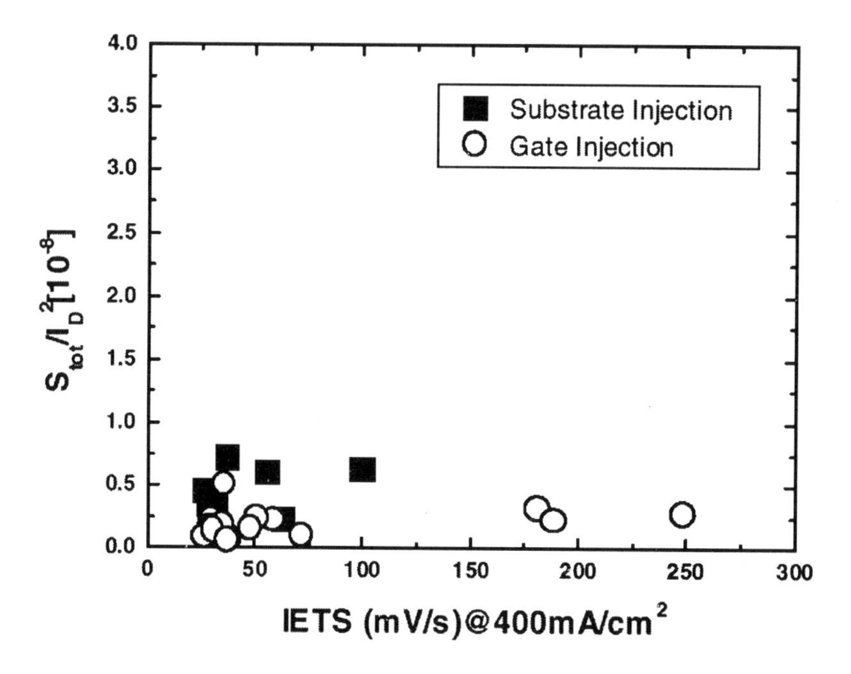

Figure 7.5. The drain current noise increase due to plasma charging damage was not observed before a stress was applied.

In general, it is rather difficult to identify the source of yield loss and reliability failure at the product level. It is far better to use highly sensitive plasma damage detection methodology during development stage to ensure that we eliminate the sources of damage or at least know the extent of damage from the processes that cannot be fixed completely.

7.2 Fixing the Damaging Process

After a plasma process is identified to be the source of charging damage, the best practice is to fix it. The method of fixing a damaging plasma process is highly process specific and cannot be generalized. For example, pressure has a dramatic effect on the level of charging damage during plasma enhanced dielectric deposition (Figure 4.15). The fix for the charging damage in a TCP metal etcher was to use a larger discharge coil and a bigger gap between the top electrode and the wafer. The fix for the charging damage in one ashing chamber was to introduce a trace amount of moisture. Tricks like these come from trial and error types of trouble-shooting. There is no systematic way of fixing plasma charging damage because most plasma systems are poorly understood. Common practice is to modify the processing parameters within the constraint of the equipment and the need to maintain a good process. Sometimes these measures are just not enough to fix a process completely. The example in Figure 7.4 is a case in point.

7.3 Use of Design Rules

When the charging damage cannot be completely eliminated, using a maximum-allowed antenna ratio design rule to limit its impact on the yield and reliability of products is the next line of defends. Figure 7.2 and 7.3 showed how smaller antenna and fewer antennae can improve the yield in the presence of charging damage. In general, the number of devices with antenna of certain ratio is not part of the design rule. In other words, if the antenna design rule allows an antenna ratio of 1000:1, the designer is free to use that ratio on as many devices as needed. This practice is reasonable when the gate-oxide is thick or when other failure mode other than gate-oxide breakdown is the failure criterion. When thin gate-oxide breakdown is the main failure mode, the number of devices associated with the antenna ratio is of significant importance.

The design rule used in the industry depends on the company and technology. It ranges from zero to infinity. Zero means that every floating node in the circuit is tied down by protection diodes. Infinity means that no rule is established. In the literature, most company targets an antenna design rule of 1000:1. In practice, the most common design rules fall between 100:1 to 1000:1.

7.4 Diode Protection

Currently, computer-aided-design (CAD) tools cannot automatically layout the circuit with the antenna rule as one of the constraints. Antenna rule violation checking is done as a separate step after the layout is completed. When antenna rule violation is found, the common method of handling the problem is to insert a minimum size diode if space is available. Diode protection in general is very effective in preventing charging damage from happening. However, diodes do represent additional circuit elements that need to be included in the circuit model. The additional load may degrade the circuit performance.

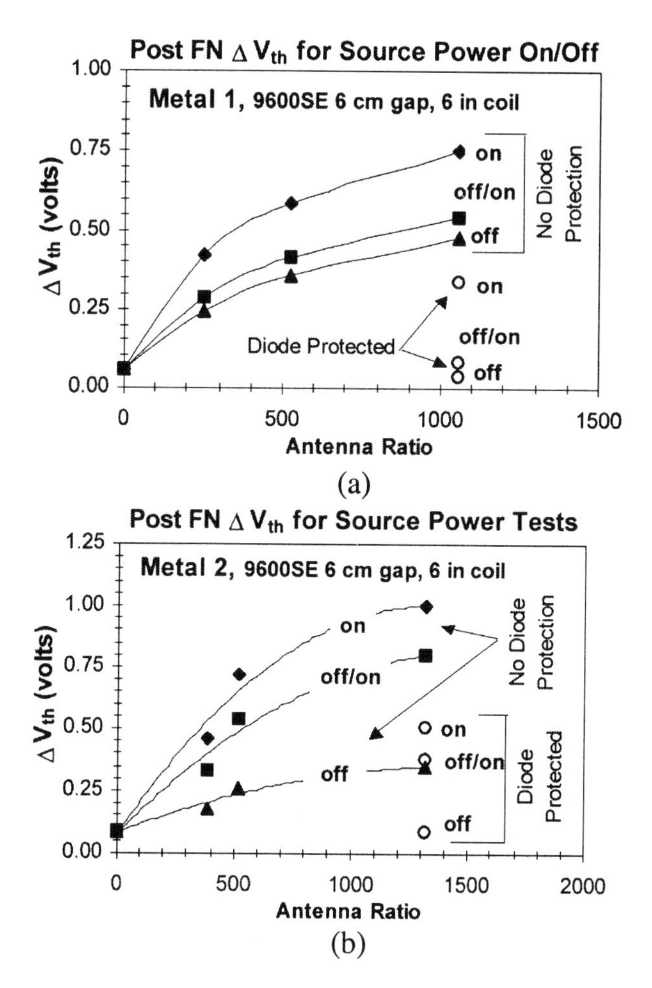

Figure 7.6. The V_t shifts are shown as a function of antenna ratio for the various combinations of power settings during metal etching. (a) Shows the results from metal 1 etching and (b) is the results from metal 2 etching. Taken from [3].

Diode protection working in the reverse bias mode relies on the photo-induced leakage current to help shunt the charging current from plasma. Although plasma always produces a large amount of photons, it is not a guarantee that enough would reach the diode to create sufficient leakage. Figure 7.6 [3] shows a case that diode protected devices are damaged by charging. The plasma process studied was metal etching using an inductively coupled plasma (ICP) source with RF bias supported by another independent source. The experiment was to study the impact of turning off the source power during the metal clearing (BE) step before the over etch (OE) step. When the source power is off, the RF source that provide the bias can keep the plasma from extinguishing and allow the etching to continue, but the plasma density is lowered by a factor of approximately 100. The off/on splits was to have the source power turned off initially during the BE step, but turned on again at half way through the BE step. The damage level for the three power settings are clearly different. The results that interests us here are that diode protection was not completely effective during metal 1 etch (Figure 7.6(a)). While the diode-protected devices have lower V_t shift for all the power settings, the power-on result is not that much better than the power-off result of the devices without diode protection. Foe the metal 2 etch, the situation is worse (Figure 7.6(b)) as less light is able to reach the diode through the metal 1 layer.

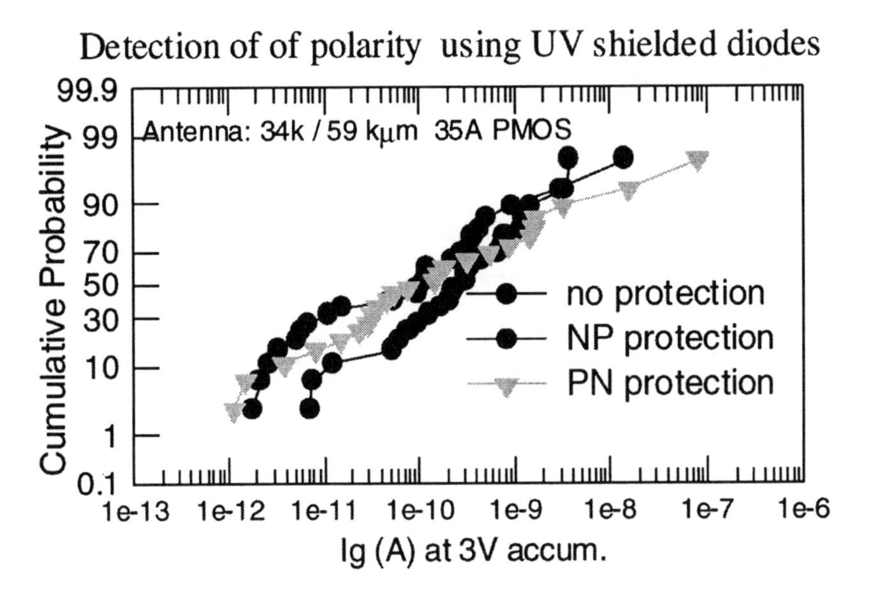

Figure 7.7. Antenna devices protected by UV-shielded diodes show the same level of damage as the unprotected antenna devices. Taken from [4].

The requirement of light in order for the diodes to protect the antenna devices is clearly demonstrated by the result reported by Krishnan *et al.* [4]. Figure 7.7 shows that if there is an UV shield over the diodes, no protection to the antenna device is seen. Figure 7.7 also demonstrates that charging damage, at least in this case, is bi-directional because devices with either np diode or pn diode protection are damaged to the same extent as the unprotected devices.

The failure of diode protection is not often reported in the literature, but they do exist. Another example can be found from Alavi *et al.*'s report [5] where they observed little protection was offered by the diodes (Figure 7.8). Although they did not discuss the question of how much light from the plasma can reach these diodes, it is still an alarming result.

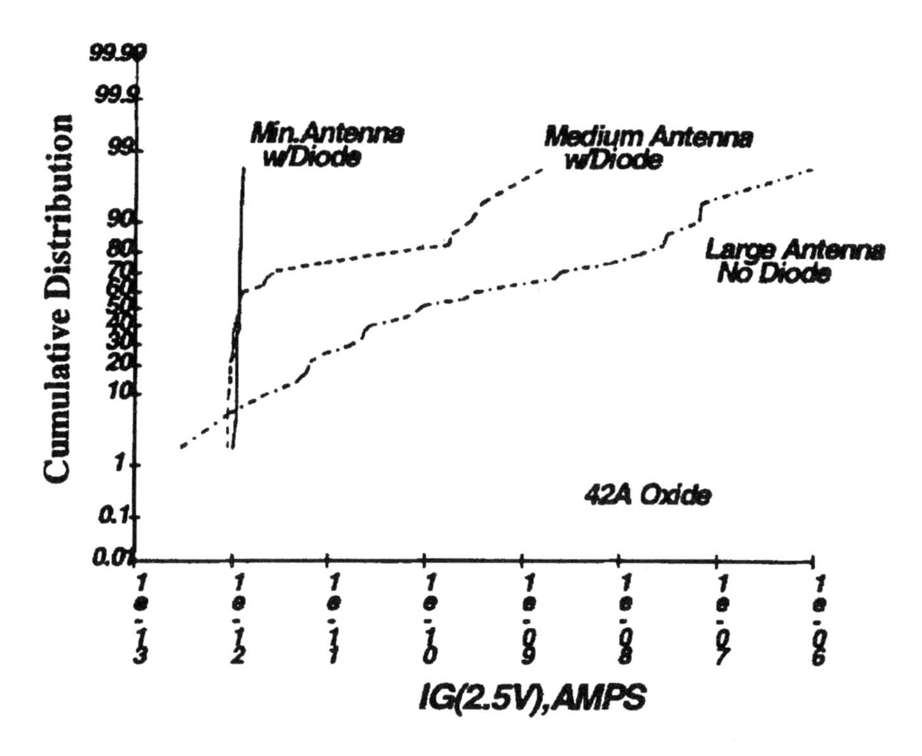

Figure 7.8. The medium antenna devices with diode protection are only marginally lower in damage than the large antenna devices without the diode protection. Taken from [5].

Despite the above examples of the failure of diode protection, diode protection does work in most cases. One just cannot count on it 100% of the times; particularly when damage is happening during back end processing where many levels of metal have already been processed. Since diode protection consumes chip space, loads down the circuit and does not guarantee success, it is best to use only as the last resort. The best strategy of dealing with plasma charging damage is to identify and eliminate. If that is not possible, at least identify and minimize, and then use design rule to limit the impact of the residual charging damage. Only when all that has failed, use diode protection.

7.5 Failure Criteria Problem

The strategy just outlined above requires the knowledge of what is the most sensitive failure criterion for the product that is impacted by the charging damage. In the literature, majorities of the reports either used the oxide breakdown or the transistor parameter shift as indicators of damage. Some included the latent interface states. Few addresses the question of how much damage is too much. Of course, if the gate-oxides are broken, it is too much. However, if only the devices with very large antenna ratios are broken, does that mean that the plasma charging damage level is acceptable? If the transistor V_t or G_m has more than 10% degradation after a brief stress, but only in devices that have very large antenna ratio, can one conclude that the charging damage level is acceptable? What is the basis for setting up the design rule? These questions are rarely addressed in the literature, yet they are the most commonly encountered problem in the industry. When a technology qualification deadline is fast approaching and that some very large antenna ratio test devices are showing charging damage, should the technology be qualified or should it not? What are the risks if the technology is qualified? If a foundry service is used to fabricate products, what data should the foundry provide to ensure that the product wouldn't fail in the field prematurely because of latent defects?

If one is making analog to digital converter or using it as part of the circuit, the requirement on the V_t and G_m matching will be extremely stringent. If one is making transceivers for cellular phones, the noise margin is extremely low. If one is making non-volatile memories, the leakage current tolerance is extremely small. Can one project the impact of plasma charging damage on the yield and reliability of such products? Very often, the performance of the product, as fabricated, is good enough to meet specifications. They may not meet the specification for very long if charging damage has created a lot of latent defects. How do one translate the damage measurement from the popular methods to these product specific situations? This is one area of research that has not been addressed until very recently. For the gate-oxide breakdown failure mode, Mason *et al.* [6] [7] did the most complete treatment of the problem while Noguchi *et al.* [8] and Van den bosch *et al.* [9] did a partial treatment of the problem. While these works are dealing only with gate-oxide breakdown failure, it is possible to extent them to the specific failure criteria of various products.

7.6 Projecting the Yield Impact to Products

When there is a sharp transition from good devices to broken devices in the antenna device yield distribution, The percent-failure or failure fraction of antenna devices can be determined accurately as shown in Figure 7.9. If the failure fraction for a number of antenna ratios are available (Figure 7.10), it is possible to project the yield of products from these measurement results.

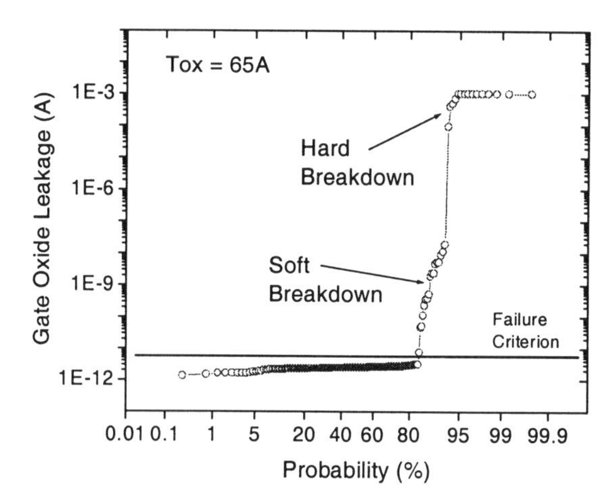

Figure 7.9. The gate leakage I_g distribution shows a sharp change going from good devices to bad (broken) devices, allowing the determination of failure fraction accurately.

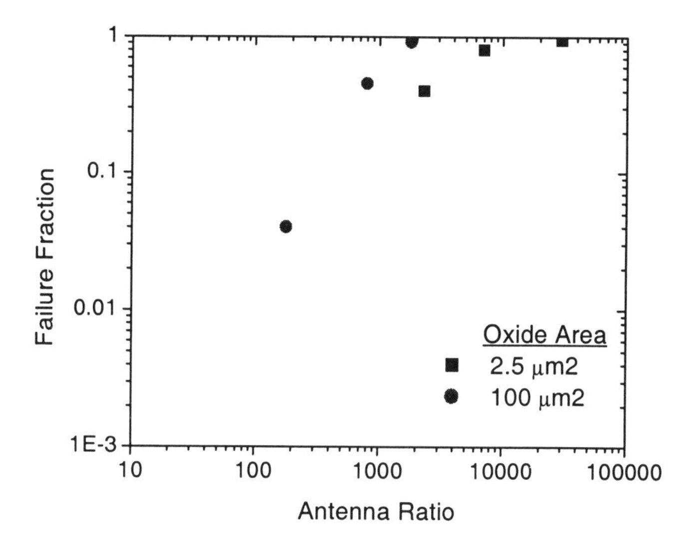

Figure 7.10. The measured failure fractions for a number of antenna ratios and two different device sizes.

To assess the impact of plasma charging damage on product yields, the antenna ratio distribution of the product is needed and can be extracted from the product layout file. (Such tools are still not commonly available, but should be soon.) Usually, the distribution would be a continuous function for a large circuit. For simplicity, the antenna ratios can be grouped into discrete value bins. Figure 7.11 shows the sample distributions of a large (total active area = $0.1cm^2$) circuit and a small (total active area of $0.0007cm^2$) circuit. The problem of finding the yield of these circuits is to find the failure fraction, F, of each antenna bin of the product.

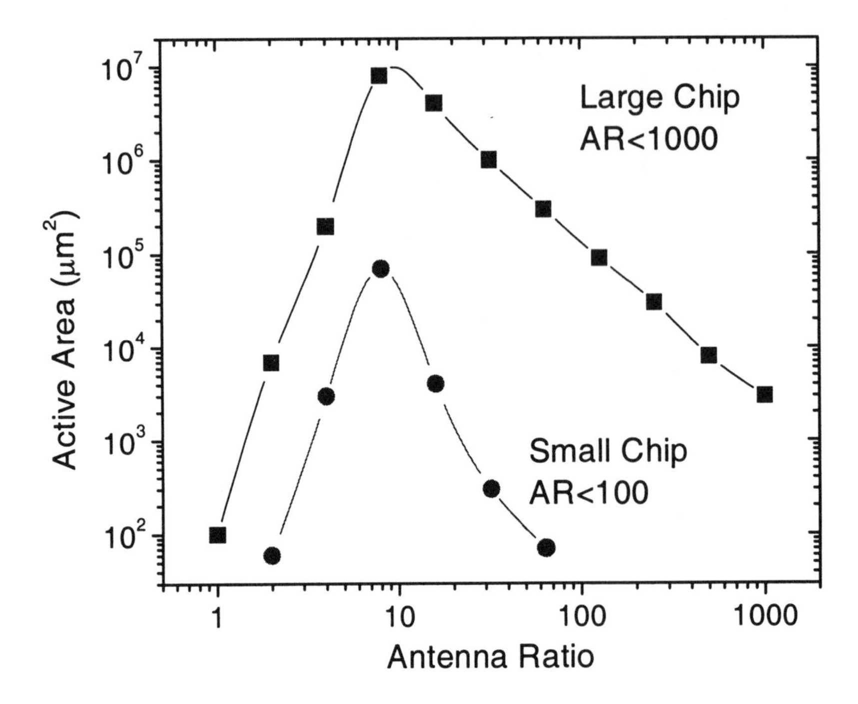

Figure 7.11. The antenna ratio distribution for a large circuit with antenna design rule of 1000:1 and a small circuit with antenna design rule of 100:1. The total active area of the large circuit is $0.1cm^2$ while the small circuit is $7x10^{-4}cm^2$.

The overall yield of a product is:

$$Y = \sum_i (1 - F_i) \tag{7.1}$$

where Y is yield and F_i is the failure fraction of the ith antenna bin.

To find the failure fraction of each antenna bin, the stress condition for each bin during plasma charging damage needs to be known. Also needed are the reliability characteristics of the gate-oxide of the technology used. During technology development, the gate-oxide reliability characteristics are routinely measured in detail. Thus all the relevant information should already be available without new measurements. Other information needed for the task is the specific

plasma process that is causing the damage. This can be obtained from the hierarchy of test structures and the damage monitoring methods. The needed information are the charging stress duration, polarity and temperature. For example, plasma enhanced dielectric deposition process caused the failure data of Figure 7.10. The stress duration was ~100 seconds, the stress polarity was substrate injection and the stress temperature was 375°C.
The starting point of making projection using the data of Figure 7.10 is a time to breakdown distribution of known-size capacitors with antenna ratio ~1:1 under a known stress voltage $V_{str.}$ Two of such distributions for the gate-oxide used in the devices of Figure 7.10 are shown in Figure 7.12. They are taken from the gate-oxide reliability database. The two distributions in Figure 7.12 are from two stress

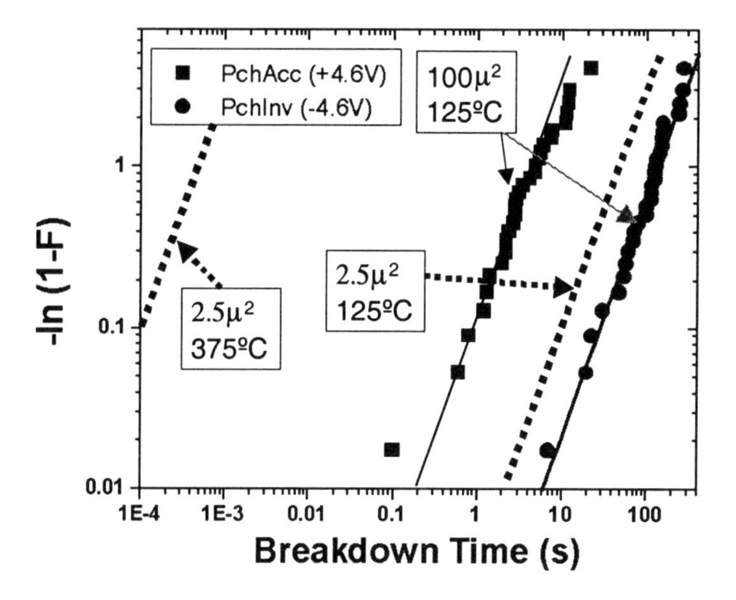

Figure 7.12. The breakdown time distributions of known-size capacitors stressed at known condition. Black solid squares are the data used as starting reference for scaling operation. Broken line on the right is due to area scaling. Broken line on the left is after temp. scaling of the broken line on the right.

polarities. Since the charging damage in question is substrate injection, the solid square distribution is the appropriate one to use. The first information to extract from this set of known breakdown data is the expected T_{BD} for the antenna devices of Figure 7.10 under the same V_{str}. The solid square distribution is for the $100\mu^2$ capacitors (Figure 7.12). The antenna devices are $2.5\mu^2$ and $100\mu^2$ (Figure 7.10). To get the breakdown time distribution of the $2.5\mu^2$ antenna devices stressed at V_{str}, the solid square distribution of Figure 7.12 needs to be translated using the area scaling relationship (Equation 1.7). The β value necessary for the calculation should already be known from the gate-oxide reliability data, or can be extract from the solid square distribution. The dotted line to the right of the solid square distribution is the result of the scaling operation. As expected, the smaller devices have longer lifetime under the same V_{str}.

To get the breakdown time distribution of the $2.5\mu^2$ antenna devices stressed at V_{str} and at 375°C instead of 125°C, this new distribution needs to be translated using the temperature scaling relationship (Equation 1.9). The temperature acceleration factor should also be part of the gate-oxide reliability database. The broken line at the far left of Figure 7.12 is the result of the scaling operation. From this line, the time required to accumulate any failure fraction of the $2.5\mu^2$ antenna devices stressed at V_{str} and 375°C can be calculated using the failure fraction scaling relationship of Chapter 1 Equation 1.5).

The failure fraction for the 6000:1 antenna ratio devices of Figure 7.10 is 0.82. The required stress time for this failure fraction can now be calculated. This calculated time is, of course, different from the charging duration of 100 seconds unless the stress voltage during charging damage happens to be exactly V_{str}. From the difference between the calculated stress time and the charging time, the actual charging voltage can be calculated using the voltage acceleration relation (Equation 1.8).

The whole procedure described above may seem complex. It can be summed up as follow: Starting from a TDDB distribution for a set of known-size capacitors stressed at known voltage and temperature. The breakdown time distribution for the antenna devices at the temperature of charging is calculated. From this new distribution, the stress time (at the known voltage) that would produce the measured failure fraction of the antenna devices is computed. The difference between the calculated stress time and charging time allows the charging voltage be determined.

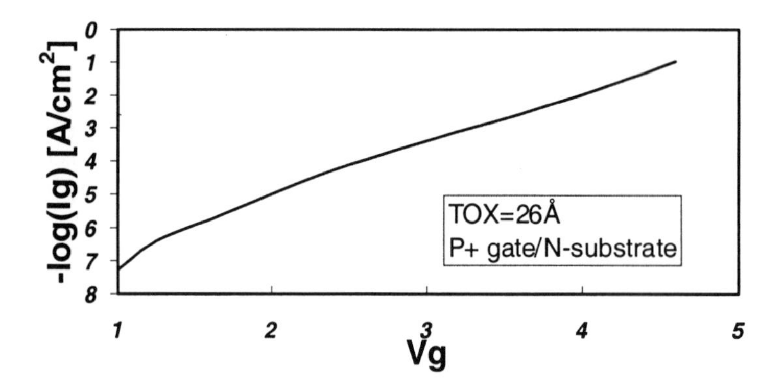

Figure 7.13. The current-voltage (I-V) curve for the 26Å gate-oxide in the devices of Figure 7.10.

Once the charging voltage during plasma damage is known, the charging current can be found directly from the gate-oxide I-V curve such as the one shown in Figure 7.13 for the gate-oxide of devices in Figure 7.10. Using this procedure, the charging current for all the antenna devices in Figure 7.10 can be obtained.

In the literature, it has been demonstrated that the plasma stress current is directly proportional to the antenna ratio [10, 11]. Had this been true in general, the stress current for all the antenna ratios can be calculated directly after the stress current of one antenna ratio is known. From there, the calculation of failure fraction for each antenna bin of a product is simply another scaling exercise. Unfortunately, the linear relationship is rarely applicable for modern plasma systems. In Section 2.6, the more general case where the stress current is not proportional to the antenna ratio was discussed. To get the charging current for any antenna ratio, the equation (Equation 2.8) that governs the charge balance between the plasma current and the tunneling current must be solved with the help of the known stress current values from a few antenna ratios.

The calculated stress voltage and current for the few antenna ratios of Figure 7.10 can be used to fit Equation 2.8. In principle, stress voltages of three antenna ratios are enough to pin down the parameters of the equation. However, since experimental data always has noise, more data points would be better. In Figure 7.10, there are 6 data points but only 4 are useful for the fitting process. Data points that have 100% yield or 100% dead are not useful for this purpose.

In Section 2.6, Equation 1.2 of Chapter 1 was used for the tunneling current. As discussed in Section 2.6, the knowledge of the precise electrical thickness of the oxide is required for Equation 1.2. While it is not difficult to do, it is very inconvenience. Furthermore, Equation 1.2 is rather complex, making the fitting tedious. In practice, a simplification of the tunneling current equation is possible. When the charging current is not linearly proportional to the antenna ratio, the range of stress voltage for any antenna ratio would not be too large. Examining Figure 7.13 reveals that, within a small range of voltage, the tunneling current in a semi-log plot is quite well approximated by a straight line. Thus Equation 1.2 can be rewritten as:

$$J_g \approx J_1 10^{-1.52(V_g - V_1)} \tag{7.2}$$

for the I-V characteristics around V_1. Here J_g is the tunneling current through the gate, J_1 and V_1 are the stress current and voltage of one point on the I-V curve that we have already determined. The value 1.52 is simply the slope of the straight line in the semi-log plot for this particular oxide and stress voltage range. Thus Equation 2.8 can be written as:

$$J_1 10^{-1.52(V_g - V_1)} = R_A L J_i \left(1 - e^{-(V_0 - V_g)/kT_e}\right) \tag{7.3}$$

Fitting the stress current of the four useful antenna ratios from Figure 7.10 to Equation 7.3, the best-fit stress current curve for all antenna ratios for the particular plasma process can be found (Figure 7.14). For comparison, the linear relationship between stress current and antenna ratio is also plotted (dotted line) in Figure 7.14. At least for the case at hand, the actual relationship between antenna ratio and stress current is very nonlinear and changes continuously as a function of antenna ratio. The failure fraction due to the same damaging plasma process for each antenna bin in Figure 7.11 can now be calculated using the various scaling relations. The result is shown in Figure 7.15. Not only is the failure fraction a strong function of antenna ratio, but also a strong function of the active area associated with the antenna ratio. Even though the 6000:1 antenna test devices have 18% yield (F=0.82), the large product with antenna rule limiting the antenna ratio to 1000:1 has zero yield. For the smaller product, on the other hand, the yield impact of the detected charging damage is only 0.03%.

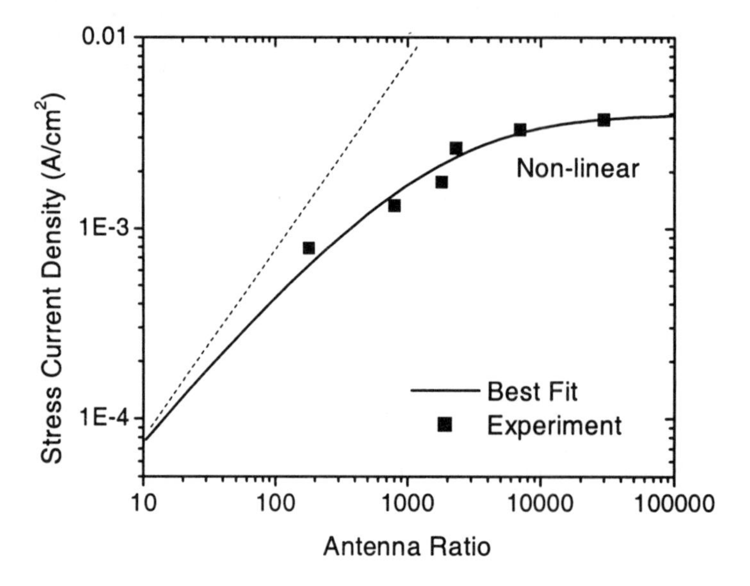

Figure 7.14. The best-fit curve of the stress current as a function of antenna ratio for the experimental data. The linear relation is shown also as dotted line.

The entire procedure described so far can be tedious and therefore should be automated. Once established, it is a powerful methodology to link the measured charging damage using small area but large antenna ratio devices to the product yield.

The assumption behind the methodology described above is that damage is uniform across the wafer. This is rarely true. Most plasma charging damage is non-uniform. To remove this shortcoming, the methodology should be used on a wafer site by wafer site basis. At the qualification phase of a technology, a few hundred wafers are processed identically to ensure the statistical robustness of the technology. These identically processed wafers allow each site to be treated independently and still have enough statistics to use the method described above. A side benefit of performing such analysis is that one can exclude regions of the wafer where the damage is more serious than acceptable from being sent to customers. Such flexibility allows the technology to be qualified in the presence of damage.

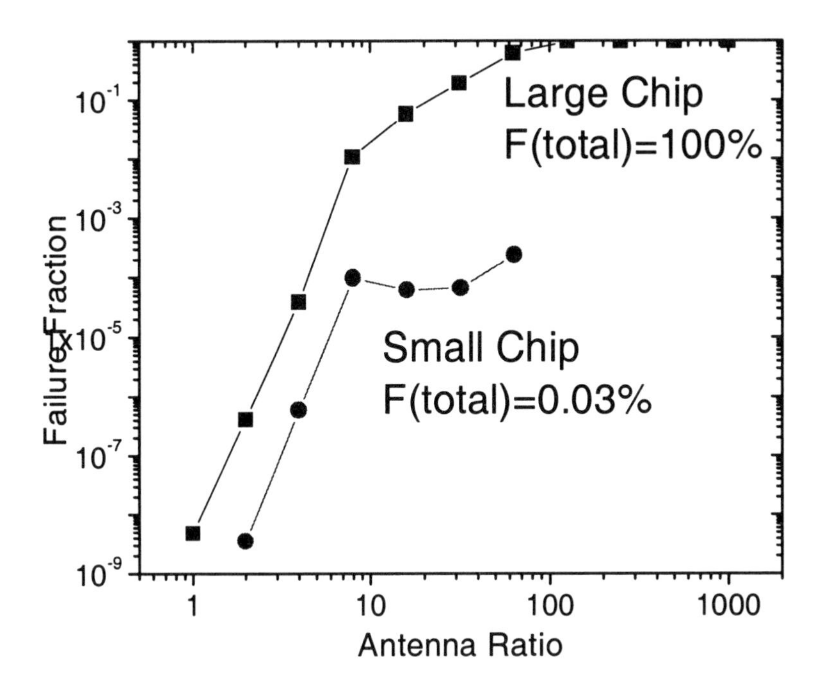

Figure 7.15. The calculated failure fractions of each antenna bin for the two products in Figure 7.11. The overall yield for the large product is zero while the overall yield for the small product is 99.97%.

The quantitative projection of yield impact due to plasma charging damage is just starting to appear in the literature. This is an important step in the development of the understanding of plasma charging damage. Only when we can discuss degradation quantitatively and specifically, can we set up meaningful criterion for tolerable damage or design rules.

The methodology discussed above deals with gate-oxide breakdown as a failure mode. This mode of failure is very important for thin oxides, but not for the thicker oxides. There has not been any reports in the literature that deal with the impact of plasma charging damage on yield quantitatively using other failure mode as criterion. The reason for that is the anneal cycles, including the forming gas anneal at the end of processing, turn most of the plasma charging induced traps into latent defects. Most other mode of degradations is not observable from wafers as processed. (Sometimes these degradations are reported due to inadequate annealing, but they can be underestimating the full impact of charging damage greatly). Since latent defects can reappear quickly upon stress, the yield impact of charging damage for other failure modes needs to be assessed only after a sort stress. For these failure modes, the above methodology is still applicable with small modification.

Since the key step of the methodology is to obtain the charging stress current for all antenna ratios during plasma exposure, the result can be used to deal with any other mode of failure as long as a calibration method is established. For example, the amount of noise increase due to latent defects as a function of stress can be established by the controlled stress-anneal-stress experiments. It is important to realize that traps generated by charging are the same intrinsic traps generated in TDDB test or during the long-term operation of the circuit. One can interpret the yield projection methodology as a method to determine the amount of traps that are generated in the gate-oxide due to plasma charging damage. Most of other failure mode can be dealt with similarly as long as the degradation is related to traps in the gate-oxide.

Unlike gate-oxide breakdown where the area and failure fraction scaling relationship are important for the calculation of failure fraction of each antenna bin, other failure modes may need their own scaling relation unique to them. In other words, the total active area associated with a particular antenna ratio may or may not matter and the time to failure distribution may not follow Weibull statistics. To calculate the failure fraction of each antenna ratio bin for failure mode other than breakdown, one need to establish the appropriate scaling relationship for that mode.

7.7 Projecting the Reliability Impact to Products

To project the impact of plasma charging damage on product reliability, the methodology of last section needs to be taken a little bit further. Figure 7.16 shows the breakdown time distribution of devices before and after charging damage. Some fraction of devices is dead (zero breakdown time). These devices are detected at wafer level test and screened off. The remaining populations are not virgin devices. Since all the devices have suffered some damage, a finite amount of traps has been generated in them. In other words, some portion of their lifetime has already been consumed. The question is how does the rest of the population fair during the burn-in test and during normal operation?

Define the gate-oxide survival distribution function as:

$$R(t) = \exp\left(-\left(t/\eta\right)^{\beta}\right) = 1 - F(t) \tag{7.4}$$

Assuming that plasma charging damage consumed a fraction ρ of the oxide lifetime:

$$\rho = \frac{t_{dam}}{\eta} \tag{7.5}$$

The survival distribution of damaged devices after the dead devices are screened off is [7]:

$$R'(t') = \frac{\exp\left(-\left(\left(t' + t_{dam}\right)/\eta\right)^{\beta}\right)}{R_0\left(t_{dam}\right)} = 1 - F(t') \tag{7.6}$$

where $R_0(t_{dam})$ is a renormalization factor reflecting that the total number of devices is smaller after screening. Figure 7.17 [7] shows the breakdown time distribution for screened devices with various level of damage expressed in terms of fractional consumed intrinsic oxide t_{63}. Note that the t_{63} here is for the product chip, not the antenna devices. Thus the t_{63} is product size sensitive and is defined at the reliability specification of 85°C. The intrinsic distribution is also shown for comparison. The oxide thickness was 55A.

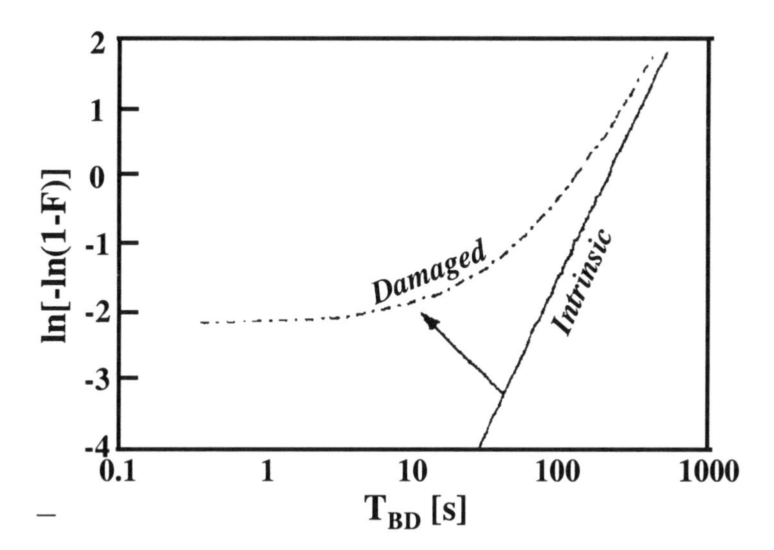

Figure 7.16. The breakdown time distribution of gate-oxide before and after the charging damage.

The solid dots in Figure 7.17 represents a burn-in condition. The burn-in tests are done at elevated temperature and voltage for a specified duration. This burn-in condition can be expressed as an equivalent consumption of intrinsic lifetime. The condition shown in the figure is 2% of the characteristic lifetime η (t_{63}). One can read directly from the family of distributions the fractional failures (the solid dots) during burn-in for each damage level. Remarkably, for the severe damage level of 50% intrinsic t_{63} consumed, the burn-in failure is only 1%. This result shows that plasma damage impact to short term reliability is small after dead devices are screened.

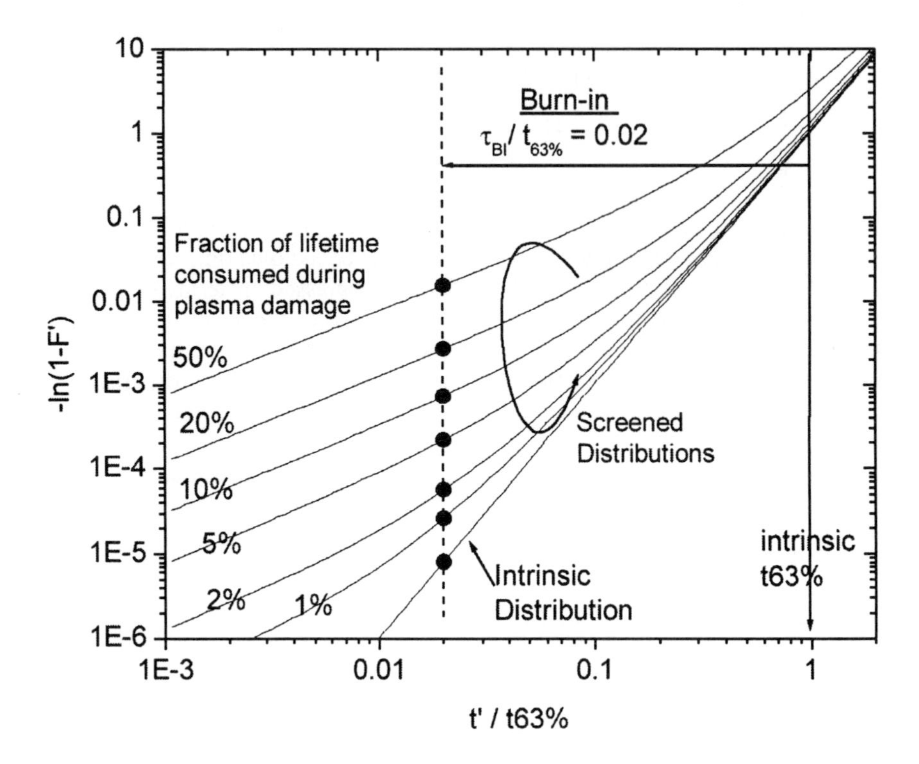

Figure 7.17. The breakdown time distribution of intrinsic devices as well as screened devices for a number of damage level expressed in terms of fraction of intrinsic lifetime consumed. Taken from [7].

The relationship between plasma damage was verified experimentally (Figure 7.18) [7]. In Figure 7.18, the burn-in failure is in arbitrary unit and is plot against yield of antenna devices. So the plot by itself does not prove that the model described here actually worked. It only shows that there is a relationship between burn-in failure and charging damage. However, Mason *et al.* had in fact gone through the whole methodology to prove that the model did work [7] quantitatively.

Since the model described here is the extension of the yield projection methodology, burn-in failure rate must correlates with yield as well. Since both are calculated quantitatively, one should be able to verify the model with this correlation. This is indeed verified (Figure 7.19) [7]. The solid curve in Figure 7.19 is the calculated relation between yield and burn-in failure based on the model. It agrees with experimental data rather well.

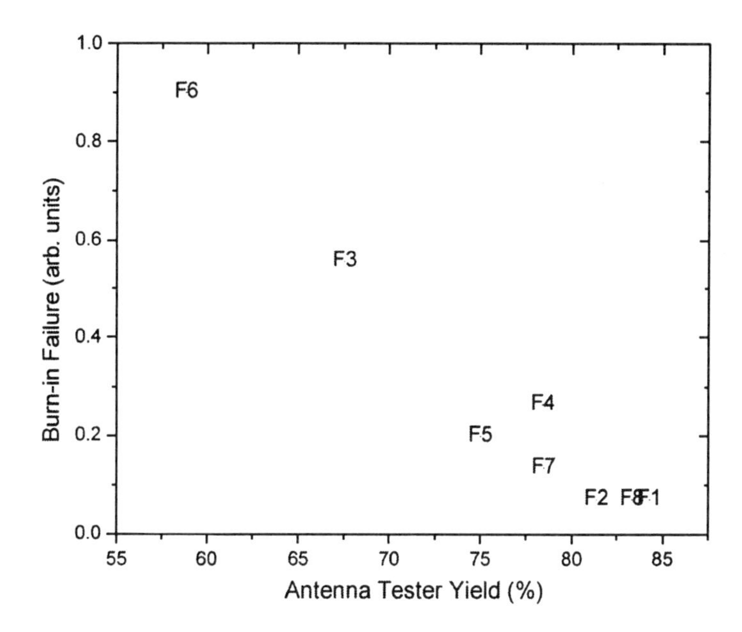

Figure 7.18. The burn-in failure rate correlates with antenna tester yield. Taken from [7].

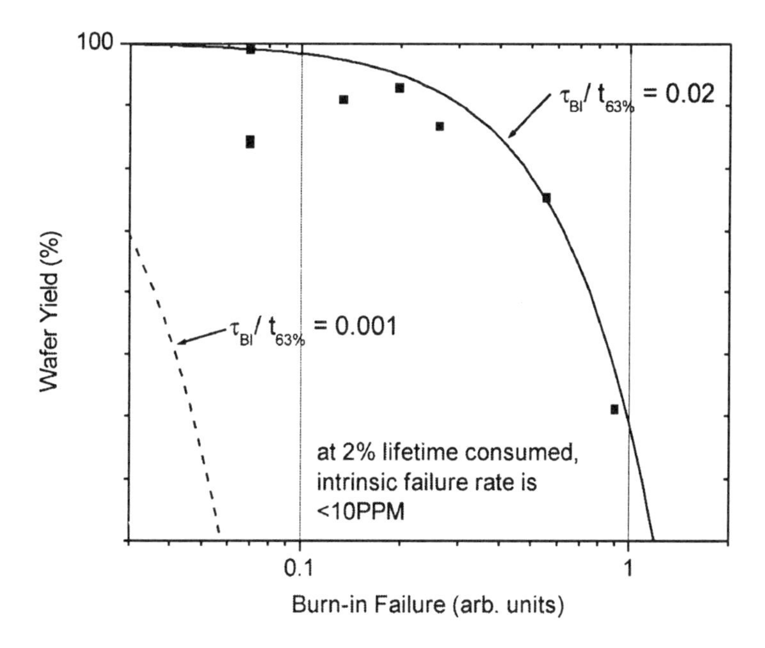

Figure 7.19. The burn-in failure rate correlated with wafer level test yield. Taken from [7].

It would not be accurate to conclude that plasma charging damage have little impact on short-term reliability. It is the screening of dead devices keep the apparent impact very low. Since the burn-in failure rate specification is usually in the 0.2% to 2% range, it is not very sensitive to plasma charging damage (after yield screening).

For long-term reliability, the specification calls for 0.01% failure in 10 years of operation. The screened population of plasma damaged devices is more likely to cause problem here.

One can use the family of breakdown time distributions in Figure 7.17 to clarify the impact of plasma charging damage on the long-term reliability of gate-oxide. Notice that the 2% consumption distribution is only a minor deviation from intrinsic distribution. One can thus assume that after the failure chip from burn-in test have been screened off, the new re-normalized distributions will remain similar to the before burn-in distributions. The time to accumulation 0.01% total failure for each distribution can be read directly from the curves. These are the lifetime of each distribution (not the characteristic lifetime t_{63}). Figure 7.20 is a reproduction of Figure 7.17 with the lifetimes marked on it. One thing immediately clear is that the lifetime decreases faster than the fractional t_{63} consumption. For example, the 20% t_{63} consumption distribution reached 0.01% cumulative failure 50 times faster than the intrinsic distribution. Another thing to notice is that as fractional t_{63} consumption increases, the time to 0.01% cumulative failure is shortening faster.

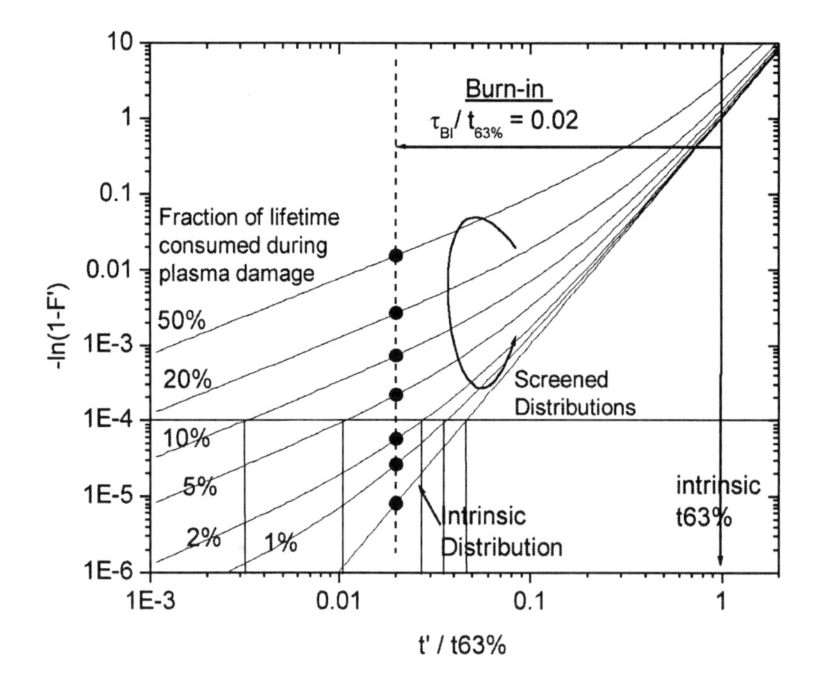

Figure 7.20. The time to reach 0.01% cumulated failure is shortened by plasma charging damage by a factor much bigger than the fractional consumption of the intrinsic lifetime.

The screened distributions lifetime shortening pace is directly related to the Weibull slope β. The shortening pace is faster when Weibull slope is shallower. The distribution shown in Figure 7.17 and 7.20 are for oxide of 55A thick, which has a β value of 3. The intrinsic lifetime of the 55A gate-oxide is about 10 times longer than the 10 years requirement at 2.5V, the tolerable level of plasma charging damage can be found from Figure 7.20 to be about the equivalent of consuming 20% of the intrinsic t_{63}. Thus from the long-term breakdown reliability point of view, the impact of plasma charging damage is minimum after screening for thick oxide unless the damage is severe.

For gate-oxides 20Å or thinner, the value of β is ~1 and the gate-oxide lifetime will barely meet the 10 years reliability requirement. For such oxides, if the reliability margin is 10% (11 years instead of exactly 10 years), the tolerable charging damage level can be calculated to be about 10^{-5} t_{63}. (Note that when β is ~1, t_{63} is ~10,000 times longer than t_{100ppm}).

It should be noted that the modeling so far treat the failure of the chip as if there is only one antenna ratio (one value of equivalent intrinsic t_{63} consumption). The real impact to chip reliability must be calculated by taking product of each individual antenna ratio bin. The chip failure is often dominated by the worst antenna ratio bin, as is in Mason *et al.*'s case [7], but it is not a general rule.

7.8 Ultra-thin Gate-oxide Issues

The reliability of ultra-thin gate-oxide is a limiting factor in deep submicron technologies. If there were no room left for premature consumption of the gate-oxide lifetime, one would naturally expect that any plasma charging damage couldn't be tolerated. Yet, many reports in the literature indicate that charging damage become less important as gate-oxide gets thinner [12-14]. The question is why? Some of the early experiments used detection methods that are not producing signal even when damage is severe and thus may not have been correct. However, more recent experiment using gate leakage method does confirm that very thin oxides can withstand charging without damage (Figure 7.21) [15].

Plasma charging damage is best described as constant current stress of gate-oxide with the current level being determined by the plasma and antenna ratio. Since processing plasma can supply ion current density up to $10mA/cm^2$ or more, the charging current in substrate injection mode can be as high as $10A/cm^2$ or more for antenna ratio of 1000:1. The current density can be two orders of magnitude higher for the gate injection mode. Majority of the charging damage cases is probably in the range of $1mA/cm^2$ to $1A/cm^2$. In this range of stress current, the thin gate-oxides appear to be able to take the stress much better than thicker gate-oxides. This is based on the observation that the Q_{BD} under constant current stress rises rapidly as oxide thickness decrease (Figure 7.22) [16]. For plasma charging, if the stress current is $200mA/cm^2$ and stress duration is 10 seconds, the total injected charge would be only 2C/cm^2, a very small fraction of the Q_{BD} of the thinnest oxide in the figure.

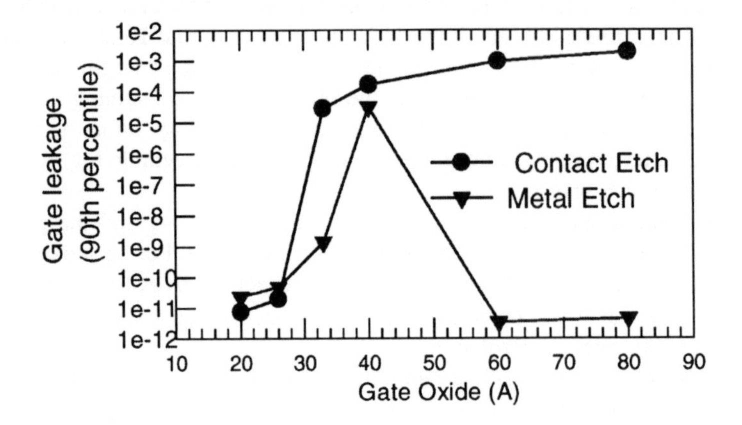

Figure 7.21. The same antenna structure and plasma process causes different levels of damage in different oxide thickness. Very thin oxides are able to withstand charging with no detectable damage. Taken from [15].

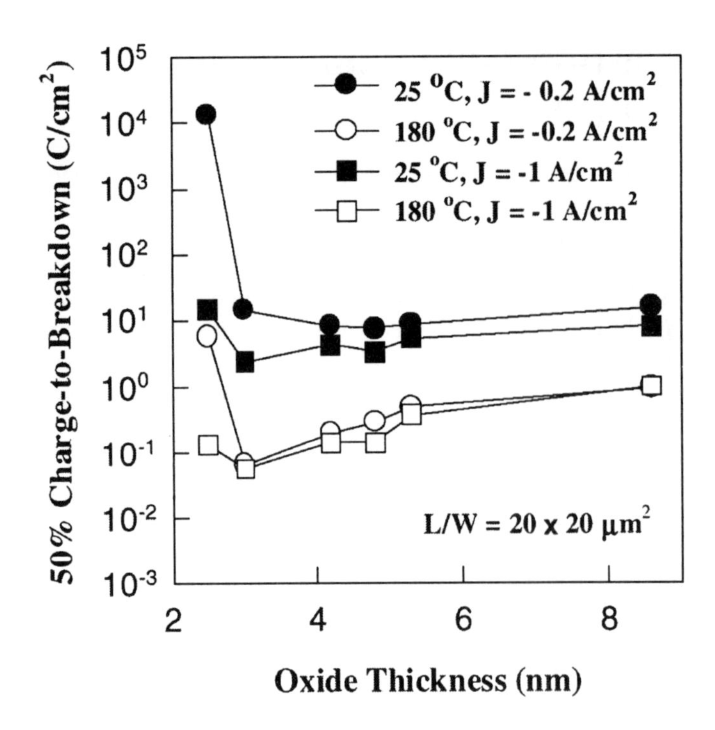

Figure 7.22. The median Q_{BD} for the ultra-thin oxide rises rapidly under constant current stress. Taken from [16].

In Figure 7.22, the Q_{BD} for the thinnest oxide does decrease to only 10C/cm^2 when the stress current is 1A/cm^2. So if the charging current is 1A/cm^2, the 10-second charging time can cause 50% of the devices to breakdown. If the process temperature were 180°C instead of room temperature, the damage is worse even for the 200mA/cm^2 case. The effects of stress current density as well as stress temperature on Q_{BD} were discussed in Chapter 1. Both are definitely much more severe for ultra-thin oxide and thus would suggest charging damage is a serious problem for ultra-thin gate-oxide if the processing temperature is high or the charging current is high, or both.

According to Figure 7.22, at least when the charging stress current and processing temperature are both at low level, charging damage will not be a problem for ultra-thin oxide. Unfortunately, the data in Figure 7.22 is the median breakdown time of 50x50μ^2 capacitors, rather than the 0.1cm^2 active area and 0.01% failure as specified by reliability requirement. To really see if plasma charging damage would be a problem or not, the Q_{BD} of scaled devices is a more appropriate quantity to compared to instead of the Q_{BD} of small devices. Figure 7.23 is the same 200mA/cm^2 constant current stress at room temperature data of Figure 7.22 after corrected for the area and failure fraction scaling. The charging stress of 2C/cm^2 is now 100 times higher than the scaled Q_{BD}. If the same scaling operation were performed for the data of the other stress conditions, the results would be far worst.

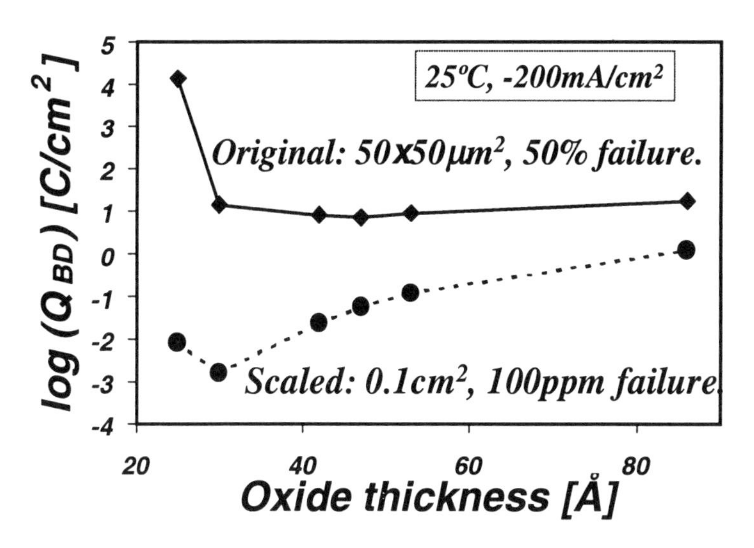

Figure 7.23. After the area and failure fraction scaling, the Q_{BD} value for the ultra-thin oxide becomes very small.

Comparing the scaled and unscaled data in Figure 7.23, measuring the 50x50μ^2 devices with 50% failure as criterion clearly has a sensitivity a million times too low to detect charging damage for the thinnest oxide in the figure. For the data in Figure 7.21, the device size used in the measurement was $5\mu^2$ and the failure fraction was 90%. The scaling correction due to area and failure fraction is ~ 10^9 for the thinnest oxides.

The scaling correction factor is a sensitive function of Weibull slope, and therefore the oxide thickness. For the same device size and failure fraction used in Figure 7.21, the scaling factor as a function of oxide thickness is shown in Figure 7.24. Thus although the small devices with the thin oxide did not breakdown after plasma processing, one cannot conclude that the damage is less severe for the thinner oxides. If one were using $0.1cm^2$ devices and 0.01% failure as criteria, the damage to the thinner oxides might be much more severe.

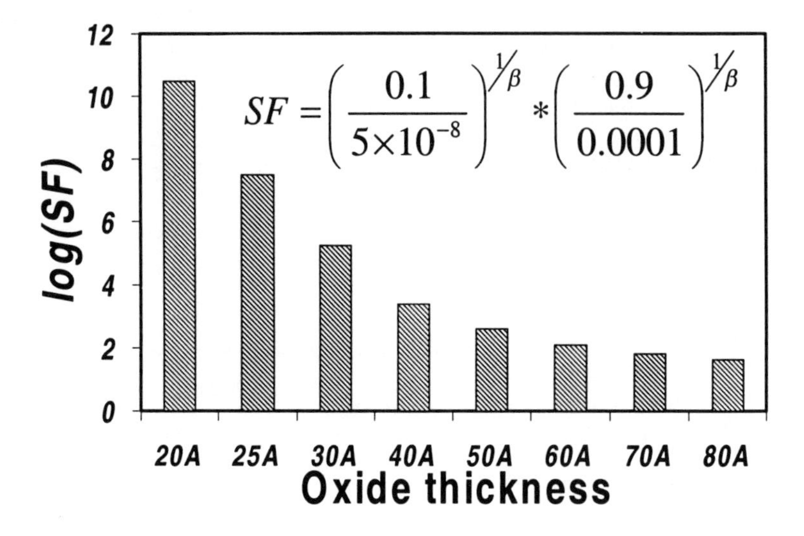

Figure 7.24. The calculated scaling factor for the device size of $5\mu^2$ and failure fraction of 0.9 as a function of oxide thickness.

The discussion above illustrated the problem of detecting damage in thin gate-oxide without taking into account the scaling effects. The same conclusion should apply to the use of design rule to limit the impact of charging damage. Conventional approach to design rule is a simple antenna ratio rule regardless of the active area associated with the antenna ratio. For example, the design rule of 100:1 allows the designer to have every active device of the circuit to have the maximum antenna ratio. Thus the rule must be set for the worse case scenario and be very conservative. For example, a design with 99% of the active device at 10:1 antenna ratio and 1% at 100:1 antenna ratio would experience a much lower impact of charging damage than one with 100% of active devices at 100:1 antenna ratio. The impact of charging damage on the two designs would be similar if the gate-oxide were 100A thick. The impact would be drastically different for 20A gate-oxide.

How can one set up antenna design rule when the gate-oxide is very thin? It is not simple, may even be impossible. Since every product will has it own unique antenna ratio distribution, it is not possible to know how to take the area effect into account in advance. To solve the problem, one must abandon the practice of using a set design rule. Instead, one should use the methodology of projecting the impact of charging damage on yield and reliability to assess each design individually. With the information on the projected yield and reliability impact for each antenna ratio bin, the designers can modify the design accordingly. The new design can then be checked again using the same methodology. Such approach would allow the designers the freedom to make antenna ratio and active area trade-off. Occasional use of large antenna ratio would also be allowed. This kind of design flexibility will be a necessity in the deep submicron technology once the full impact of plasma charging damage on yield and reliability is properly measured.

7.9 The Damage Measurement Problem for Ultra-thin Gate-oxide

In order to use the methodology described in this chapter to project the impact of plasma charging damage on product yield and reliability, one must be able to accurately measure the failure fraction of antenna devices as a function of antenna ratios. If the measured failure fractions for the large antenna ratio, small area devices contain errors, these errors will be amplified during the many scaling operations and the resulting projection may have too large an uncertainty. Thus any charging damage measurement method must be able to capture the percentage of broken down devices accurately for it to be useful.

In Chapter 6, a number of plasma charging damage measurement techniques was discussed. Most of the commonly used measurement methods are unsuitable for use in ultra-thin gate-oxides. Currently, the most popular measurement method for ultra-tin gate-oxide damage measurement is the initial gate-leakage current. As was shown in Chapter 6, this method cannot produce an accurate failure fraction for ultra-thin oxide (Figure 6.33).

One of the ways to improve the measured failure fraction accuracy of the initial gate-leakage measurement is to use extremely small devices to limit the direct tunneling current. However, using a smaller device will make the scaling factor bigger. One way to compensate for the sensitivity gap (the scaling factor) is to use larger antenna ratio test devices. The idea is that when a larger antenna ratio is used, the stress current will increase. Higher stress current leads to higher stress voltage. It is possible to calculate, using the voltage acceleration scaling, the needed voltage increase to compensate for the sensitivity gap.

For example, consider a circuit designed with a 1000:1 antenna rule and the largest antenna ratio bin (1000:1) has an associated active area of $1000\mu^2$. To calculate the failure fraction of this antenna ratio bin, failure fraction data from antenna devices that are much smaller but with large antenna ratios are needed. Note that for the yield and reliability projection methodology, the antenna test devices must have a failure fraction that is neither zero nor 1. In other words, the all failure and no failure results are not useful signals for the calculation. Antenna test devices that has too low a sensitivity may produce zero failure fraction (100% yield) while the product is suffering from charging damage. Antenna test devices that has too high a sensitivity may produce failure fraction of 1 (zero yield) while the product may not be impacted. Thus, the antenna devices must be designed to be at the right sensitivity range so that useful failure fraction signal can be obtained for the calculation. To design such antenna test devices, the scaling factors involved must be known.

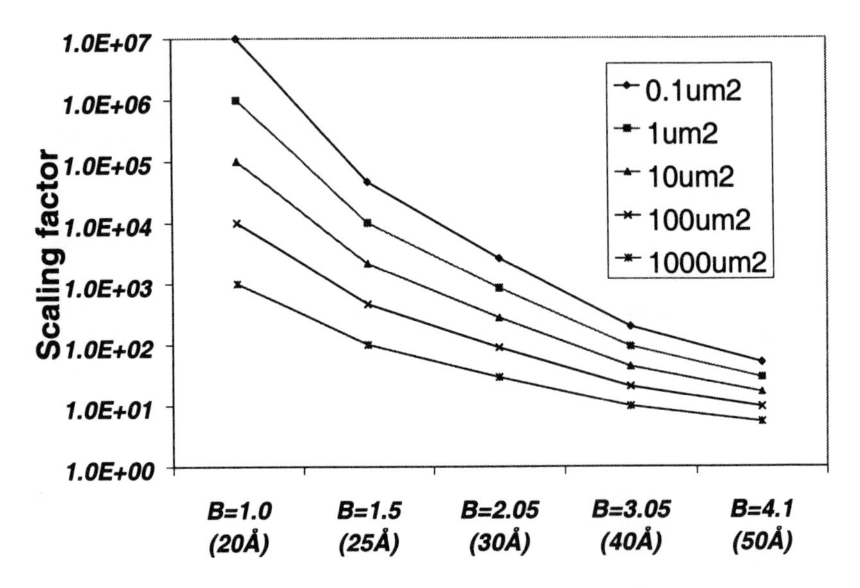

Figure 7.25. The scaling factor for a $1000\mu^2$ active area as a function of oxide thickness for a variety of antenna device size.

Figure 7.25 shows the scaling factor for this antenna bin for various antenna device sizes. As expected, both the smaller devices and thinner oxides have larger scaling factors. Assuming for the moment that the charging current for the 1000:1 antenna is in the $10mA/cm^2$ to $100mA/cm^2$ range, the corresponding stress voltage range during charging damage can be found from the I-V curve of each oxide thickness. Once the stress voltage range is known, the mid-point of the range can serve as the voltage reference and the voltage acceleration factor at the reference voltage can be found from Figure 1.27 of Chapter 1. The scaling factor can then be converted into need voltage increase using the voltage acceleration factor. Figure 7.26 shows the calculated stress voltage increases needed for the scaling factors in Figure 7.25. These stress voltage increases are what the antenna devices needed to experience during plasma charging in order to provide the correct sensitivity. Note that although the scaling factors are very large, the stress voltage increase needed is not. This is because the voltage acceleration factor is extremely large for thin oxides.

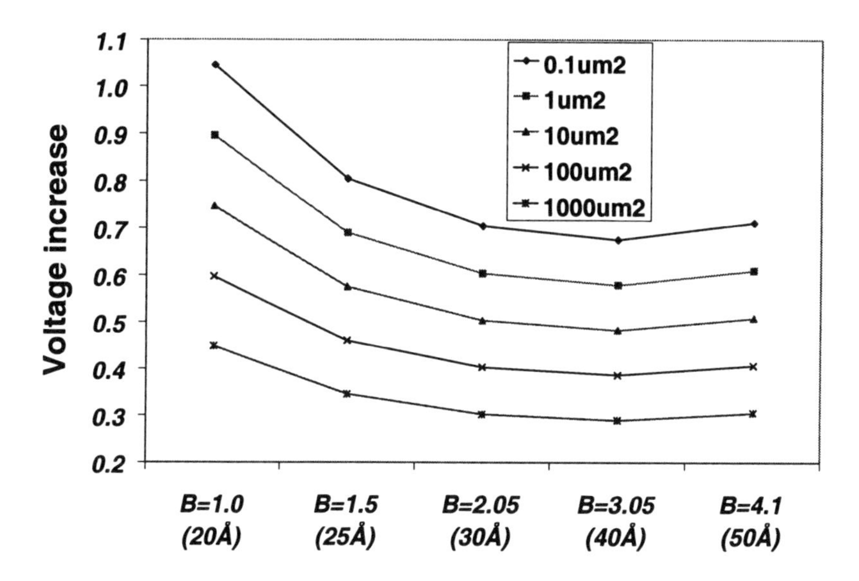

Figure 7.26. The amount of stress voltage increase that can account for the scaling factor in Figure 7.25.

From the calculated stress voltage increase, the needed stress current increases can be read from the I-V curve of each oxide thickness. The results are shown in Figure 7.27. For the example considered, the needed increase in stress current is modest. The highest factor is only about 200. The problem is, how can one choose an antenna ratio that will guarantee the stress current be increased by that much during plasma charging? If the charging stress current is linearly proportional to antenna ratio, the answer is simple. Unfortunately, the relationship between charging current and antenna ratio is non-linearly and cannot be known in advance. It can also saturate like in Figure 7.14. When that happen, no amount of antenna ratio increase can produce the desired stress current increase.

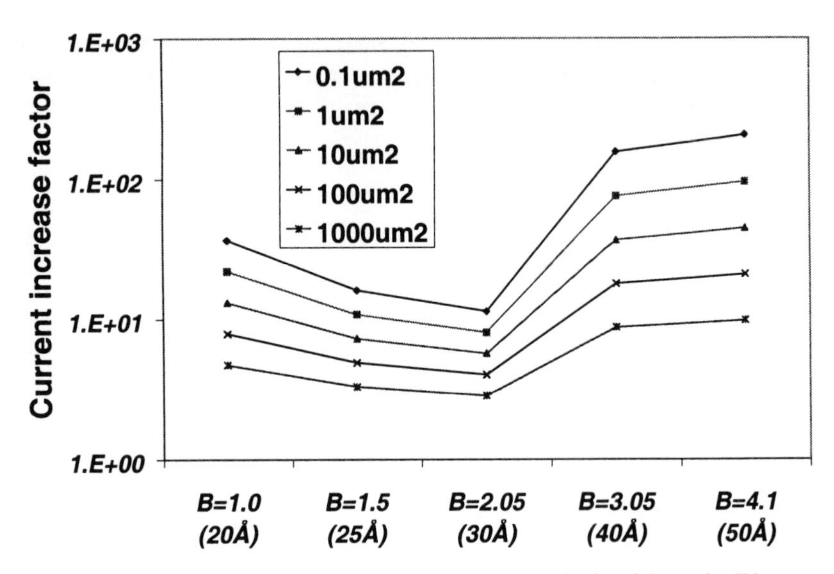

Figure 7.27. The corresponding stress current increases as calculated from the IV curves of various oxides.

On the one hand, direct tunneling forces the use of ever-smaller devices to improve the accuracy of the measured percentage of broken devices. On the other hand, smaller device requires a large scaling factor that is not possible to compensate. This measurement dilemma [17] is currently an unsolved problem.

References

1. Watanabe, T. and Y. Yoshida, *Dielectric Breakdown of Gate Insulator Due to Reactive Ion Etching*. Solid State Technol., 1984(April): p. 263.
2. Luchies, J., P. Simon, and F.K.W. Maly, *Relation Between Product yield and Plasma Process Induced Damage*. in *International Symp. Plasma Process Induced Damage*. 1998. p. 7.
3. Siu, S., R. Patrick, and V. Vahedi, *Effect of Plasma density and Uniformity, Electron temperature, Process Gas, and Chamber on Electron-shading Damage*. in *International Symp. Plasma Process Induced Damage (P2ID)*. 1999. p. 25.
4. Krishnan, S., *et al.*, *Inductively coupled plasma (ICP) metal etch dmage to 35-60A gate oxide*. in *International Electron Devices Meeting (IEDM)*. 1996. p. 731.
5. Alavi, M., *et al.*, *Effect of MOS Device Scaling on Process Induced Gate Charging*. in *International Symp. Plasma Process Induced Damage (P2ID)*. 1997. p. 7.

5. Alavi, M., *et al.*, *Effect of MOS Device Scaling on Process Induced Gate Charging*. in *International Symp. Plasma Process Induced Damage (P2ID)*. 1997. p. 7.
6. Mason, P., *et al.*, *Quantitative Yield and Reliability Projection from Antenna Test Results - A Case Study*. in *Symp. VLSI Technology*. 2000. p. 96.
7. Mason, P., *et al.*, *Relationship Between Yield and Reliability Impact of Plasma Damage to Gate Oxide*. in *International Symp. Plasma Process Induced Damage (P2ID)*. 2000. p. 2.
8. Noguchi, K., A. Matsumoto, and N. Oda, *A Model for Evaluating Cumulative Oxide Damage from Multiple Plasma Processes*. in *International Reliability Physics Symp. (IRPS)*. 2000. p. 364.
9. bosch, G.V.d., *et al.*, *Anomalously weak Antenna Ratio dependence of Plasma Process-Induced Damage*. in *International Symp. Plasma Process Induced Damage (P2ID)*. 2000. p. 6.
10. Ma, S. and J.P. McVittie, *Prediction of Plasma charging Induced Gate Oxide Tunneling Current and Antenna Dependence by Plasma Charging Probe*. in *International Symp. Plasma Process Induced Damage (P2ID)*. 1996. p. 20.
11. Shin, H. and C. Hu, *Dependence of Plasma-Induced Oxide Charging Current in Al Antenna Geometry*. IEEE Electron Dev. Lett., 1992. **13**(12): p. 600.
12. Hu, C., *Gate Oxide Scaling Limits and Projection*. in *International Electron Device Meeting (IEDM)*. 1996. p. 319.
13. Park, D. and C. Hu, *Plasma Charging Damage on Ultra-Thin Gate Oxides*. in *International Symp. Plasma Process Induced Damage (P2ID)*. 1997. p. 15.
14. Bersuker, G., J. Werking, and S. Kim, *Process Induced Charging Damage in Thin Gate Oxides*. in *International Symp. Plasma Process Induced Damage (P2ID)*. 1997. p. 21.
15. Krishnan, S., *et al.*, *Antenna Device Reliability for ULSI Processing*. in *International Electron Device Meeting (IEDM)*. 1998. p. 601.
16. Lin, H.-C., *et al.*, *Characterization of Plasma Charging Damage in Ultrathin Gate Oxide*. in *International Reliability Physics Symposium (IRPS)*. 1998. p. 312.
17. Cheung, K.P., P. Mason, and D. Hwang, *Plasma charging damage of ultra-thin gate-oxide -- The measurement dilemma*. in *International Symp. Plasma Process Induced Damage (P2ID)*. 2000. p. 10..

Index